Springer Monographs in Mathematics

Springer
Berlin
Heidelberg
New York
Barcelona
Hong Kong
London
Milan
Paris
Tokyo

Lluís Puig
路易斯•步驰

有限群的块 Blocks of Finite Groups

块的超聚焦子代数 The Hyperfocal Subalgebra of a Block

Springer

Lluís Puig
Université de Paris 7 - Denis Diderot
Institut de Mathématiques de Jussieu
175, rue du Chevaleret
75013 Paris, France

e-mail: puig@math.jussieu.fr

Catalog-in-Publication Data applied for

Die Deutsche Bibliothek - CIP-Einheitsaufnahme

Puig, Lluís: Blocks of finite groups : the hyperfocal subalgebra of a block
/ Lluís Puig. - Berlin ; Heidelberg ; New York ; Barcelona ; Hong Kong ;
London ; Milan ; Paris ; Tokyo : Springer, 2002
(Springer monographs in mathematics)

Mathematics Subject Classification (2000): 20C11

ISBN 978-3-642-07802-6

Springer-Verlag Berlin Heidelberg New York
a member of BertelsmannSpringer Science+Business Media GmbH

http://www.springer.de

Printed in Germany

目录

Contents

1. 引言

1.1. 本教程来源于我在武汉大学的一个系列讲演，可是教程的内容比讲演的内容扩充了很多. 我当时的主要目的是把研读群论的学生引导到能阅读和理解文章 "The hyperfocal subalgebra of a block" [10].

1.2. 这篇文章的主要结果，即超聚焦子代数的存在性与唯一性，在 Brauer 块理论里可能是基本的；特别地，块论中的另一个重要结果，所谓幂零块的源代数的结构定理 (文章 [9] 的主要结果)，可作为它的直接推论. 后来我终于把证明简化了，可以在一本不长的教程中完整地把它叙述出来. 这样，读者可以看到事情的整个思想和面貌.

1.3. 差不多六十年前 Richard Brauer 引进了块理论；他的目的是研究特征为非零素数 p 的域 k 上的有限群 G 的群代数 kG：任意 kG 的不可分解直和项的双边理想决定一个 G-块. 可是 Brauer 的主要发现可能是：存在无限个彼此不同构的群它们有一个 "共同块". 当然，我应该明确 "共同块" 这个词汇的精确含义；事实上，可以有不同的含义.

1.4. 我的意见：最好方式是考虑所谓 G-块的源代数. 我自己在文章 "Pointed groups and construction of characters" [7] 里引进了源代数的定义；原来的目的只是研究上面

1. Introduction

1.1. The present course comes from a series of lectures we gave at Wuhan University, but its contents are significantly larger. Our aim was to provide an introduction for the understanding of the paper "The hyperfocal subalgebra of a block" [10], to students in group theory.

1.2. The subject of that paper, namely the existence and the uniqueness of the *hyperfocal subalgebra of a block*, seems fundamental in Brauer block theory; for instance, an important result in block theory, *the structure of the source algebra of a nilpotent block* (the main result in [9]), can be obtained as a corollary. We subsequently succeeded in shortening the proof and decided to issue a complete and self-contained account which will allow, we hope, a better understanding of the subject.

1.3. About sixty years ago, Richard Brauer introduced the *block theory*; his purpose was to study the group algebra kG of a finite group G over a field k of nonzero characteristic p: any indecomposable two-sided ideal which is also a direct summand of kG determines a G-block. But Brauer's main discovery was perhaps the existence of families of infinitely many nonisomorphic groups having "a block in common". Of course, the expression "a block in common" demands a definition; actually, more than one reasonable definition might exist.

1.4. Our point of view is to consider the so-called *source algebra* of a G-block, which we introduced in "Pointed groups and construction of characters" [7]; the original purpose was just to study the nilpotent blocks mentioned above, which were

提到的幂零块, 当时它已经由文章 "A Frobenius theorem for blocks" [2] 引进. 幂零块的源代数在文章 "Nilpotent blocks and their source algebras" [9] 里解决以后, 如下事实是越来越清楚了: 有关 G-块的一般结构由 G-块的源代数决定. 所以, 对我来说, 两个群有一个 "共同块" 的含义是它们各自有一个块其对应的源代数是相互同构的.

already introduced in "A Frobenius theorem for blocks" [2]. Once the *source algebra* of a nilpotent block was determined in "Nilpotent blocks and their source algebras" [9], it became more and more clear that the source algebra of a G-block determines everything concerning the G-block. Thus, for us, to say that two groups have "a block in common" means that we find a block in each one in such a way that the corresponding source algebras are isomorphic.

1.5. 在这本教程里, 上述技术性词汇都是细心引进的. 为了阅读它只是需要例如书 [12] 的知识, 以及 Wedderburn 定理, Nakayama 引理一类的基本代数知识; 也就是说, 在这个基础上, 证明都是完备的.

1.5. In this course, all the concepts mentioned above are carefully introduced. To follow the exposition just requires familiarity with the contents of [12], Wedderburn's Theorem, Nakayama's Lemma and other basic algebraic topics; in other words, in this text all the proofs are complete.

1.6. 为了研究 kG 与复数域上的群代数 $\mathbb{C}G$ 的关系, Brauer 考虑了在满足 $\mathcal{O}/J(\mathcal{O}) \cong k$ 的特征为零的局部环 $\mathcal{O}$ 上的群代数 $\mathcal{O}G$; 为了研究 kG 与 $\mathcal{O}G$ 的关系, 基本结果是其幂等元的关系. 所以, 对一般的 $\mathcal{O}$-代数 A, 为了从 $A/J(A)$ 到 A 提升幂等元, 在§2 里开始研究能够在 $\mathcal{O}$ 上做这件事的充分条件; 我的方法来源于 [13,II,§4].

1.6. In order to study the relationship between kG and the group algebra $\mathbb{C}G$ over the complex numbers, Brauer considers the group algebra $\mathcal{O}G$ over a local ring $\mathcal{O}$ of characteristic zero fulfilling $\mathcal{O}/J(\mathcal{O}) \cong k$; then, to relate kG and $\mathcal{O}G$, the basic result concerns the relationship between the idempotents. Thus, we start by considering in §2 sufficient conditions on $\mathcal{O}$ to lift idempotents from $A/J(A)$ to A, for any $\mathcal{O}$-algebra A; our method comes from [13,II,§4].

1.7. 一般情况下, 不需要在 k 上附加其它条件; 只有在某些论断里 需要 k 是 "充分大" 的, 才假定 k 是代数闭的. 从 §3 以后, 我假定 $\mathcal{O}$ 是完备的离散赋值环, 这就能够提升幂等元; 一般来说允许 $\mathcal{O} = k$, 可是在最后三节里 $\mathcal{O}$ 的特征应该为零.

1.7. In general, we do not need more hypotheses on k, and in those cases where we do need k be "big enough", we simply assume that k is algebraically closed. From §3 on, we assume that $\mathcal{O}$ is a complete discrete valuation ring, which is enough for lifting idempotents; we allow the possibility $\mathcal{O} = k$, except in the last three sections where the characteristic of $\mathcal{O}$ has to be zero.

1.8. 在 §3 里, 我引进所谓 $\mathcal{O}$-代数的点和 $\mathcal{O}$-代数的除子; 事实上, 对任意 $\mathcal{O}$-代数 A, 其除子和投射 A-模的同构类基本上是一样的, 可是在考虑某些 $\mathcal{O}$-代数和某些同态的, 除子的观点比较容易. 以前, 只是考虑 A 的点, 没有考虑 A 的除子; 现在从除子的观点研究, 用所谓"可嵌入同构"代替了原来的"embeddings" (见 [7]).

1.9. 从 §4 开始考虑一个有限群 G, 和带有 G-作用的 $\mathcal{O}$-代数 A, 这种结构称为 G-代数; 事实上, 那里考虑更一般的情况. 对 G 的任意子群 H 可考虑所有 H-稳定元素构成的 $\mathcal{O}$-子代数 A^H 和它的除子, 称为 A 上的 H-除子. 那么, 重要工具是与 A 的 H-除子 ω 相关的 H-代数 A_ω. 事实上, A_ω 只是在"可嵌入同构"意义下定义的; 关于这个事实我作了细心分析.

1.10. 在 §5 可以看到考虑除子比仅仅考虑点的更有利之处; 确实, 在 A 上的 H-除子之间, 这里 H 跑遍 G 的子群, "限制"和"诱导"可以妥善处置. 关于 H-除子的诱导需要构造适当的 G-代数; 特别是, 我们得到一个 Higman 投射性条件的推广. 请注意, 文章 "Induction, restriction and G-algebras" [1] 已经引进 A 的 H-除子以及它们的限制和它们的诱导, 可是关于诱导的定义, 他的和我的不一样.

1.8. In §3, we introduce the *points* and the *divisors* on an $\mathcal{O}$-algebra A; actually, the *divisors* on A and the isomorphism classes of projective A-modules basically coincide, but when considering several $\mathcal{O}$-algebras together with several algebra homomorphisms, the *divisor* point of view is easier. Previously we only considered *points* on A; in the present course, the *divisor* point of view allows us to replace the previous "embeddings" (see [7]) by the so-called *embeddable isomorphisms*.

1.9. In §4 we already consider a finite group G, and an $\mathcal{O}$-algebra A endowed with a G-action, called a *G-algebra*; in fact, we will consider a more general situation. Thus, for any subgroup H of G, we can consider the subalgebra A^H of elements fixed by H, and the divisors of A^H, called *H-divisors* on A. Then, an important construction is the H-algebra A_ω relative to an H-divisor ω on A. Actually, A_ω is only defined up to an *embeddable isomorphism*; this is a question that we will analyse with care.

1.10. In §5, the advantage in considering *divisors* instead of merely considering *points* can already be understood; it is in the set of all the H-divisors on A, when H runs over the set of all the subgroups of G, that "restriction" and "induction" can be handled. For the induction of H-divisors, we have to construct a suitable G-algebra; in particular, we get a generalization of the Higman Projectivity Condition. The H-divisors on A, and their restrictions and inductions, have already been introduced in "Induction, restriction and G-algebras" [1], but note that the definition of the induction there does not coincide with ours.

1.11. 在§6里，我引进 A 上的所谓局部点群，以及对应于 A 上的 H-点的 H-代数 A_β 的源代数. 在 §7 里，我证明了 Green 不可分解性定理的推广；这个推广是已经知道的 (见 [6] 和 [14,§5])，可是我这里的证明可能是新的. 在 §8 里，开始了超聚焦子代数的存在性的证明的准备工作；那里，我在比文章 [8] 更一般的情况下引进了所谓 A-融合.

1.11. In §6, we introduce the *local pointed groups* on A, together with the *source algebra* of the H-algebra A_β associated with an H-point on A. In §7, we prove a generalization of the Green Indecomposability Theorem; this generalization is already known (see [6] and [14,§5]), but our proof here is possibly new. In §8, we begin the preparatory work for the proof of the existence of the *hyperfocal subalgebra*; here, in a more general context than in [8], we introduce the so-called *A-fusions*.

1.12. 在 A 上的 K-除子之间的 A-融合，其中 K 跑遍 G 的子群，是很重要的工具，这是因为如果 A 上的某些 K-除子"包含"在 A 的 H-除子 ω 里，那么其 A-融合在对应 H-代数 A_ω 是保存着的. 事实上，本书 A-融合的定义比在文章 [10] 里的好多了. 由于樊恽的注意，在 §9 里的结果比文章 [10] 的对应的结果更具一般性；在 §9 里，对于证明超聚焦子代数的存在性来说，我得到一个关键步骤.

1.12. The *A-fusions* between K-divisors on A, where K runs over the set of subgroups of G, is a very important tool because if some K-divisors on A are "contained" in an H-divisor ω on A then their A-fusions are preserved in the corresponding H-algebra A_ω. As a matter of fact, we give a definition of the A-fusions which improves on the formulation given in [10]. Following a remark of Fan Yun, we obtain results in §9 more general than the corresponding ones in [10]; in §9, we reach a key step in the proof for the existence of the *hyperfocal subalgebra*.

1.13. 在 §10 里，我把所有上述的定义和结果应用到群代数；例如，$\mathcal{O}G$ 上的 G-点就是 G-块. 一般来说，在群代数里上述的结果都有特种内容；这样，我这里再一次证明文章 [8] 里关于 $\mathcal{O}G$ 上的局部点群之间的 $\mathcal{O}G$-融合的主要结果. 事实上，$\mathcal{O}G$ 上的局部点群之间的 $\mathcal{O}G$-融合仅依赖于 $\mathcal{O}G$ 上的"少数"局部点群的 $\mathcal{O}G$-自融合，即 $\mathcal{O}G$ 上的极大局部点群和所谓关键点群.

1.13. In §10, we apply all the definitions and results above to the case of the *group algebra;* for instance, the G-points on $\mathcal{O}G$ are just the G-blocks. In general, all the results above have a particular form in the group algebra; thus, here we prove again the main result in [8] on the $\mathcal{O}G$-fusions between the local pointed groups on $\mathcal{O}G$. Actually, the $\mathcal{O}G$-fusions between the local pointed groups on $\mathcal{O}G$ only depend on the $\mathcal{O}G$-fusions of "few" local pointed groups to themselves, like the maximal ones and the so-called *essential* local pointed groups.

1.14. 在 §11 里, 为了表示这个事实, 我引进 $\mathcal{O}G$ 上的关键点群, 以及适当的定义; $\mathcal{O}G$-融合只是依赖某些 $\mathcal{O}G$-自融合, 这个事实对于证明超聚焦子代数的存在性来说也是一个关键步骤. 在 §12 里, 我研究 G-块的源代数的特殊性; 而且, 在一个例子里我展示无限个相互不同构的群有一个"共同块"; 更确切地, 它们的所谓主块的源代数都是相互同构的.

1.15. 在 §13 里, 我在任意 G-块的源代数里正式引进超聚焦子代数; 一般来说, 在 G-块的代数里 没有对应的结构: 这个事实也表现了考虑源代数的好处. 那里, 我刻画了超聚焦子代数上的局部点群, 以及它们之间的融合; 而且, 已经证明了从超聚焦子代数的存在性可推出上述的幂零块的源代数的结构定理.

1.16. 在 §14 和 §15 里, 我需要 $\mathcal{O}$-代数理论中关于 p-进分析的某些结果; 它们基本依赖于交换 $\mathcal{O}$-代数上的指数函数和对数函数的存在性. 为了不打断研究思路, 我把这些辅助的结果集中 在 §16 里.

1.17. 最后, 在 §14 和 §15 里, 我证明超聚焦子代数的存在性和唯一性; 事实上, 由于用归纳法, 为了证明其存在性, 我不仅需要超聚焦子代数的唯一性, 而且需要一个更精细的结果. 在 §14 里, 我给出了这个更苛刻的唯一性; 它的证明应用了 §16 里的关于非交换上同调的一个结果: 它相当于说某个 1-上循环就是 1-上边缘.

1.14. In order to formulate this statement we introduce in §11 the *essential* pointed groups along with suitable definitions; the fact that the $\mathcal{O}G$-fusions only depend on some *$\mathcal{O}G$-autofusions* is also a key point in the proof of existence of the *hyperfocal subalgebra*. In §12, we study the particularities of the source algebra of a G-block; we exhibit an example with infinitely many pairwise nonisomorphic groups having "a block in common": the source algebras of their *principal* blocks are pairwise isomorphic.

1.15. In §13, we introduce the *hyperfocal subalgebra* in the source algebra of a G-block; in general, in the algebra of a G-block there is no such structure: this fact shows the advantage in considering the source algebra. There, we describe the local pointed groups on the *hyperfocal subalgebra* and the fusions between them; moreover, from the existence of the *hyperfocal subalgebra* we can already prove the theorem on the structure of the source algebras of nilpotent blocks.

1.16. In §14 and §15, we need some results on *p-adic* analysis; they basically depend on the existence of the *exponential* and the *logarithmic* functions in commutative $\mathcal{O}$-algebras. In order to avoid any interruption in the exposition, we collect these results in §16.

1.17. Finally, in §14 and §15, we prove the existence and the uniqueness of the *hyperfocal subalgebra*; actually, because we argue by induction, in order to prove existence we need to frame the uniqueness in a stronger way. In §14, we prove this stronger form of uniqueness employing a result of §16 which can be considered a result on noncommutative cohomology: roughly speaking it says that some 1-*cocycle* is a 1-*coboundary*.

1.18. 在§15里，我证明超聚焦子代数的存在性; 其关键是适当的提升的存在性 (见推论 15.9): 樊恽在 [4] 里已经得到了这种结果. 这里, 我给出的证明部分不同于在文章 [10] 里的证明; 特别是, 文章 [10] 里的证明要引用文章 [11] 的结果而这里不再需要了.

1.19. 本教程的准备和写作得到了很多人的关心与帮助: **武汉大学** 邀请. **樊恽** 教授安排了我在数学系的系列讲演; 并且他细心评阅这本教程, 既从中文的观点也从数学的观点评阅; 事实上, 是他鼓动我在本书给出超聚焦子代数的存在性和唯一性的一个完整证明. **张继平** 教授准备了一个英中数学词汇. **Joe庄** 细心评阅引言的英文的方面. **Alberto Arabia** 教授制作了计算机的特种中文软件, 让我在我的计算机里能够使用汉字; 为了得到这双栏目排版, 他还制作了另一个计算机的软件. 我的汉语老师 **茹小雷** 和我的夫人 **Isabel** 帮助我做七千多汉字的编码以及录入计算机的工作. 我真诚感谢所有这些朋友和亲人.

1.18. In §15, we prove the existence of the *hyperfocal subalgebra*; the key to the proof is the existence of a suitable lifting (see Corollary 15.9): in [4], Fan Yun already obtained a result of this type. The proof we give here is partially different from the proof in [10]; in particular, the proof in [10] employs some results of [11] which are no longer necessary here.

1.19. The elaboration of this course has been made possible thanks to the collaboration of many people: The invitation by **Wuhan University**. The organization of the series of lectures by **Fan Yun**, who revised both the Chinese writing and the mathematical content in this course with great care, and encouraged me to issue a complete account on the *hyperfocal subalgebra*. **Zhang Ji Ping** prepared for me a mathematical English-Chinese vocabulary. **Joe Chuang** carefully revised the English part of this introduction. **Alberto Arabia** created a computer program allowing me to employ Chinese characters in my usual TeX program, and a second one for the two-column pattern. My Chinese language teacher, **Ru Xiao Lei**, and my wife **Isabel** helped me to implement the codes of about seven thousand Chinese characters on the computer. I deeply thank them all.

2. 幂等元的提升

2. Lifting Idempotents

2.1. 为了从特征为素数 p 提升一个幂等元到特征为零, 有效的方法是使用一个特征为零的完备的离散赋值环使得其根上的商环是特征为 p 的域. 事实上, 完备化条件是允分的但不必要的. 下面, 我考虑这些问题.

2.1. In order to lift idempotents from characteristic p to characteristic zero, the safest method is to work over a complete discrete valuation ring of characteristic zero with a residue field of characteristic p. Yet, as a matter of fact, the completeness is a sufficient but not necessary condition. In this section, we will discuss on this question.

2.2. 设 $\mathcal{O}$ 是一个离散赋值环使得 $k = \mathcal{O}/J(\mathcal{O})$ 是特征为 p 的域, 其中 $J(\mathcal{O})$ 是 $\mathcal{O}$ 的根基, 并且 $\mathcal{O}$ 的分式域 $\mathcal{K}$ 是为特征零的域. 也就是说, 有一个满射的群同态 $\vartheta: \mathcal{K}^* \to \mathbb{Z}$ 使得

2.2. Let $\mathcal{O}$ be a discrete valuation ring such that $k = \mathcal{O}/J(\mathcal{O})$ is a field of characteristic p, where $J(\mathcal{O})$ denotes the radical of $\mathcal{O}$, and that its field of quotients $\mathcal{K}$ has characteristic zero. That is to say, we have a surjective group homomorphism $\vartheta: \mathcal{K}^* \to \mathbb{Z}$ fulfilling

$$2.2.1 \qquad \{\lambda \in \mathcal{K}^* \mid \vartheta(\lambda) \geq 0\} = \mathcal{O} - \{0\};$$

特别是, 有 $\pi \in \mathcal{O}$ 使得 $J(\mathcal{O}) = \pi\mathcal{O}$. 请注意, 如果 $\mathcal{K}'$ 是 $\mathcal{K}$ 的 Galois 扩张, 那么从范数映射 $\mathfrak{N}: \mathcal{K}'^* \to \mathcal{K}^*$ 不难定义一个满射的群同态 $\vartheta': \mathcal{K}'^* \to \mathbb{Z}$; 进一步, 不难证明下面的集合

in particular, there is $\pi \in \mathcal{O}$ such that $J(\mathcal{O}) = \pi\mathcal{O}$. Note that, if $\mathcal{K}'$ is a Galois extension of $\mathcal{K}$, then from the norm map $\mathfrak{N}: \mathcal{K}'^* \to \mathcal{K}^*$ it is not difficult to define a surjective group homomorphism $\vartheta': \mathcal{K}'^* \to \mathbb{Z}$; moreover, it is easy to check that the set

$$2.2.2 \qquad \{\lambda' \in \mathcal{K}'^* \mid \mathfrak{N}(\lambda') \in \mathcal{O}\} \cup \{0\}$$

就是 $\mathcal{K}'$ 中 $\mathcal{O}$ 上的整元构成的环 $\mathcal{O}'$; 也就是说, $\mathcal{O}'$ 也是离散赋值环.

coincides with the integral closure $\mathcal{O}'$ of $\mathcal{O}$ in $\mathcal{K}'$; in other words, $\mathcal{O}'$ still is a discrete valuation ring.

2.3. 设 $\{\lambda_n\}_{n\in\mathbb{N}}$ 是一个 $\mathcal{K}$ 的元素的序列; 如果存在 $\lambda \in \mathcal{K}$ 使得对任意 $n \in \mathbb{N}$ 有 $\lambda - \lambda_n \in \pi^{n+1}\mathcal{O}$, 那么我们说 $\{\lambda_n\}_{n\in\mathbb{N}}$ 有极限 λ, 并且我们记 $\lambda = \lim_{n\to\infty}\{\lambda_n\}$ (请注意, 这个条件比通常的收敛条件更强). 例子: 如果对任意 $n \in \mathbb{N}$ 有 $\lambda_n = \lambda$, 这个序列当然有极限 λ. 有极限的序列满足所谓 Cauchy 条件;

2.3. Let $\{\lambda_n\}_{n\in\mathbb{N}}$ be a sequence of elements of $\mathcal{K}$; whenever there is $\lambda \in \mathcal{K}$ such that we have $\lambda - \lambda_n \in \pi^{n+1}\mathcal{O}$ for any $n \in \mathbb{N}$, we say that the sequence $\{\lambda_n\}_{n\in\mathbb{N}}$ has the *limit* λ, and we write $\lambda = \lim_{n\to\infty}\{\lambda_n\}$ (note that this condition is stronger than the usual one). For instance, if for any $n \in \mathbb{N}$ we have $\lambda_n = \lambda$, this sequence obviously has the limit λ. The sequences having a limit fulfill the so-called Cauchy condition; in particular, for any

特别是, 对任意 $n \in \mathbb{N}$ 有

$n \in \mathbb{N}$, we have

2.3.1 $$\lambda_{n+1} - \lambda_n \in \pi^{n+1}\mathcal{O}$$

(事实上, 上面的条件比 Cauchy 条件更强). 反过来, 如果满足这个条件的 $\mathcal{K}$-序列都有极限, 那么我们说 $\mathcal{K}$ 和 $\mathcal{O}$ 是完备的.

(actually, this condition is stronger than Cauchy's condition). Conversely, if all the $\mathcal{K}$-sequences fulfilling this condition have a limit then we say that $\mathcal{K}$ and $\mathcal{O}$ are *complete*.

2.4. 不难造一个包含 $\mathcal{K}$ 的完备域 $\hat{\mathcal{K}}$. 我们在所有满足条件 2.3.1 的 $\mathcal{K}$-序列里考虑下面的等价关系: 如果该集合里的两个序列 $\{\lambda_n\}_{n\in\mathbb{N}}$ 和 $\{\mu_n\}_{n\in\mathbb{N}}$ 满足

2.4. It is not difficult to construct a complete field $\hat{\mathcal{K}}$ containing $\mathcal{K}$. We consider the following equivalence relationship in the set of all the sequences fulfilling condition 2.3.1: if two sequences $\{\lambda_n\}_{n\in\mathbb{N}}$ and $\{\mu_n\}_{n\in\mathbb{N}}$ of this set fulfill

2.4.1 $$\mu_n - \lambda_n \in \pi^{n+1}\mathcal{O}$$

就说这两个序列是等价的. 例子: 固定 $\ell \in \mathbb{N}$, 如果 $\mu_n = \lambda_\ell$ 或 λ_n 当 $n \le \ell$ 或 $n \ge \ell$, 那么这两个序列是等价的; 又, $\{\lambda_{\ell+n}\}_{n\in\mathbb{N}}$ 也满足条件 2.3.1 并 $\{\lambda_n\}_{n\in\mathbb{N}}$ 和它是等价的. 请注意, 如果 $\{\lambda_n\}_{n\in\mathbb{N}}$ 没有极限 0, 那么对适当的 ℓ, λ_ℓ 不属于 $\pi^{\ell+1}\mathcal{O}$; 此时, 从条件 2.3.1 可推出 $\vartheta(\lambda_{\ell+n}) = \vartheta(\lambda_\ell)$, 其中 $n \in \mathbb{N}$ (cf. 2.2.1).

we say that they are equivalent. For instance, fixing $\ell \in \mathbb{N}$, if $\mu_n = \lambda_\ell$ or λ_n according to $n \le \ell$ or $n \ge \ell$ then these sequences are equivalent; moreover, $\{\lambda_{\ell+n}\}_{n\in\mathbb{N}}$ also fulfills condition 2.3.1 and clearly is equivalent to $\{\lambda_n\}_{n\in\mathbb{N}}$. Note that, if $\{\lambda_n\}_{n\in\mathbb{N}}$ has not the limit 0 then, for a suitable ℓ, λ_ℓ does not belong to $\pi^{\ell+1}\mathcal{O}$; thus, from condition 2.3.1 we get $\vartheta(\lambda_{\ell+n}) = \vartheta(\lambda_\ell)$ for any $n \in \mathbb{N}$ (cf. 2.2.1).

2.5. 设 $\hat{\mathcal{K}}$ 是所有满足条件 2.3.1 的 $\mathcal{K}$-序列的等价类的集合; 显然, 序列的加法在 $\hat{\mathcal{K}}$ 里决定一个交换群结构. 而且, 设 $\hat{\lambda} = \widetilde{\{\lambda_n\}}_{n\in\mathbb{N}}$ 与 $\hat{\mu} = \widetilde{\{\mu_n\}}_{n\in\mathbb{N}}$ 是 $\hat{\mathcal{K}}$ 的两个非等于零元素; 我们能假定对任意 $n \in \mathbb{N}$ 有 $\vartheta(\lambda_n) = \vartheta(\lambda_\circ)$ 与 $\vartheta(\mu_n) = \vartheta(\mu_\circ)$; 取 $\ell \in \mathbb{N}$ 满足 $-\vartheta(\lambda_\circ) \le \ell$ 与 $-\vartheta(\mu_\circ) \le \ell$; 此时, 因为下面的差

2.5. Let $\hat{\mathcal{K}}$ be the set of equivalent classes of sequences of $\mathcal{K}$ which fulfill condition 2.3.1; clearly, the usual sum of sequences determines a structure of commutative group in $\hat{\mathcal{K}}$. Moreover, let $\hat{\lambda} = \widetilde{\{\lambda_n\}}_{n\in\mathbb{N}}$ and $\hat{\mu} = \widetilde{\{\mu_n\}}_{n\in\mathbb{N}}$ be two nonzero elements of $\hat{\mathcal{K}}$; by the remarks above, we may assume that, for any $n \in \mathbb{N}$, we have $\vartheta(\lambda_n) = \vartheta(\lambda_\circ)$ and $\vartheta(\mu_n) = \vartheta(\mu_\circ)$; choose ℓ fulfilling $-\vartheta(\lambda_\circ) \le \ell$ and $-\vartheta(\mu_\circ) \le \ell$; at that point, since the difference

2.5.1 $$\lambda_{n+1}\mu_{n+1} - \lambda_n\mu_n = \lambda_{n+1}(\mu_{n+1} - \mu_n) + (\lambda_{n+1} - \lambda_n)\mu_n\,,$$

显然属于 $\pi^{n+1-\ell}\mathcal{O}$, 所以序列 $\{\lambda_{\ell+n}\,\mu_{\ell+n}\}_{n\in\mathbb{N}}$ 满足2.3.1; 那么, 不难验证 $\{\lambda_{\ell+n}\,\mu_{\ell+n}\}_{n\in\mathbb{N}}$ 的等价类,记 $\hat{\lambda}\hat{\mu}$,不依赖我们的选择.

belongs to $\pi^{n+1-\ell}\mathcal{O}$, $\{\lambda_{\ell+n}\,\mu_{\ell+n}\}_{n\in\mathbb{N}}$ fufills 2.3.1; then, it is not difficut to check that the equivalent class of $\{\lambda_{\ell+n}\,\mu_{\ell+n}\}_{n\in\mathbb{N}}$, noted $\hat{\lambda}\hat{\mu}$, does not depend on our choice.

2.6. 这个乘法运算在 $\hat{\mathcal{K}}-\{0\}$ 里决定一个交换群结构; 确实, 不难验证结合律. 另一方面, 令 v 记 $\vartheta(\lambda_\circ)$ 与 0 中较大的数; 不难验证有

2.6. This operation determines a commutative group structure in $\hat{\mathcal{K}}-\{0\}$; indeed, the associativity is easily checked. On the other hand, denote by v the biggest of $\vartheta(\lambda_\circ)$ and 0; we clearly have

2.6.1 $$\lambda_{v+n} = \lambda_v\Big(1+\sum_{i=1}^{n}\delta_i\pi^i\Big)$$

其中 $n\in\mathbb{N}$ 与 $\delta_i\in\mathcal{O}$; 用归纳法我们能定义下面的序列 $\{\mu_n\}_{n\in\mathbb{N}}$

where $n\in\mathbb{N}$ and $\delta_i\in\mathcal{O}$; arguing by induction, we can define the following sequence $\{\mu_n\}_{n\in\mathbb{N}}$

2.6.2 $$\mu_n = 1+\sum_{i=1}^{n}\varepsilon_i\pi^i\,,\quad \varepsilon_i = -\delta_i-\sum_{j=1}^{i-1}\delta_j\varepsilon_{i-j}\,;$$

因为对任意 $i\in\mathbb{N}$ 有 $\varepsilon_i\in\mathcal{O}$, 所以它满足条件 2.3.1; 此时, 不难验证其等价类乘以 λ_v^{-1} 常数序列的等价类就是 $\hat{\lambda}$ 的逆元素.

since we have $\varepsilon_i\in\mathcal{O}$ for any $i\in\mathbb{N}$, this sequence fulfills condition 2.3.1; thus, it is easy to check that its equivalent class multiplied by the equivalent class of the constant sequence λ_v^{-1} is the inverse of $\hat{\lambda}$.

2.7. 当然, 我定义 $0\hat{\lambda}=0=\hat{\lambda}0$; 那么不难验证分配律. 这样, 就证明了 $\hat{\mathcal{K}}$ 是一个域, 而只要把 $\lambda\in\mathcal{K}$ 与 λ 常数序列的等价类等同一致, $\mathcal{K}$ 就是一个 $\hat{\mathcal{K}}$ 的子域. 在 $\hat{\mathcal{K}}$ 里, 设 $\hat{\mathcal{O}}$ 是所有 $\mathcal{O}$-序列的等价类的集合; 显然, $\hat{\mathcal{O}}$ 是 $\hat{\mathcal{K}}$ 的子环, 并且它包含 $\mathcal{O}$.

2.7. Obviously, we set $0\hat{\lambda}=0=\hat{\lambda}0$; now, it is not difficult to check the distributivity. Hence, we have proved that $\hat{\mathcal{K}}$ is a field, and it suffices to identify $\lambda\in\mathcal{K}$ with the equivalent class of the constant sequence λ to get $\mathcal{K}$ as a subfield of $\hat{\mathcal{K}}$. Let $\hat{\mathcal{O}}$ be the subset of $\hat{\mathcal{K}}$ of all the equivalent classes of sequences in $\mathcal{O}$; clearly, $\hat{\mathcal{O}}$ is a subring of $\hat{\mathcal{K}}$ which contains $\mathcal{O}$.

2.8. 而且, 我们能把 $\mathcal{K}$ 的离散赋值 $\vartheta\colon\mathcal{K}^*\to\mathbb{Z}$ 扩张为一个 $\hat{\mathcal{K}}$ 的离散赋值 $\hat{\vartheta}$ 如下:设 $\hat{\lambda}=\widetilde{\{\lambda_n\}}_{n\in\mathbb{N}}$ 是一个 $\hat{\mathcal{K}}^*$ 的元素; 别忘了, 我们能假定对任

2.8. Moreover, we can extend the discrete valuation $\vartheta\colon\mathcal{K}^*\to\mathbb{Z}$ of $\mathcal{K}$ to a discrete valuation $\hat{\vartheta}$ of $\hat{\mathcal{K}}$ as follows: let $\hat{\lambda}=\widetilde{\{\lambda_n\}}_{n\in\mathbb{N}}$ be an element of $\hat{\mathcal{K}}^*$; recall that we may assume that we have $\vartheta(\lambda_n)=\vartheta(\lambda_\circ)$ for any

意 $n \in \mathbb{N}$ 有 $\vartheta(\lambda_n) = \vartheta(\lambda_\circ)$; 那么就令 $\hat{\theta}(\hat{\lambda}) = \theta(\lambda_\circ)$; 特别是, $\hat{\theta}(\hat{\lambda}) \geq 0$ 当且仅当 $\hat{\lambda} \in \hat{\mathcal{O}}$.

$n \in \mathbb{N}$; in this case, we just set $\hat{\theta}(\hat{\lambda}) = \theta(\lambda_\circ)$; in particular, note that we have $\hat{\theta}(\hat{\lambda}) \geq 0$ if and only if $\hat{\lambda} \in \hat{\mathcal{O}}$.

2.9. 最后, 我断言 $\hat{\mathcal{K}}$ 和 $\hat{\mathcal{O}}$ 是完备的. 设 $\{\hat{\lambda}_n\}_{n\in\mathbb{N}}$ 是 $\hat{\mathcal{K}}$ 的一个元素的序列使得对任意 $n \in \mathbb{N}$ 有

2.9. Finally, we claim that $\hat{\mathcal{K}}$ and $\hat{\mathcal{O}}$ are complete. Let $\{\hat{\lambda}_n\}_{n\in\mathbb{N}}$ be a sequence of elements of $\hat{\mathcal{K}}$ such that, for any $n \in \mathbb{N}$, we have

2.9.1 $$\hat{\lambda}_{n+1} - \hat{\lambda}_n \in \pi^{n+1}\hat{\mathcal{O}};$$

这样, 如果 $\hat{\lambda}_n = \widetilde{\{\lambda_{n,m}\}}_{m\in\mathbb{N}}$, 对适当的 $\mu_{n,m} \in \mathcal{O}$, 其中 $n, m \in \mathbb{N}$, 还有

thus, if $\hat{\lambda}_n = \widetilde{\{\lambda_{n,m}\}}_{m\in\mathbb{N}}$ then, for a suitable choice of $\mu_{n,m} \in \mathcal{O}$, where $n, m \in \mathbb{N}$, we still have

2.9.2 $$(\lambda_{n+1,m} - \lambda_{n,m}) - \pi^{n+1}\mu_{n,m} \in \pi^{m+1}\mathcal{O};$$

别忘了, 由 2.4 的子里, 能假定对任意 $m \leq n$ 有 $\lambda_{n,m} = \lambda_{n,n+1}$; 此时, 元素 $\lambda_{n+1,m} - \lambda_{n,m}$ 属于 $\pi^{n+1}\mathcal{O}$, 其中 $m, n \in \mathbb{N}$; 因为这些序列满足条件 2.3.1, 所以对任意 $n \in \mathbb{N}$ 下面的差

recall that, according to the example in 2.4, we may assume that $\lambda_{n,m} = \lambda_{n,n+1}$ for any $m \leq n$; then, for any $m, n \in \mathbb{N}$, we obtain $\lambda_{n+1,m} - \lambda_{n,m} \in \pi^{n+1}\mathcal{O}$; since all these sequences fulfill condition 2.3.1, for any $n \in \mathbb{N}$, the difference

2.9.3 $$\lambda_{n+1,n+1} - \lambda_{n,n} = \lambda_{n+1,n+1} - \lambda_{n+1,n} + \lambda_{n+1,n} - \lambda_{n,n};$$

属于 $\pi^{n+1}\mathcal{O}$; 这样, $\{\lambda_{n,n}\}_{n\in\mathbb{N}}$ 满足 2.3.1, 从而这个序列决定 $\hat{\mathcal{K}}$ 的一个元素 $\hat{\lambda}$.

belongs to $\pi^{n+1}\mathcal{O}$; thus, $\{\lambda_{n,n}\}_{n\in\mathbb{N}}$ fulfills condition 2.3.1 and therefore determines an element $\hat{\lambda}$ of $\hat{\mathcal{K}}$.

2.10. 进一步, 对任意 $m, n \in \mathbb{N}$ 只要 $m \geq n$ 就有

2.10. Moreover, for any $m, n \in \mathbb{N}$, it suffices that $m \geq n$ to get

2.10.1 $$\lambda_{m,m} - \lambda_{n,m} = \sum_{i=n}^{m-1} (\lambda_{i+1,m} - \lambda_{i,m}) \in \pi^{n+1}\mathcal{O};$$

所以, 只要把上面的例子应用到序列 $\{\lambda_{m,m}\}_{m\in\mathbb{N}}$, 就 $\hat{\lambda} - \hat{\lambda}_n$ 属于 $\pi^{n+1}\hat{\mathcal{O}}$, 其中 $n \in \mathbb{N}$; 也就是说, 序列 $\{\hat{\lambda}_n\}_{n\in\mathbb{N}}$ 有极限 $\hat{\lambda}$. 请注意, 如果 $\{\lambda_n\}_{n\in\mathbb{N}}$ 在 $\mathcal{K}$ 中是满足条件 2.3.1 的序列, 那么 $\{\lambda_n\}_{n\in\mathbb{N}}$ 的等价类在 $\hat{\mathcal{K}}$ 中就是这个序列的极限; 所以只要 $\mathcal{K}$ 是完备的, 就得到 $\hat{\mathcal{K}} = \mathcal{K}$.

hence, it suffices to apply the example above to the sequence $\{\lambda_{m,m}\}_{m\in\mathbb{N}}$ to get that $\hat{\lambda} - \hat{\lambda}_n$ belongs to $\pi^{n+1}\hat{\mathcal{O}}$ for any $n \in \mathbb{N}$; that is, the sequence $\{\hat{\lambda}_n\}_{n\in\mathbb{N}}$ has the limit $\hat{\lambda}$. Note that, if $\{\lambda_n\}_{n\in\mathbb{N}}$ is a sequence in $\mathcal{K}$ which fulfills condition 2.3.1, then the element of $\hat{\mathcal{K}}$ determined by the equivalent class of $\{\lambda_n\}_{n\in\mathbb{N}}$ actually is the limit of this sequence; hence, if $\mathcal{K}$ is complete then we simply get $\hat{\mathcal{K}} = \mathcal{K}$.

2.11. **我们总是假定$\mathcal{O}$-代数都是有限秩的自由$\mathcal{O}$-模.** 设 A 是一个 $\mathcal{O}$-代数; 令 J 记 A 的根. 已经知道, 如果 $\mathcal{O}$ 是完备的, 那么 A-序列 $\{a_n\}_{n\in\mathbb{N}}$ 只要满足 $a_{n+1}-a_n \in J^{n+1}$, 其中 $n\in\mathbb{N}$, 它就有极限; 也就是说, 存在 $a\in A$ 使得对任意 $n\in\mathbb{N}$ 有 $a-a_n\in J^{n+1}$. 确实, 对适当的 $r\in\mathbb{N}$ 有 $J^r\subset\pi.A$, 从而对任意 $n\in\mathbb{N}$ 下面的差

2.11. *We always assume that the $\mathcal{O}$-algebras are $\mathcal{O}$-free $\mathcal{O}$-modules of finite rank.* Let A be an $\mathcal{O}$-algebra; denote by J the radical of A. It is well-known that if $\mathcal{O}$ is complete then it suffices that a sequence $\{a_n\}_{n\in\mathbb{N}}$ in A fulfills $a_{n+1}-a_n\in J^{n+1}$ for any $n\in\mathbb{N}$, to guarantee that it has a limit; precisely, in that case there is $a\in A$ such that $a-a_n\in J^{n+1}$ for any $n\in\mathbb{N}$. Indeed, we have $J^r\subset\pi.A$ for a suitable $r\in\mathbb{N}$ and therefore, for any $n\in\mathbb{N}$, the difference

2.11.1 $$a_{(r+1)(n+1)}-a_{(r+1)n}=\sum_{i=0}^{r}(a_{(r+1)n+i+1}-a_{(r+1)n+i})$$

属于 $J^{(r+1)n+1}$; 因为 A 是有限秩的自由 $\mathcal{O}$-模并对任意 $n\geq r$ 有 $J^{(r+1)n+1}\subset\pi^{n+1}\cdot A$, 所以存在 $a\in A$ 使得对任意 $n\geq r$ 有

belongs to $J^{(r+1)n+1}$; since A is an $\mathcal{O}$-free $\mathcal{O}$-module of finite rank and, for any $n\geq r$, we have $J^{(r+1)n+1}\subset\pi^{n+1}\cdot A$, there is $a\in A$ such that, for any $n\geq r$, we have

2.11.2 $$a-a_{(r+1)n}\in\pi^{n+1}.A\subset J^{n+1},$$

从而还有

and therefore we also have

2.11.3 $$a-a_n=a-a_{(r+1)n}+\sum_{i=n+1}^{(r+1)n}(a_i-a_{i-1})\in J^{n+1};$$

进一步, 对任意 $n\leq r$ 也得到

moreover, for any $n\in\mathbb{N}$, we still obtain

2.11.4 $$a-a_n=a-a_r+\sum_{i=n}^{r-1}(a_{i+1}-a_i)\in J^{n+1}.$$

2.12. 特别是, 如果 A' 是 $\mathcal{O}$-代数并 f 是 $\mathcal{O}$-代数同态从 A 到 A' 使得对任意 $n\in\mathbb{N}$ 有 $f(a_{n+1})-f(a_n)\in J'^{n+1}$ 其中 J' 是 A' 的根基, 那么 $f(a)-f(a_{rn})$ 属于 $\pi^{n+1}.A'\subset J'^{n+1}$, 从而得到 $f(a)-f(a_n)\in J'^{n+1}$; 也就是说, $\{f(a_n)\}_{n\in\mathbb{N}}$ 有极限 $f(a)$. 而且如果 $r\in J$ 那么 $\{\sum_{\ell=0}^n r^\ell\}_{n\in\mathbb{N}}$ 有极限, 记为 $\sum_{\ell\in\mathbb{N}} r^\ell$; 这个极限就是 $1-r$ 的逆元素; 从而, A^* 包含 $1+J$.

2.12. In particular, if A' is an $\mathcal{O}$-algebra and f an $\mathcal{O}$-algebra homomorphism from A to A' such that, for any $n\in\mathbb{N}$, we have $f(a_{n+1})-f(a_n)\in J'^{n+1}$, where J' is the radical of A', then $f(a)-f(a_{rn})$ belongs to $\pi^{n+1}.A'\subset J'^{n+1}$ and therefore we get $f(a)-f(a_n)\in J'^{n+1}$; in other terms, $\{f(a_n)\}_{n\in\mathbb{N}}$ has the limit $f(a)$. Moreover, if $r\in J$ then the sequence $\{\sum_{\ell=0}^n r^\ell\}_{n\in\mathbb{N}}$ has a limit, noted $\sum_{\ell\in\mathbb{N}} r^\ell$; this limit actually coincides with the inverse of $1-r$; consequently, A^* contains $1+J$. Note that,

请注意，如果$\mathcal{K}'$是$\mathcal{K}$的Galois扩张，并$\mathcal{O}'$是$\mathcal{K}'$中$\mathcal{O}$上的整元的环$\mathcal{O}'$，那么$\mathcal{O}'$与$\mathcal{K}'$也是完备的.

if $\mathcal{K}'$ is a Galois extension of $\mathcal{K}$, and $\mathcal{O}'$ is the integral closure of $\mathcal{O}$ in $\mathcal{K}'$, then $\mathcal{O}'$ and $\mathcal{K}'$ are complete too.

定理 2.13. **假定$\mathcal{O}$是完备的. 设A是一个交换$\mathcal{O}$-代数并设I是A的理想. 令J记A的根基并$s: A \to A/J$记自然的映射，再令**

Theorem 2.13. *Assume that $\mathcal{O}$ is complete. Let A be a commutative $\mathcal{O}$-algebra and I an ideal of A. Denote by J the radical of A and by $s: A \to A/J$ the canonical map, and set* †

2.13.1
$$\text{幂}^{p^{\mathbb{N}}}(I) = \bigcap_{n\in\mathbb{N}} \left(\{a^{p^n} \mid a \in I\} + J^{n+1}\right),$$
$$\text{幂}^{p^{\mathbb{N}}}\big(s(I)\big) = \bigcap_{n\in\mathbb{N}} \{s(a)^{p^n} \mid a \in I\}.$$

那么s决定一个从$\text{幂}^{p^{\mathbb{N}}}(I)$到$\text{幂}^{p^{\mathbb{N}}}\big(s(I)\big)$的双射；而且，有

Then, s determines a bijection between $\text{幂}^{p^{\mathbb{N}}}(I)$ and $\text{幂}^{p^{\mathbb{N}}}\big(s(I)\big)$; moreover, we have

2.13.2
$$\text{幂}^{p^{\mathbb{N}}}(I) \cdot \text{幂}^{p^{\mathbb{N}}}(I) \subset \text{幂}^{p^{\mathbb{N}}}(I).$$

证明: 请注意，如果$a, b \in A$满足$s(a) = s(b)$，那么对任意$n \in \mathbb{N}$有$a^{p^n} - b^{p^n} \in J^{n+1}$；确实，只要使用归纳法，我们就能假定$n > 1$并且$c = a^{p^{n-1}} - b^{p^{n-1}}$属于$J^n$；显然有

Proof: First of all, note that if $a, b \in A$ fulfill the equality $s(a) = s(b)$ then, for any $n \in \mathbb{N}$, we have $a^{p^n} - b^{p^n} \in J^{n+1}$; indeed, we argue by induction on n and may assume that $n > 1$ and that the element $c = a^{p^{n-1}} - b^{p^{n-1}}$ belongs to J^n; clearly, we have

2.13.3
$$a^{p^n} = (b^{p^{n-1}} + c)^p \in b^{p^n} + pJ^n + J^{np} \subset b^{p^n} + J^{n+1}.$$

现在，如果$a, b \in \text{幂}^{p^{\mathbb{N}}}(I)$那么对任意$n \in \mathbb{N}$存在$a_n, b_n \in I$与$r_n, s_n \in J^{n+1}$使得

Now, if $a, b \in \text{幂}^{p^{\mathbb{N}}}(I)$ then for any $n \in \mathbb{N}$, we can find $a_n, b_n \in I$ and $r_n, s_n \in J^{n+1}$ such that

2.13.4
$$a = (a_n)^{p^n} + r_n, \quad b = (b_n)^{p^n} + s_n;$$

特别是，ab属于下面的交

in particular, ab belongs to the intersection

2.13.5
$$\bigcap_{n\in\mathbb{N}} \left((a_n b_n)^{p^n} + J^{n+1}\right);$$

从而ab也属于$\text{幂}^{p^{\mathbb{N}}}(I)$. 而且，如果有$s(a) = s(b)$那么还有

hence, ab still belongs to $\text{幂}^{p^{\mathbb{N}}}(I)$. Moreover, if we have $s(a) = s(b)$ then $s(a_n)^{p^n} = s(b_n)^{p^n}$

† The Chinese character 幂 is pronounced "mi" as in "middle" and means "power" or "exponent".

$s(a_n)^{p^n} = s(b_n)^{p^n}$, 从而 $0 = s(a_n - b_n)^{p^n}$, 这是因为 A/J 是特征 p 的. 另一方面, 因为 A/J 是域的直积所以 $s(a_n) = s(b_n)$; 这样, 得到

and therefore we still have $s(a_n - b_n)^{p^n} = 0$ since A/J has characteristic p. On the other hand, since A/J is a direct product of fields, we also get $s(a_n) = s(b_n)$; thus, we obtain

2.13.6 $$a - b \in \bigcap_{m\in\mathbb{N}} J^m \subset \bigcap_{m\in\mathbb{N}} \pi^m \cdot A = \{0\}.$$

所以 s 决定一个单映射从 幂$^{p^{\mathbb{N}}}(I)$ 到 幂$^{p^{\mathbb{N}}}(s(I))$; 我们要证明这个映射也是满射. 设 $\bar{a}$ 是一个 幂$^{p^{\mathbb{N}}}(s(I))$ 的元素; 也就是说, 对任意 $n \in \mathbb{N}$ 存在 $a_n \in I$ 使得 $s(a_n)^{p^n} = \bar{a}$; 特别是, 有

Hence, s determines an injective map from 幂$^{p^{\mathbb{N}}}(I)$ to 幂$^{p^{\mathbb{N}}}(s(I))$; we will prove that this map is surjective too. Let $\bar{a}$ be an element of 幂$^{p^{\mathbb{N}}}(s(I))$; explicitly, this means that, for any $n \in \mathbb{N}$, there is $a_n \in I$ such that $s(a_n)^{p^n} = \bar{a}$; in particular, we get

2.13.7 $$s\big((a_{n+1})^p\big)^{p^n} = \bar{a} = s(a_n)^{p^n}$$

仍然因为 A/J 就是特征为 p 的域的直积, 所以我们得到 $s\big((a_{n+1})^p\big) = s(a_n)$; 从而, 也得到

and therefore, since A/J is a direct product of fields of characteristic p, we have $s\big((a_{n+1})^p\big) = s(a_n)$; consequently, we also obtain

2.13.8 $$(a_{n+1})^{p^{n+1}} - (a_n)^{p^n} = \big((a_{n+1})^p\big)^{p^n} - (a_n)^{p^n} \in J^{n+1}.$$

那么存在 $a \in I$ 使得对任意 $n > 0$ 有 $a - (a_n)^{p^n} \in J^{n+1}$ (见 2.11), 从而得到

Then, there is $a \in I$ such that, for any $n \in \mathbb{N}$, we have $a - (a_n)^{p^n} \in J^{n+1}$ (see 2.11), and therefore we get

2.13.9 $$s(a) = \big(s(a_n)\big)^{p^n} = \bar{a};$$

也就是说, $s(a)$ 属于 幂$^{p^{\mathbb{N}}}(s(I))$.

that is to say, $s(a)$ belongs to 幂$^{p^{\mathbb{N}}}(s(I))$.

推论 2.14. **假定 $\mathcal{O}$ 是完备的. 设 A 是 $\mathcal{O}$-代数并设 I 是一个 A 的理想. 令 J 记 A 的根, 再令 $s: A \to \bar{A}$ 为自然的映射, 其中 $\bar{A} = A/J$. 对任意幂等元 $\bar{\imath} \in \bar{I}$, 其中 $\bar{I} = s(I)$, 存在一个幂等元 $i \in A$ 使得 $s(i) = \bar{\imath}$. 而且, 如果 $i' \in A$ 是一个幂等元使得 $s(i') = \bar{\imath}$ 有 $a \in A^*$ 使得 $i' = i^a$.**

Corollary 2.14. *Assume that $\mathcal{O}$ is complete. Let A be an $\mathcal{O}$-algebra and I an ideal of A. Denote by J the radical of A and by $s: A \longrightarrow \bar{A}$ the canonical map, where $\bar{A} = A/J$. For any idempotent $\bar{\imath} \in \bar{I}$, where $\bar{I} = s(I)$, there exists an idempotent $i \in A$ such that $s(i) = \bar{\imath}$. Moreover, if $i' \in A$ is an idempotent such that $s(i') = \bar{\imath}$, then there exists $a \in A^*$ such that $i' = i^a$.*

证明: 显然有 $a \in A$ 使得 $s(a) = \bar{\imath}$; 令 $B = \sum_{n\in\mathbb{N}} \mathcal{O}\cdot a^n$;

Proof: Clearly there is $a \in A$ such that $s(a) = \bar{\imath}$; set $B = \sum_{n\in\mathbb{N}} \mathcal{O}\cdot a^n$; then B is a

那么 B 是 A 的子代数, 而有 $B \cap J = J(B)$; 所以, 只要用 $B, B \cap I$ 分别代替 A, I 就能假定 A 是交换的. 那么, 由定理 2.13, 既然 $\bar{\imath}$ 属于 幂$^{p^{\mathbb{N}}}(s(I))$ 所以存在 $i \in$ 幂$^{p^{\mathbb{N}}}(I)$ 满足 $s(i) = \bar{\imath}$, 还满足 $s(i^2) = \bar{\imath}^2 = \bar{\imath}$; 仍使用定理2.13, 既然 i^2 也属于幂$^{p^{\mathbb{N}}}(I)$, 所以 $i^2 = i$. 最后, 如果 $i' \in A$ 是一个幂等元使得 $s(i') = \bar{\imath}$, 那么令 $a = ii' + (1-i)(1-i')$; 一方面, 显然 $ia = ai'$, 另一方面有

subalgebra of A and we have $B \cap J = J(B)$; consequently, up to the replacement of A and I by B and $B \cap I$, we may assume that A is commutative. Then, according to Theorem 2.13, since $\bar{\imath}$ belongs to 幂$^{p^{\mathbb{N}}}(s(I))$, there exists $i \in$ 幂$^{p^{\mathbb{N}}}(I)$ fulfilling $s(i) = \bar{\imath}$, and in particular $s(i^2) = \bar{\imath}^2 = \bar{\imath}$; according to Theorem 2.13 again, since i^2 also belongs to 幂$^{p^{\mathbb{N}}}(I)$, we get $i^2 = i$. Finally, if $i' \in A$ is an idempotent fulfilling $s(i') = \bar{\imath}$, then consider the element $a = ii' + (1-i)(1-i')$; on the one hand, we clearly have $ia = ai'$; on the other hand, we get

2.14.1 $$s(a) = \bar{\imath}^2 + (1-\bar{\imath})^2 = 1\,,$$

从而, 还有 $a \in 1 + J \subset A^*$ (见 2.11).

and therefore we still get $a \in 1 + J \subset A^*$ (see 2.11).

2.15. 一般的说, 因为一个 $\mathcal{O}$-代数 A 是自由 $\mathcal{O}$-模, 我们能把 $a \in A$ 与 $1 \otimes a \in \hat{\mathcal{O}} \otimes_{\mathcal{O}} A$ 或者 $1 \otimes a \in \mathcal{K} \otimes_{\mathcal{O}} A$ 或者 $1 \otimes a \in \hat{\mathcal{K}} \otimes_{\mathcal{O}} A$ 都等同一致. 而且, 既然 $\mathcal{K} \cap \hat{\mathcal{O}} = \mathcal{O}$ 显然

2.15. As a general rule, since any $\mathcal{O}$-algebra A is an $\mathcal{O}$-free $\mathcal{O}$-module, we can identify $a \in A$ with $1 \otimes a \in \hat{\mathcal{O}} \otimes_{\mathcal{O}} A$, $1 \otimes a \in \mathcal{K} \otimes_{\mathcal{O}} A$ or $1 \otimes a \in \hat{\mathcal{K}} \otimes_{\mathcal{O}} A$. Moreover, since we have $\mathcal{K} \cap \hat{\mathcal{O}} = \mathcal{O}$, we clearly get

2.15.1 $$(\mathcal{K} \otimes_{\mathcal{O}} A) \cap (\hat{\mathcal{O}} \otimes_{\mathcal{O}} A) = A\,.$$

命题 2.16. 设 A 是一个 $\mathcal{O}$-代数. 如果每一个 $\mathcal{K} \otimes_{\mathcal{O}} A$ 的本原幂等元在 $\hat{\mathcal{K}} \otimes_{\mathcal{O}} A$ 里也是本原的, 那么每一个 A 的本原幂等元在 $\hat{\mathcal{O}} \otimes_{\mathcal{O}} A$ 里也是本原的.

Proposition 2.16. *Let A be an $\mathcal{O}$-algebra. If any primitive idempotent of $\mathcal{K} \otimes_{\mathcal{O}} A$ remains primitive in $\hat{\mathcal{K}} \otimes_{\mathcal{O}} A$, then any primitive idempotent of A still remains primitive in $\hat{\mathcal{O}} \otimes_{\mathcal{O}} A$.*

证明: 设 i 是一个 A 的本原幂等元; 只要用 iAi 代替 A 就能假定 A 的单元素是本原的. 那么, 设 $\hat{\imath}$ 是一个 $\hat{A} = \hat{\mathcal{O}} \otimes_{\mathcal{O}} A$ 的不等于零幂等元并且设 $\hat{J}'$ 和 $\hat{J}''$ 是两个 $\hat{\mathcal{K}} \otimes_{\hat{\mathcal{O}}} \hat{A}$ 的相互正交本原幂等元的集合使得

Proof: Let i be a primitive idempotent of A; up to the replacement of A by iAi, we may assume that the unity element of A is primitive. Then, let $\hat{\imath}$ be a nonzero idempotent of $\hat{A} = \hat{\mathcal{O}} \otimes_{\mathcal{O}} A$, and consider two sets $\hat{J}'$ and $\hat{J}''$ of pairwise orthogonal primitive idempotents such that

2.16.1 $$\sum_{\hat{\jmath} \in \hat{J}'} \hat{\jmath} = \hat{\imath}\,, \qquad \sum_{\hat{\jmath} \in \hat{J}''} \hat{\jmath} = 1 - \hat{\imath}\,.$$

另一方面，由我们的假设，一个满足 $\sum_{j\in J} j = 1$ 的 $\mathcal{K}\otimes_{\mathcal{O}}A$ 的相互正交本原幂等元的集合 J 在 $\hat{\mathcal{K}}\otimes_{\mathcal{O}}A$ 里也是本原幂等元的集合. 从而，由下面的引理 2.17，存在一个双映射 $\tau\colon J\to\hat{J}$ 其中 $\hat{J}=\hat{J}'\cup\hat{J}''$，和 $(\hat{\mathcal{K}}\otimes_{\mathcal{O}}A)^*$ 的元素 $\hat{a}$ 使得对任意 $j\in J$ 有 $\tau(j)=j^{\hat{a}}$；令 $J'=(\tau)^{-1}(\hat{J}')$ 与 $i=\sum_{j\in J'} j$，就特别有 $\hat{\imath}=i^{\hat{a}}$. 事实上，我们能假定 $\hat{a}\in\hat{A}$；那么选择 $h,\ell\in\mathbb{N}$ 使得

On the other hand, from our hypothesis, a set of pairwise orthogonal primitive idempotents of $\mathcal{K}\otimes_{\mathcal{O}}A$ such that $\sum_{j\in J} j = 1$ remains a set of primitive idempotents in $\hat{\mathcal{K}}\otimes_{\mathcal{O}}A$. Moreover, according to Lemma 2.17 below, there are a bijective map $\tau\colon J\to\hat{J}$, where $\hat{J}=\hat{J}'\cup\hat{J}''$, and an element $\hat{a}$ in $(\hat{\mathcal{K}}\otimes_{\mathcal{O}}A)^*$ such that $\tau(j)=j^{\hat{a}}$ for any $j\in J$; in particular, setting $J'=(\tau)^{-1}(\hat{J}')$ and $i=\sum_{j\in J'} j$, we get $\hat{\imath}=i^{\hat{a}}$. Obviously, we may assume that $\hat{a}$ belongs to $\hat{A}$; then, consider $h,\ell\in\mathbb{N}$ such that

2.16.2 $$\hat{a}^{-1}\in\pi^{-h}\cdot\hat{A}\,,\quad i\in\pi^{-\ell}\cdot A\,.$$

设 $\{a_n\}_{n\in\mathbb{N}}$ 与 $\{b_n\}_{n\in\mathbb{N}}$ 是两个 A 的序列使得

Choose two sequences $\{a_n\}_{n\in\mathbb{N}}$ and $\{b_n\}_{n\in\mathbb{N}}$ of elements of A such that

2.16.3 $$\hat{a}=\lim_{n\to\infty}\{a_n\}\,,\quad \pi^h\cdot\hat{a}^{-1}=\lim_{n\to\infty}\{b_n\}\,;$$

也就是说，对任意 $n\in\mathbb{N}$ 元素 $\hat{a}-a_n$ 和 $\pi^h\cdot\hat{a}^{-1}-b_n$ 都属于 $\pi^{n+1}\cdot\hat{A}$. 特别是，使 $m\geq 2h+\ell$ 固定，显然有

in other terms, for any $n\in\mathbb{N}$, the elements $\hat{a}-a_n$ and $\pi^h\cdot\hat{a}^{-1}-b_n$ belong to $\pi^{n+1}\cdot\hat{A}$. In particular, choosing $m\geq 2h+\ell$, we have

2.16.4 $$\begin{aligned}1&=\big(a_m+(\hat{a}-a_m)\big)\big(\pi^{-h}\cdot b_m+(\hat{a}^{-1}-\pi^{-h}\cdot b_m)\big)\\&=\pi^{-h}\cdot a_m b_m+c\end{aligned}$$

其中 c 属于 $\pi^{m+1-h}\cdot\hat{A}$，还属于 $\mathcal{K}\otimes_{\mathcal{O}}A$，这是因为 $a_m b_m$ 属于 A；从而有

where c belongs to $\pi^{m+1-h}\cdot\hat{A}$ and, at the same time, belongs to $\mathcal{K}\otimes_{\mathcal{O}}A$ since $a_m b_m$ belongs to A; consequently, we have

2.16.5 $$c\in\pi^{m+1-h}\cdot\hat{A}\cap(\mathcal{K}\otimes_{\mathcal{O}}A)=\pi^{m+1-h}\cdot A$$

并且 A 的元素 $1-c$ 在 A 中是可逆的；所以 a_m 在 $\mathcal{K}\otimes_{\mathcal{O}}A$ 中也是可逆的，更精确有

and therefore the element $1-c$ is inversible in A; hence, a_m is inversible in $\mathcal{K}\otimes_{\mathcal{O}}A$ and explicitly we have

2.16.6 $$\begin{aligned}(a_m)^{-1}&=\pi^{-h}\cdot b_m(1-c)^{-1}\\&=\pi^{-h}\cdot b_m+\sum_{n=1}^{\infty}\pi^{-h}\cdot b_m c^n\,;\end{aligned}$$

这样 $a_m\, i\,(a_m)^{-1}$ 属于 $\mathcal{K}\otimes_{\mathcal{O}} A$，同时也属于 $\hat{A}$，这是因为

thus, $a_m\, i\,(a_m)^{-1}$ belongs to $\mathcal{K}\otimes_{\mathcal{O}} A$ and, simultaneously, it belongs to $\hat{A}$ since

2.16.7 $$\hat{\imath} - a_m\, i\,(a_m)^{-1} = (\hat{a} - a_m)\, i\,(\hat{a}^{-1}) + a_m\, i\,(\hat{a}^{-1} - (a_m)^{-1})\,;$$

仍由于等式 $\hat{A}\cap(\mathcal{K}\otimes_{\mathcal{O}} A) = A$，这个幂等元 $a_m\, i\,(a_m)^{-1}$ 属于 A，从而它是单位元；所以 $\hat{\imath}$ 也是单位元.

hence, since $\hat{A}\cap(\mathcal{K}\otimes_{\mathcal{O}} A) = A$, the idempotent $a_m\, i\,(a_m)^{-1}$ belongs to A, and therefore it coincides with the unity element; thus, $\hat{\imath}$ coincides with unity element too.

引理 2.17. **设 $\mathcal{L}$ 是一个域并设 B 是一个有限维数的 $\mathcal{L}$-代数. 如果 J 与 J' 都是两个相互正交本原幂等元的集合使得 $\sum_{j\in J} j = 1 = \sum_{j'\in J} j'$，那么存在 $b\in B^*$ 使得 $J' = J^b$.**

Lemma 2.17. *Let $\mathcal{L}$ be a field and B an $\mathcal{L}$-algebra of finite dimension. If J and J' are two sets of pairwise orthogonal primitive idempotents of B such that $\sum_{j\in J} j = 1 = \sum_{j'\in J} j'$, then there exists $b\in B^*$ such that $J' = J^b$.*

证明: 如果 B 的根基就是零那么，只要使用 Wedderburn 定理，就完成证明. 这样，能假定 B 的根基非零，那么有一个 B 的非零理想 N 使得 $N^2 = \{0\}$；令 $\bar{B} = B/N$，再令 $\bar{b}$ 记 $b\in B$ 的像；首先指出，如果 i 是 B 的本原幂等元，那么 $\bar{\imath}$ 也是本原的；这是因为，如果 $0\neq \ell\in iBi$ 满足 $\bar{\ell}^2 = \bar{\ell}$，那么得到 $0 = (\ell^2-\ell)^2 = \ell^4 - 2\ell^3 + \ell^2$，而不难验证这个式子可推出

Proof: If the radical of B is zero, it suffices to apply the Wedderburn Theorem to prove the statement. Thus, we may assume that the radical of B is not zero, and then B has a nonzero ideal N such that $N^2 = \{0\}$; set $\bar{B} = B/N$ and denote by $\bar{b}$ the image of $b\in B$; first of all, we claim that if i is a primitive idempotent of B then $\bar{\imath}$ still is primitive; indeed, if we choose $0\neq \ell\in iBi$ fulfilling $\bar{\ell}^2 = \bar{\ell}$, then we get $0 = (\ell^2-\ell)^2 = \ell^4 - 2\ell^3 + \ell^2$ according to our choice of N and, from this equality, it is not difficult to check that we have

2.17.1 $$\ell^n = (n-2)\ell^3 - (n-3)\ell^2$$

其中 $n\geq 2$，还可推出 $3\ell^2 - 2\ell^3$ 是一个幂等元，从而有 $i = 3\ell^2 - 2\ell^3$. 因为 $\bar{\imath} = 3\bar{\ell}^2 - 2\bar{\ell}^3 = \bar{\ell}$ 所以 $\bar{\imath}$ 也是本原的.

for any $n\geq 2$, which easily implies that $3\ell^2 - 2\ell^3$ is an idempotent and therefore we get $3\ell^2 - 2\ell^3 = i$; since we have $\bar{\imath} = 3\bar{\ell}^2 - 2\bar{\ell}^3 = \bar{\ell}$, $\bar{\imath}$ is a primitive idempotent.

现在，对维数 $\dim_{\mathcal{L}}(B)$ 使用归纳法；由上面的结果，$\bar{J}$ 与 $\bar{J}'$ 是两个 $\bar{B}$ 的相互正交本原幂等元的集合使得 $\sum_{\bar{\jmath}\in\bar{J}} \bar{\jmath} = 1 = \sum_{\bar{\jmath}'\in\bar{J}} \bar{\jmath}'$；而且，$B^*$ 显然包含 $1+N$；所以，自然映射 $B^*\to\bar{B}^*$ 是满射；

Now, we argue by induction on $\dim_{\mathcal{L}}(B)$; according to the previous argument, $\bar{J}$ and $\bar{J}'$ are two sets of pairwise orthogonal primitive idempotents of $\bar{B}$ such that $\sum_{\bar{\jmath}\in\bar{J}} \bar{\jmath} = 1 = \sum_{\bar{\jmath}'\in\bar{J}} \bar{\jmath}'$; moreover, B^* clearly contains $1+N$; consequently, the canonical map $B^*\to\bar{B}^*$ is surjective;

这样，存在 $b \in B^*$ 和一个双射 $\tau: J \to J'$ 使得对任意 $j \in J$ 有 $\overline{\tau(j)} = \overline{j^b}$. 那么，考虑 $c = \sum_{j\in J} jb\tau(j)$；一方面可得到

thus, there are $b \in B^*$ and a bijective map $\tau: J \to J'$ such that, for any $j \in J$, we have $\overline{\tau(j)} = \overline{j^b}$. Then, consider the element $c = \sum_{j\in J} jb\tau(j)$; on the one hand, we have

2.17.2 $$\bar{c} = \sum_{j\in J} \bar{\jmath}\bar{b}\overline{j^b} = \bar{b},$$

从而 c 是可逆的；另一方面对任意 $j \in J$ 有 $jc = jb\tau(j) = c\tau(j)$.

and therefore c is inversible; on the other hand, we have $jc = jb\tau(j) = c\tau(j)$ for any $j \in J$.

推论 2.18. **设 A 是 $\mathcal{O}$-代数. 令 J 记 A 的根基并 $s: A \to A/J = \bar{A}$ 记自然的映射. 假定 $\mathcal{K} \otimes_{\mathcal{O}} A$ 的每一个本原幂等元在 $\hat{\mathcal{K}} \otimes_{\mathcal{O}} A$ 里也是本原的. 那么对任意 $\bar{A}$ 的幂等元 $\bar{\imath}$ 存在一个幂等元 $i \in A$ 使得 $s(i) = \bar{\imath}$. 而且，如果 $i' \in A$ 也是一个幂等元使得 $s(i') = \bar{\imath}$，就存在 $a \in A^*$ 使得 $i' = i^a$**

Corollary 2.18. *Let A be an $\mathcal{O}$-algebra. Denote by J the radical of A and by $s: A \to \bar{A} = A/J$ the canonical map. Assume that any primitive idempotent of $\mathcal{K} \otimes_{\mathcal{O}} A$ remains primitive in $\hat{\mathcal{K}} \otimes_{\mathcal{O}} A$. Then, for any idempotent $\bar{\imath}$ of $\bar{A}$, there is an idempotent $i \in A$ such that $s(i) = \bar{\imath}$. Moreover, if $i' \in A$ is also an idempotent such that $s(i') = \bar{\imath}$, then there is $a \in A^*$ such that $i' = i^a$.*

证明: 设 J 是一个 A 的相互正交本原幂等元的集合使得 $\sum_{j\in J} j = 1$ 并且设 $\bar{J}'$ 和 $\bar{J}''$ 是两个 $\bar{A}$ 的相互正交本原幂等元的集合使得

Proof: Let J be a set of pairwise orthogonal primitive idempotents of A such that $\sum_{j\in J} j = 1$, and choose two sets $\bar{J}'$ and $\bar{J}''$ of pairwise orthogonal primitive idempotents of $\bar{A}$ such that

2.18.1 $$\sum_{\bar{\jmath}\in\bar{J}'} \bar{\jmath} = \bar{\imath}, \quad \sum_{\bar{\jmath}\in\bar{J}''} \bar{\jmath} = 1 - \bar{\imath}.$$

由命题 2.16，对任意 $j \in J$，幂等元 $s(j)$ 在 $\bar{A}$ 里也是本原的；那么，由引理 2.17，存在双映射 $\tau: J \to \bar{J}' \cup \bar{J}''$ 和 $\bar{a} \in \bar{A}^*$ 使得对任意 $j \in J$ 有 $\tau(j) = s(j)^{\bar{a}}$. 可是，任意满足 $s(a) = \bar{a}$ 的元素 $a \in A$ 是可逆的，这是因为，由 Nakayama 引理，从 $\bar{A}\bar{a} = \bar{A}$ 可推出 $Aa = A$. 所以有

According to Proposition 2.16, the idempotent $s(j)$ is also primitive in $\bar{A}$ for any $j \in J$; then, according to Lemma 2.17, there are an element $\bar{a} \in \bar{A}^*$ and a bijective map $\tau: J \to \bar{J}' \cup \bar{J}''$ such that we have $\tau(j) = s(j)^{\bar{a}}$ for any $j \in J$. But any element $a \in A$ such that $s(a) = \bar{a}$ is inversible since, according to the Nakayama Lemma, from $\bar{A}\bar{a} = \bar{A}$, we can deduce that $Aa = A$. Hence, we have

2.18.2 $$s\Big(\sum_{j\in(\tau)^{-1}(\bar{J}')} j^a\Big) = \bar{\imath}.$$

最后, 如果 i 与 i' 是两个 A 的幂等元使得 $s(i)=\bar{\imath}=s(i')$, 那么一方面 $c=ii'+(1-i)(1-i')$ 也是可逆的, 因为从 $s(c)=1$ 可推出 $Ac=A$; 另一方面, 显然有 $ic=ii'=ci'$.

Finally, if i and i' are two idempotents of A such that $s(i)=\bar{\imath}=s(i')$ then, on the one hand, $c=ii'+(1-i)(1-i')$ is inversible since $s(c)=1$ implies that $Ac=A$; on the other hand, clearly we have $ic=ii'=ci'$.

3. $\mathcal{O}$-代数的点和点的重数

3. Points of the $\mathcal{O}$-algebras and Multiplicity of the Points

3.1. 虽然幂等元的提升不需要 $\mathcal{O}$ 完备性，但是今后总假定 $\mathcal{O}$ 是一个完备的离散赋值环使得其根上的商环是特征为 p 的域，记为 k. 而且，不用假定 $\mathcal{O}$ 的分式域 $\mathcal{K}$ 特征为零，即允许 $\mathcal{K} = \mathcal{O} = k$.

3.1. Although the lifting of idempotents need not the completeness of $\mathcal{O}$, from now on we assume that $\mathcal{O}$ is a *complete discrete valuation ring* with a *residue field* k of characteristic p. We need not assume that the *field of quotients* $\mathcal{K}$ of $\mathcal{O}$ has characteristic zero, allowing $\mathcal{K} = \mathcal{O} = k$.

3.2. 设 A 是一个 $\mathcal{O}$-代数，令 J 记 A 的根并令 $s: A \to A/J$ (或者 s_A) 记自然的同态; 这样 $\bar{A} = A/J$ 就是单代数的直积; 别忘了, $s^{-1}(\bar{A}^*) = A^*$ (见 2.15) 并记 $J^* = 1 + J \subset A^*$. 我们要考虑 A 的本原幂等元的共轭类; 也就是说，对任意 A 的本原幂等元 i，我们考虑

3.2. Let A be an $\mathcal{O}$-algebra and denote by J the radical of A and by $s: A \to A/J$ (or by s_A) the canonical homomorphism; thus, $\bar{A} = A/J$ is just a direct product of simple algebras; recall that $s^{-1}(\bar{A}^*) = A^*$ (see 2.15) and set $J^* = 1 + J \subset A^*$. We will consider the conjugacy classes of primitive idempotents of A; that is to say, for any primitive idempotent i of A, we consider

3.2.1 $$\alpha = \{i^a \mid a \in A^*\}$$

并且称 α 是一个 A 的点. 令 $\mathcal{P}(A)$ 记 A 的点的集合. 请注意，由推论 2.14, s 决定一个双映射 $\mathcal{P}(A) \leftrightarrow \mathcal{P}(\bar{A})$.

and say that α is a *point* of A. Denote by $\mathcal{P}(A)$ the set of points of A. Note that, according to Corollary 2.14, s determines a bijective map $\mathcal{P}(A) \to \mathcal{P}(\bar{A})$.

3.3. 而且，因为在单代数中本原幂等元都是彼此共轭的，所以 s 还决定一个从 $\mathcal{P}(A)$ 到 $\bar{A}$ 的单直积项的集合双映射; 令 $A(\alpha)$ 记对应 $\alpha \in \mathcal{P}(A)$ 的 $\bar{A}$ 的单直和项，再令 $s_\alpha: A \to A(\alpha)$ 记自然的同态; 也就是说，由 Wedderburn 定理，有 $A(\alpha) \cong \mathrm{End}_{D^\alpha}(V_\alpha)$，其中 V_α 是单 $A(\alpha)$-模与 $D^\alpha = \mathrm{End}_{A(\alpha)}(V_\alpha)$ 是可除 k-代数; 请注意，如果 k 是代数闭的，那么 $D^\alpha = k$.

3.3. Moreover, since in a simple algebra all the primitive idempotents are mutually conjugate, s also determines a bijective map between $\mathcal{P}(A)$ and the set of simple direct factors of $\bar{A}$; denote by $A(\alpha)$ the direct factor of $\bar{A}$ corresponding to $\alpha \in \mathcal{P}(A)$ and by $s_\alpha: A \to A(\alpha)$ the canonical algebra homomorphism; that is to say, by Wedderburn Theorem, we have $A(\alpha) \cong \mathrm{End}_{D^\alpha}(V_\alpha)$, where V_α is a simple $A(\alpha)$-module and $D^\alpha = \mathrm{End}_{A(\alpha)}(V_\alpha)$ is a division k-algebra; note that, whenever k is algebraically closed, we have $D^\alpha = k$.

命题 3.4. 如果I与I'是A的两个相互正交本原幂等元的集合使得$\sum_{i\in I} i = 1 = \sum_{i'\in I'} i'$, 那么存在$a \in A^*$使得$I' = I^a$.

Proposition 3.4. *If I and I' are two sets of pairwise orthogonal primitive idempotents such that $\sum_{i\in I} i = 1 = \sum_{i'\in I'} i'$, then there exists $a \in A^*$ such that $I' = I^a$.*

注意 3.5. 这样, $|I| = \mathrm{irk}(A)$ 不依赖I的选择. 设$\hat{\mathcal{K}}$是$\mathcal{K}$的有限扩张, 和令$\hat{\mathcal{O}}$在$\hat{\mathcal{K}}$中记$\mathcal{O}$上的整元的环, 从而它也是完备的离散赋值环; 一般来说, 对任意$i \in I$幂等元$1 \otimes i$在$\hat{A} = \hat{\mathcal{O}} \otimes_{\mathcal{O}} A$中可能不是本原的; 可是, 如果$k$是代数闭的, 那么有$\mathrm{irk}(A) = \mathrm{irk}(\hat{A})$, 这是因为对任意$\alpha \in \mathcal{P}(A)$, $A(\alpha)$是一个k上的矩阵代数(见3.3). 在任意情况里, 不难证明对适当的$\hat{\mathcal{K}}$, $\hat{A}(\hat{\alpha})$是一个$\hat{k}$上的矩阵代数, 其中$\hat{\alpha} \in \mathcal{P}(\hat{A})$与$\hat{k} = \hat{\mathcal{O}}/J(\hat{\mathcal{O}})$.

Remark 3.5. In particular, $|I| = \mathrm{irk}(A)$ does not depend on the choice of I. Let $\hat{\mathcal{K}}$ be a finite extension of $\mathcal{K}$ and denote by $\hat{\mathcal{O}}$ the integral closure of $\mathcal{O}$ in $\hat{\mathcal{K}}$, so that $\hat{\mathcal{O}}$ still is a complete discrete valuation ring; generally speaking, for any $i \in I$, the idempotent $1 \otimes i$ of $\hat{A} = \hat{\mathcal{O}} \otimes_{\mathcal{O}} A$ could be not primitive; but, if k is algebraically closed, then we have $\mathrm{irk}(A) = \mathrm{irk}(\hat{A})$ since, for any $\alpha \in \mathcal{P}(A)$, $A(\alpha)$ is a matrix algebra over k (see 3.3). In any case, it is not difficult to prove that, for a suitable finite extension $\hat{\mathcal{K}}$ and for any $\hat{\alpha} \in \mathcal{P}(\hat{A})$, $\hat{A}(\hat{\alpha})$ is a matrix algebra over $\hat{k} = \hat{\mathcal{O}}/J(\mathcal{O})$.

证明: 由推论 2.18, 如果ℓ属于$I \cup I'$, $s(\ell)$在$\bar{A}$里也是本原的; 所以, 由引理 2.17, 存在一个双映射$\tau: I \to I'$和$\bar{a} \in \bar{A}^*$使得对任意$i \in I$有$s(\tau(i)) = s(i)^{\bar{a}}$. 那么, 选择$a \in A$使得$s(a) = \bar{a}$, 并且考虑$c = \sum_{i\in I} ia\tau(i)$; 一方面可得到

Proof: According to Corollary 2.18, if ℓ belongs to $I \cup I'$ then $s(\ell)$ still is primitive in $\bar{A}$; consequently, by Lemma 2.17, there are a bijective map $\tau: I \to I'$ and an element $\bar{a} \in \bar{A}^*$ such that, for any $i \in I$, we have $s(\tau(i)) = s(i)^{\bar{a}}$. Then, choose $a \in A$ such that $s(a) = \bar{a}$, and consider $c = \sum_{i\in I} ia\tau(i)$; on the one hand , we obtain

3.4.1 $$s(c) = \sum_{i\in I} s(i)\bar{a}s(i)^{\bar{a}} = \bar{a}\,,$$

从而c是可逆的; 另一方面对任意$i \in I$有$ic = ia\tau(i) = c\tau(i)$.

and thus, c is inversible; on the other hand, for any $i \in I$ we get $ic = ia\tau(i) = c\tau(i)$.

3.6. 这个命题还可以用模论表述; 事实上, 这个命题和下述推论是相互等价的. **这里的A-模都是有限秩的自由$\mathcal{O}$-模.**

3.6. This proposition can be applied in module theory; actually, it is equivalent to the corollary below. *All the A-modules here are $\mathcal{O}$-free $\mathcal{O}$-modules of finite rank.*

推论 3.7. 设 M 是 A-模, 并且设 $\{M_i\}_{i\in I}$ 与 $\{M_{i'}\}_{i'\in I'}$ 是两个不可分解 A-模的集合使得

Corollary 3.7. *Let M be an A-module and consider two sets $\{M_i\}_{i\in I}$ and $\{M_{i'}\}_{i'\in I'}$ of indecomposable A-modules fulfilling*

3.7.1 $$\bigoplus_{i\in I} M_i \cong M \cong \bigoplus_{i'\in I'} M_{i'} .$$

存在一个双映射 $\tau: I \to I'$ 使得对任意 $i\in I$ 有 $M_{\tau(i)} \cong M_i$.

Then, there exists a bijective map $\tau: I \to I'$ such that we have $M_{\tau(i)} \cong M_i$ for any $i \in I$.

证明: 考虑 $B = \mathrm{End}_A(M)$; 那么, 对任意 $i \in I \cup I'$, 自然的映射 $M \to M_i \to M$ 决定 B 的元素 e_i ; 显然 $\{e_i\}_{i\in I}$ 与 $\{e_{i'}\}_{i'\in I'}$ 是两个 B 的相互正交本原幂等元的集合使得

Proof: Consider $B = \mathrm{End}_A(M)$; then, for any $i \in I \cup I'$, the composition of the canonical maps $M \to M_i$ and $M_i \to M$ determines an element e_i of B ; clearly, $\{e_i\}_{i\in I}$ and $\{e_{i'}\}_{i'\in I'}$ are two sets of pairwise orthogonal primitive idempotents of B fulfilling

3.7.2 $$\sum_{i\in I} e_i = \mathrm{id}_M = \sum_{i'\in I'} e_{i'} ;$$

那么, 由命题 3.4 存在双映射 $\tau: I \to I'$ 与 $f \in B^*$ 使得对任意 $i \in I$ 有 $e_{\tau(i)} = f{\cdot}e_i{\cdot}f^{-1}$, 从而 $f(M_i) = M_{\tau(i)}$.

hence, according to Proposition 3.4, there exists a bijective map $\tau: I \to I'$ such that, for any $i \in I$, we have $e_{\tau(i)} = f{\cdot}e_i{\cdot}f^{-1}$, which is easily translated to $f(M_i) = M_{\tau(i)}$.

推论 3.8. 两个 A 的幂等元 i 与 j 是相互共轭当且仅当存在 $a, b \in A$ 使得

Corollary 3.8. *Two idempotent i and j in A are conjugate if and only if there are $a, b \in A$ fulfilling*

3.8.1 $$i = ab , \quad j = ba .$$

注意 3.9. 那么还得到

Remark 3.9. Then, we also get

3.9.1 $$j = (jbi)(iaj) , \quad i = iabababi = (iaj)(jbi) ;$$

特别是, 如果 i 与 j 属于 A 的一个理想 I , 它们在 $\mathcal{O}{\cdot}1_A + I$ 里也是相互共轭的.

in particular, whenever i and j belong to an ideal I of A , they are still conjugate in the $\mathcal{O}$-subalgebra $\mathcal{O}{\cdot}1_A + I$.

证明: 如果存在 $c \in A^*$ 使得 $j = i^c$, 显然 $j = c^{-1}(ic)$ 并 $i = (ic)c^{-1}$. 假定存在 $a, b\in A$ 使得 $i = ab$ 与 $j = ba$; 那么, 考虑 A-模 Ai 与 Aj , 以及被下面的式子决定的 A-模的同态 $m_a: Ai \to Aj$ 与 $m_b: Aj \to Ai$

Proof: Whenever there is $c \in A^*$ such that $j = i^c$, clearly we have $j = c^{-1}(ic)$ and $i = (ic)c^{-1}$. Assume that there are $a, b \in A$ such that $i = ab$ and $j = ba$; then, consider the A-modules Ai and Aj , together with the A-module homomorphisms $m_a: Ai \to Aj$ and $m_b: Aj \to Ai$ determined by the formulæ

3.8.2 $$m_a(di) = dia = daj , \quad m_b(dj) = djb = dbi ,$$

其中 $d \in A$; 显然 $m_b \circ m_a = \mathrm{id}_{Ai}$ 并 $m_a \circ m_b = \mathrm{id}_{Aj}$, 这样得到 $Ai \cong Aj$. 从而, 由推论 3.7, 也得到 $A(1-i) \cong A(1-j)$; 也就是说, 存在一个 A-模的同构 $f: A \to A$ 使得

for any $d \in A$; clearly, we have $m_b \circ m_a = \mathrm{id}_{Ai}$ and $m_a \circ m_b = \mathrm{id}_{Aj}$; thus, we have $Ai \cong Aj$. Moreover, according to Corollary 3.7, we also have $A(1-i) \cong A(1-j)$; that is to say, there exists an A-module isomorphism $f: A \to A$ such that

3.8.3 $$f(Ai) = Aj\,, \quad f\big(A(1-i)\big) = A(1-j)\,;$$

对任意 $d \in A$ 必然有 $f(d) = df(1)$; 所以 A 的元素 $c = f(1)$ 是可逆的, 并且 $i^c \in Aj$ 与 $1-i^c \in A(1-j)$; 从而, 由 $1 = i^c + (1-i^c)$ 就得到 $i^c = j$.

for any $d \in A$, we obviously have $f(d) = df(1)$; consequently, the element $c = f(1)$ of A is inversible, and moreover, $i^c \in Aj$ and $1-i^c \in A(1-j)$; hence, since $1 = i^c + (1-i^c)$, we get $i^c = j$.

推论 3.10. 设 j 是一个 A 的幂等元, 并且设 I 与 I' 是两个 A 的相互正交本原幂等元的集合使得 $\sum_{i\in I} i = j = \sum_{i'\in I'} i'$. 那么对任意 $\alpha \in \mathcal{P}(A)$ 有

Corollary 3.10. *Let j be an idempotent of A, and consider two subsets I and I' of pairwise orthogonal primitive idempotents such that $\sum_{i\in I} i = j = \sum_{i'\in I'} i'$. Then, for any $\alpha \in \mathcal{P}(A)$, we have*

3.10.1 $$|\alpha \cap I| = |\alpha \cap I'|\,.$$

证明: 显然 jAj 是一个 A 的子 $\mathcal{O}$-代数 (可是它们的单位元素不一样); 在 jAj 里使用命题 3.4, 就能找到 $a_j \in (jAj)^*$ 使得 $I' = I^{a_j}$; 而且, 显然元素 $a = a_j + (1-j)$ 在 A 中是可逆的, 也得到 $I^a = I^{a_j} = I'$. 所以, $(\alpha \cap I)^a = \alpha \cap I'$.

Proof: Clearly, jAj is an $\mathcal{O}$-subalgebra of A (but they unity elements do not necessarily coincide); by applying Proposition 3.4 to jAj, we can find an element $a_j \in (jAj)^*$ such that $i' = I^{a_j}$; moreover, the element $a = a_j + (1-j)$ of A is clearly inversible, and we still have $I^a = I^{a_j} = I'$. Consequently, we get $(\alpha \cap I)^a = \alpha \cap I'$.

3.11. 对任意 A 的幂等元 j 与 A 的点 α, 如果 I 是相互正交本原幂等元的集合 I 使得 $\sum_{i\in I} i = j$, 那么令

3.11. For any idempotent j and any point α of A, choosing a subset I of pairwise orthogonal primitive idempotents such that $\sum_{i\in I} i = j$, we set

3.11.1 $$\mathrm{m}_\alpha^j = |\alpha \cap I|$$

并且说 m_α^j 是 j 上的 α 的重数.

and say that m_α^j is the *multiplicity* of α in j.

推论 3.12. 两个 A 的幂等元 j 与 j' 是相互共轭当且仅当对任意 $\alpha \in \mathcal{P}(A)$ 有 $\mathrm{m}_\alpha^j = \mathrm{m}_\alpha^{j'}$.

Corollary 3.12. *Two idempotents j and j' of A are conjugate to each other if and only if we have $\mathrm{m}_\alpha^j = \mathrm{m}_\alpha^{j'}$ for any $\alpha \in \mathcal{P}(A)$.*

证明: 设 I 与 I' 是两个 A 的相互正交本原幂等元的集合使得 $\sum_{i\in I} i = j$ 与 $\sum_{i'\in I'} i' = j'$. 如果对任意 $\alpha \in \mathcal{P}(A)$ 有 $\mathrm{m}_\alpha^j = \mathrm{m}_\alpha^{j'}$, 显然存在一个双映射 $\tau: I \to I'$ 使得每一个 $\alpha \in \mathcal{P}(A)$ 满足 $\tau(\alpha \cap I) = \alpha \cap I'$; 特别是, 对任意 $i \in I$ 存在 $a_i \in A^*$ 使得 $\tau(i) = i^{a_i}$. 令 $a = \sum_{i\in I} i a_i \tau(i)$ 与 $b = \sum_{i\in I} \tau(i)(a_i)^{-1} i$; 那么得到 $ab = i$ 并 $ba = j$; 所以, 由推论 3.7, j 与 j' 是相互共轭的.

Proof: Let I and I' be two sets of pairwise orthogonal primitive idempotents of A such that $\sum_{i\in I} i = j$ and $\sum_{i'\in I'} i' = j'$. If we have $\mathrm{m}_\alpha^j = \mathrm{m}_\alpha^{j'}$ for any $\alpha \in \mathcal{P}(A)$, then we clearly have a bijective map $\tau: I \to I'$ fulfilling $\tau(\alpha \cap I) = \alpha \cap I'$ for any $\alpha \in \mathcal{P}(A)$; in particular, for any $i \in I$ we have $a_i \in A^*$ such that $\tau(i) = i^{a_i}$. Consider $a = \sum_{i\in I} i a_i \tau(i)$ and $b = \sum_{i\in I} \tau(i)(a_i)^{-1} i$; then, we get $ab = i$ and $ba = j$; consequently, according to Corollary 3.7, j and j' are conjugate to each other.

3.13. 这样, 任意 A 的幂等元的共轭类 κ 决定一个映射

3.13. In this way, any conjugacy class of idempotents of A determines a map

3.13.1 $$\mathrm{m}: \mathcal{P}(A) \longrightarrow \mathbb{N}$$

并且 m 也决定 κ. 更一般的说, 我们把任意一个这样的映射都考虑和称为 A 的除子; 以后, 如果 ω 与 ω' 是两个 A 的除子使得 $\omega'(\alpha) \le \omega(\alpha)$, 其中 $\alpha \in \mathcal{P}(A)$, 那么我们说 ω 包含 ω', 并且我们记为 $\omega' \subset \omega$; 类似地可定义 $\omega \cup \omega'$, $\omega \cap \omega'$, 它们分别是既包含 ω 与 ω' 的最低 A-除子, 和既包含在 ω 与 ω' 中的最高 A-除子; 进一步, 记

and m determines κ too. More generally, we consider all these maps and we call each of them a *divisor* of A; moreover, if ω and ω' are two divisors of A such that $\omega'(\alpha) \le \omega(\alpha)$ for any $\alpha \in \mathcal{P}(A)$, then we say that ω *contains* ω' and, in this case, we write $\omega' \subset \omega$; similarly, we can define $\omega \cup \omega'$ and $\omega \cap \omega'$, they respectively are the smallest divisor of A containing ω and ω', and the biggest divisor of A contained in ω and ω'; furthermore, we set

3.13.2 $$\langle \omega, \omega' \rangle = \sum_{\alpha\in\mathcal{P}(A)} \omega(\alpha)\omega'(\alpha).$$

3.14. 令 $\mathcal{D}(A)$ 记 A 的除子的集合; 认为每一个 $\alpha \in \mathcal{P}(A)$ 与其除子一致; 而且对任意 A 的幂等元 i, 令 μ_A^i 记被 i 决定的 A-除子; 这样, 就得到

3.14. Denote by $\mathcal{D}(A)$ the set of divisors of A; we identify any $\alpha \in \mathcal{P}(A)$ with the divisor determined by α; moreover, for any idempotent i of A, we denote by μ_A^i the A-divisor determined by i; thus, we get

3.14.1 $$\mu_A^i = \sum_{\alpha\in\mathcal{P}(A)} \mathrm{m}_\alpha^i \cdot \alpha.$$

请注意,在A中用元素右乘可以诱导下面的$\mathcal{O}$-代数同构

Note that the multiplication on the right in A induces the following $\mathcal{O}$-algebra isomorphism

3.14.2 $$iAi \cong \mathrm{End}_A(Ai)^\circ \,;$$

而且,只要在每一个 A 的点 α 中选择一个元素 i_α 就得到下面的 $\mathcal{O}$-模同构

furthermore, the choice of an element i_α in any point α of A determines the following A-module isomorphism

3.14.3 $$Ai \cong \bigoplus_{\alpha\in\mathcal{P}(A)} (Ai_\alpha)^{\mathrm{m}_\alpha^i} \,.$$

3.15. 一般的说,对任意 A 的除子 ω, 我们考虑下面的 A-模与 $\mathcal{O}$-代数

3.15. In general, for any divisor ω of A, we consider the following A-module and $\mathcal{O}$-algebra

3.15.1 $$M_\omega = \bigoplus_{\alpha\in\mathcal{P}(A)} (Ai_\alpha)^{\omega(\alpha)} \,, \quad A_\omega = \mathrm{End}_A(M_\omega)^\circ \,;$$

当然,不同的选择 $\{i'_\alpha\}_{\alpha\in\mathcal{P}:(A)}$ 产生不同的A-模

of course, a different choice $\{i'_\alpha\}_{\alpha\in\mathcal{P}:(A)}$ produces a different A-module

3.15.2 $$M'_\omega = \bigoplus_{\alpha\in\mathcal{P}(A)} (Ai'_\alpha)^{\omega(\alpha)}$$

可是它们是彼此同构的,并且任意 A-模同构 $f\colon M_\omega \cong M_{\omega'}$ 决定一个 $\mathcal{O}$-代数同构

but they are mutually isomorphic, and any A-module isomorphism $f\colon M_\omega \cong M_{\omega'}$ determines an $\mathcal{O}$-algebra isomorphism

3.15.3 $$\mathrm{End}_A(M_\omega) \cong \mathrm{End}_A(M'_\omega)$$

称为可嵌入同构. 特别是,对任意 A 的幂等元 ℓ, 同构 3.14.2 和 3.14.3 决定下面的 $\mathcal{O}$-代数同构

called an *embeddable isomorphism*. In particular, for any idempotent ℓ of A, isomorphisms 3.14.2 and 3.14.3 determine the following $\mathcal{O}$-algebra isomorphism

3.15.4 $$A_{\mu_A^\ell} \cong \ell A\ell \,.$$

3.16. 请注意,如果 A 的点 α 满足 $\omega(\alpha)$ 不零,它就决定一个 A_ω 的点;这是因为,如果 $i\in\alpha$, 那么自然的 A-模同态

3.16. Note that, any point α of A such that $\omega(\alpha)$ is not zero determines a point of A_ω; this comes from the fact that, if $i\in\alpha$, the natural A-module homomorphisms

3.16.1 $$e_i : M_\omega \longrightarrow Ai \longrightarrow M_\omega$$

都决定 A_ω 的相互共轭本原幂等元; 而且, 因为 $e_i(M_\omega) = Ai$ 所以不难验证 A-模同态的限制以及 iAj 的右边乘法可以决定下面的同构

determine primitive idempotents mutually conjugated in A_ω; moreover, since $e_i(M_\omega) = Ai$, it is easy to check that the restriction of A-module homomorphisms and the multiplication on the right by iAj induce

3.16.2 $$e_j A_\omega e_i \cong \mathrm{Hom}_A(Ai, Aj) \cong iAj$$

其中$j \in \beta$ 并β 是一个 A 上的点. 这样, 把 $\mathcal{P}(A_\omega)$ 与满足 $0 \neq \omega(\alpha)$ 的 $\alpha \in \mathcal{P}(A)$ 的集合看作一致; 进一步, 再把在 $\mathcal{P}(A) - \mathcal{P}(A_\omega)$ 中取值零的 A-除子的集合与 $\mathcal{D}(A_\omega)$ 看作一致; 特别是, 任意满足 $\omega' \subset \omega$ 的 $\omega' \in \mathcal{D}(A)$ 都属于 $\mathcal{D}(A_\omega)$. 只要通过可嵌入同构所有这两个等同一致就不依赖 A_ω 的选择. 请注意, 已经知道有 $\mathcal{P}(A_\omega) = \mathcal{P}(A)$ 当且仅当 A_ω 是 Morita 等价于 A.

where $j \in \beta$, β being a point of A. Thus, we may identify $\mathcal{P}(A_\omega)$ with the set of $\alpha \in \mathcal{P}(A)$ such that $\omega(\alpha) \neq 0$; more generally, we can also identify the set of divisors of A which vanish over $\mathcal{P}(A) - \mathcal{P}(A_\omega)$ with $\mathcal{D}(A_\omega)$; in particular, all the divisors ω' of A such that $\omega' \subset \omega$ belong to $\mathcal{D}(A_\omega)$. Both identifications do not depend on the choice of A_ω provided, from one choice to another, we go throughout some *embeddable* isomorphism. Note that, as it is well-known, we have $\mathcal{P}(A_\omega) = \mathcal{P}(A)$ if and only if A_ω and A are Morita equivalent.

3.17. 这样, 对任意除子 $\omega \in \mathcal{D}(A)$ 考虑基数为 $\omega(\alpha)$ 的有限集合 Ω_α, 而且所有从 $\Omega_\alpha \times \Omega_\beta$ 到 $i_\alpha A i_\beta$ 的映射的 $\mathcal{O}$-模 $(i_\alpha A i_\beta)^{\Omega_\alpha \times \Omega_\beta}$, 其中 α, β 跑遍 $\mathcal{P}(A)$; 那么 A_ω 就成为

3.17. That is to say, for any divisor $\omega \in \mathcal{D}(A)$ consider finite sets Ω_α of cardinal $\omega(\alpha)$, and moreover the $\mathcal{O}$-module $(i_\alpha A i_\beta)^{\Omega_\alpha \times \Omega_\beta}$ of all the maps from the set $\Omega_\alpha \times \Omega_\beta$ to the $\mathcal{O}$-algebra $i_\alpha A i_\beta$, where α and β run over $\mathcal{P}(A)$; then, A_ω becomes

3.17.1 $$A_\omega = \bigoplus_{\alpha,\beta \in \mathcal{P}(A)} (i_\alpha A i_\beta)^{\Omega_\alpha \times \Omega_\beta},$$

这里乘法定义如下: 对任意 $(i_\alpha A i_\beta)^{\Omega_\alpha \times \Omega_\beta}$, $(i_\alpha A i_\beta)^{\Omega_\alpha \times \Omega_\beta}$ 的分别元素 $a_{\alpha,\beta}$, $b_{\delta,\gamma}$, 如果 $\beta \neq \delta$ 那么规定 $a_{\alpha,\beta} b_{\delta,\gamma} = 0$, 而且如果 $\beta = \delta$ 那么 $a_{\alpha,\beta} b_{\beta,\gamma}$ 在 $(i_\alpha A i_\gamma)^{\Omega_\alpha \times \Omega_\gamma}$ 中对任意 $(u, w) \in \Omega_\alpha \times \Omega_\gamma$ 的取值是

where the product is defined as follows: for any $a_{\alpha,\beta}$ and $b_{\alpha,\beta}$ respectively belonging to $(i_\alpha A i_\beta)^{\Omega_\alpha \times \Omega_\beta}$ and $(i_\delta A i_\gamma)^{\Omega_\delta \times \Omega_\gamma}$, if $\beta \neq \delta$ then we just set $a_{\alpha,\beta} b_{\delta,\gamma} = 0$, whereas if $\beta = \delta$ then $a_{\alpha,\beta} b_{\delta,\gamma}$ is the element of $(i_\alpha A i_\gamma)^{\Omega_\alpha \times \Omega_\gamma}$ which, for any $u \in \Omega_\alpha$ and any $w \in \Omega_\gamma$, takes the value

3.17.2 $$(a_{\alpha,\beta} b_{\beta,\gamma})(u, w) = \sum_{v \in \Omega_\beta} a_{\alpha,\beta}(u, v) b_{\beta,\gamma}(v, w)$$

其中 $\alpha,\beta,\gamma,\delta \in \mathcal{P}(A)$. 这样, 如果 $\theta \in \mathcal{D}(A_\omega)$ 那么同构 3.16.2 显然决定一个 $\mathcal{O}$-代数同构如下, 也称可嵌入的同构

where $\alpha,\beta,\gamma,\delta \in \mathcal{P}(A)$. Consequently, if $\theta \in \mathcal{D}(A)$ then isomorphism 3.16.2 clearly determines the following $\mathcal{O}$-algebra isomorphism, still called *embeddable isomorphism*

3.17.3 $$(A_\omega)_\theta \cong A_\theta\,.$$

注意 3.18. 选择 $n \in \mathbb{N}$ 使得 $n\mu_A^1$ 包含 ω, 并且令 $E = \mathrm{End}_\mathcal{O}(\mathcal{O}^n)$; 如果 e 是一个 E 的本原幂等元, 可见 $eEe \cong \mathcal{O}$; 所以得到 (见 3.15.4)

Remark 3.18. Choose $n \in \mathbb{N}$ such that $n\mu_A^1$ contains ω, and consider $E = \mathrm{End}_\mathcal{O}(\mathcal{O}^n)$; if e is a primitive idempotent of E, obviously we have $eEe \cong \mathcal{O}$; consequently, we get (see 3.15.4)

3.18.1 $$A \cong (1\otimes e)(A\otimes_\mathcal{O} E)(1\otimes e) \cong (A\otimes_\mathcal{O} E)_{\mu^{1\otimes e}_{A\otimes_\mathcal{O} E}}\quad ;$$

那么 ω 与 μ_A^1 成为 $A\otimes_\mathcal{O} E$ 的除子 (见 3.15), 并且显然有 $n\mu_A^1 = \mu^{1\otimes 1}_{A\otimes_\mathcal{O} E}$; 所以存在一个 $A\otimes_\mathcal{O} E$ 的幂等元 ℓ 使得 $\omega = \mu^\ell_{A\otimes_\mathcal{O} E}$, 从而同构 3.15.4 和 3.17.3 决定可嵌入同构如下

then, ω and μ_A^1 become divisors of $A\otimes_\mathcal{O} E$ (see 3.15), and moreover clearly we have $n\mu_A^1 = \mu^{1\otimes 1}_{A\otimes_\mathcal{O} E}$; hence, there exists an idempotent ℓ of $A\otimes_\mathcal{O} E$ such that $\omega = \mu^\ell_{A\otimes_\mathcal{O} E}$ and therefore isomorphisms 3.15.4 and 3.17.3 determine the embeddable isomorphism

3.18.2 $$A_\omega \cong \ell(A\otimes_\mathcal{O} E)\ell\,.$$

这样, 考虑一个 A 的除子 ω 的时候, 只要用造当的 $A\otimes_\mathcal{O} E$ 代替 A, 我们就能假定 $\omega = \mu_A^i$, 其中 i 是一个 A 的幂等元.

Thus, when considering a divisor ω of A, we may assume that we have $\omega = \mu_A^i$ for a suitable idempotent i of A; indeed, as above it suffices to replace A by $A\otimes_\mathcal{O} E$.

命题 3.19. **设 i 是一个 A 的幂等元. 任意 A 的幂等元的共轭类包含一个元素 j 使得 $ij = ji$ 并且 $\mu_A^{ij} = \mu_A^i \cap \mu_A^j$. 那么, $i - ij$ 与 $j - ij$ 也是幂等元, 并且有**

Proposition 3.19. *Let i be an idempotent of A. Any conjugacy class of idempotents of A contains an element j such that $ij = ji$, and moreover we have $\mu_A^{ij} = \mu_A^i \cap \mu_A^j$. Then $i - ij$ and $j - ij$ are idempotents too and we have*

3.19.1 $$\mu_A^i = \mu_A^{i-ij} + \mu_A^{ij}\,,\quad \mu_A^j = \mu_A^{j-ij} + \mu_A^{ij}\,.$$

证明: 设 I' 与 I'' 是两个 A 的相互正交本原幂等元的集合使得

Proof: Let I' and I'' be two subsets of pairwise orthogonal primitive idempotents of A such that

3.19.2 $$i = \sum_{i'\in I'} i'\,,\quad 1 - i = \sum_{i''\in I''} i''$$

并令$I=I'\cup I''$；设j'是A的幂等元；对任意$\alpha\in\mathcal{P}(A)$显然存在$J_\alpha\subset\alpha\cap I$使得$|J_\alpha|=\mu_A^{j'}(\alpha)$与$|J_\alpha\cap I'|=(\mu_A^i\cap\mu_A^{j'})(\alpha)$；令

and set $I = I'\cup I''$; let j' be an idempotent of A; for any $\alpha\in\mathcal{P}(A)$ clearly we can find a subset $J_\alpha\subset\alpha\cap I$ such that $|J_\alpha| = \mu_A^{j'}(\alpha)$ and $|J_\alpha\cap I'| = (\mu_A^i\cap\mu_A^{j'})(\alpha)$; consider

3.19.3 $$J=\bigcup_{\alpha\in\mathcal{P}(A)} J_\alpha\,,\quad j=\sum_{\ell\in J}\ell\,.$$

由推论 3.12, j 和 j' 是相互共轭, 而且显然

According to Corollary 3.12, j and j' are conjugate to each other, and moreover we have

3.19.4 $$ij=\sum_{\ell\in I'\cap J}\ell=ji\,,\quad i-ij=\sum_{\ell\in I'-J}\ell\,,\quad j-ij=\sum_{\ell\in J-I'}\ell\,.$$

现在, 对任意 $\alpha\in\mathcal{P}(A)$, 只要使用式子如下, 就可证明式子 3.19.1

Now, for any $\alpha\in\mathcal{P}(A)$, it suffices to apply next equalities to get equalities 3.19.1.

3.19.5 $$\begin{aligned}|\alpha\cap I'| &= |\alpha\cap(I'-J)|+|\alpha\cap(I'\cap J)|\\ |\alpha\cap J| &= |\alpha\cap(J-I')|+|\alpha\cap(I'\cap J)|\,.\end{aligned}$$

3.20. 设 B 是 $\mathcal{O}$-代数; 当然, 对任意幂等元 $i\in A$ 与幂等元 $j\in B$, 张量积 $i\otimes j$ 是 $A\otimes_{\mathcal{O}}B$ 的幂等元, 而且有

3.20. Let B be an $\mathcal{O}$-algebra; obviously, for any idempotents $i\in A$ and $j\in B$, the tensor product $i\otimes j$ is an idempotent of $A\otimes_{\mathcal{O}}B$, and moreover we have

3.20.1 $$(i\otimes j)(A\otimes_{\mathcal{O}}B)(i\otimes j)\cong iAi\otimes_{\mathcal{O}}jBj\,;$$

进一步, 如果 i',j' 分别是 i,j 的共轭, 那么$i'\otimes j'$ 在$A\otimes_{\mathcal{O}}B$中也是 $i\otimes j$ 的共轭; 也就是说, 存在唯一双线性映射

furthermore, if i' and j' respectively are conjugate to i and j, then $i'\otimes j'$ is conjugate to $i\otimes j$; that is to say, there exists a unique bilinear map

3.20.2 $$t_{A,B}:\mathcal{D}(A)\times\mathcal{D}(B)\longrightarrow\mathcal{D}(A\otimes_{\mathcal{O}}B)$$

使得 $t_{A,B}(\mu_A^i,\mu_B^j)=\mu_{A\otimes_{\mathcal{O}}B}^{i\otimes j}$; 令 $\omega\otimes\theta$ 记 $\mathcal{D}(A)\times\mathcal{D}(B)$ 的元素 (ω,θ) 的像. 那么, 同构式 3.20.1 决定下面的可嵌入同构 (见注意 3.18)

such that $t_{A,B}(\mu_A^i,\mu_B^j)=\mu_{A\otimes_{\mathcal{O}}B}^{i\otimes j}$; we denote by $\omega\otimes\theta$ the image of $(\omega,\theta)\in\mathcal{D}(A)\times\mathcal{D}(B)$. Then, isomorphism 3.20.1 determines the following *embeddable isomorphism* (see Remark 3.18)

3.20.3 $$(A\otimes_{\mathcal{O}}B)_{\omega\otimes\theta}\cong A_\omega\otimes_{\mathcal{O}}B_\theta\,.$$

3.21. 另一方面设$f{:}A\to B$是 $\mathcal{O}$-代数同态; 再次, 对任意幂等元 $i\in A$, 当然 $f(i)$ 是 B

3.21. On the other hand, let $f{:}\,A\to B$ be an $\mathcal{O}$-algebra homomorphism; as above, for any idempotent i of A, $f(i)$ is an idem-

的幂等元, 而且 f 决定 $\mathcal{O}$-代数同态

potent of B; moreover, f determines an $\mathcal{O}$-algebra homomorphism

3.21.1 $$iAi \longrightarrow f(i)Bf(i);$$

如果 $i' = i^a$, 其中 a 是一个 A 的可逆元素, 那么下面的元素

if we have $i' = i^a$ for some inversible element a of A, then the element

3.21.2 $$b = f(a) + \big(1_B - f(1_A)\big)$$

在 B 中也是可逆的, 并且显然 $f(i') = f(i)^b$; 也就是说, f 从 A 的幂等元的共轭类的集合到 B 的幂等元的共轭类的集合决定一个映射. 事实上, f 显然决定唯一线性映射

still is inversible in B, and moreover we get $f(i') = f(i)^b$; in other terms, f determines a map from the set of conjugacy classes of idempotents of A to the set of conjugacy classes of idempotents of B. In fact, f clearly determines a unique linear map

3.21.3 $$\mathrm{res}_f : \mathcal{D}(A) \longrightarrow \mathcal{D}(B)$$

使得 $\mathrm{res}_f(\mu_A^i) = \mu_B^{f(i)}$, 其中 i 是 A 的幂等元.

such that we have $\mathrm{res}_f(\mu_A^i) = \mu_B^{f(i)}$ for any idempotent i of A.

3.22. 那么, 由注意 3.18, 对任意 $\omega \in \mathcal{D}(A)$ 式子 3.21.1 决定下面的自然幺同态

3.22. Then, by Remark 3.18, for any $\omega \in \mathcal{D}(A)$, homomorphism 3.21.1 determines a natural unitary homomorphism

3.22.1 $$f_\omega : A_\omega \longrightarrow B_{\mathrm{res}_f(\omega)},$$

并且对任意 $\theta \in \mathcal{D}(B)$ 令 (见 3.13.2)

and, for any $\theta \in \mathcal{D}(B)$, we set (see 3.13.2)

3.22.2 $$\mathrm{m}(f)_\omega^\theta = \langle \mathrm{res}_f(\omega), \theta \rangle,$$

或者 f 已经知道的时候记 m_ω^θ; 特别是, 只要选择 $B = A(\alpha)$ 并选择 $f = s_\alpha$, 其中 $\alpha \in \mathcal{P}(A)$ 使得 $\alpha \subset \omega$, 这个同态就显然决定下面的 k-代数同构

or simply m_ω^θ whenever f is clearly known; in particular, for any $\alpha \in \mathcal{P}(A)$ such that $\alpha \subset \omega$, choosing $B = A(\alpha)$ and $f = s_\alpha$, clearly the above homomorphism determines the following k-algebra homomorphism

3.22.3 $$A_\omega(\alpha) \cong A(\alpha)_{\omega_\alpha},$$

其中 $\omega_\alpha = \mathrm{res}_{s_\alpha}(\omega)$. 最后, 如果 C 是另一个 $\mathcal{O}$-代数并且 $g{:}\,B \to C$ 是另一个 $\mathcal{O}$-代数同态, 就有

where we set $\omega_\alpha = \mathrm{res}_{s_\alpha}(\omega)$. Finally, if C is another $\mathcal{O}$-algebra and $g{:}\,B \to C$ another $\mathcal{O}$-algebra homomorphism, we have

3.22.4 $$\mathrm{res}_g \circ \mathrm{res}_f = \mathrm{res}_{g\circ f}.$$

命题 3.23. 设 $f: A \to B$ 是一个 $\mathcal{O}$-代数满同态, 并设 I 是一个 A 的理想. 如果 $\alpha \in \mathcal{P}(A)$ 满足 $\alpha \not\subset \operatorname{Ker}(f)$, 那么 $f(\alpha)$ 是一个 B 的点, 并且 f 决定一个 k-代数同构

Proposition 3.23. *Let $f: A \to B$ be a surjective $\mathcal{O}$-algebra homomorphism and I an ideal of A. If we have a point $\alpha \in \mathcal{P}(A)$ such that $\alpha \not\subset \operatorname{Ker}(f)$, then $f(\alpha)$ is a point of B and f determines a k-algebra isomorphism*

3.23.1 $$f^\alpha: A(\alpha) \cong B\big(f(\alpha)\big)$$

使得 $f^\alpha \circ s_\alpha = s_{f(\alpha)} \circ f$. 而且, 有 $f(\alpha) \subset f(I)$ 当且仅当有 $\alpha \subset I$. 进一步, 这个对应决定下面的单射

such that $f^\alpha \circ s_\alpha = s_{f(\alpha)} \circ f$. Moreover, we have $f(\alpha) \subset f(I)$ if and only if we have $\alpha \subset I$. Furthermore, this correspondence determines an injective map

3.23.2 $$f^{\mathcal{P}}: \mathcal{P}(B) \longrightarrow \mathcal{P}(A).$$

证明: 因为 $s_B \circ f$ 也是满射的, 所以得到 $f\big(J(A)\big) \subset J(B)$, 从而 f 也决定一个 $\mathcal{O}$-代数满同态 $\bar{f}: \bar{A} \to \bar{B}$ 其中 $\bar{B} = B/J(B)$; 那么, 只要使用定理 2.13 并且使用 $\bar{A}$ 与 $\bar{B}$ 是两个单代数的直积的事实, 就可完证明.

Proof: Since $s_B \circ f$ is also surjective, we clearly get $f\big(J(A)\big) \subset J(B)$, an therefore f also determines a surjective $\mathcal{O}$-algebra homomorphism $\bar{f}: \bar{A} \to \bar{B}$ where we set $\bar{B} = B/J(B)$; then, to complete the proof, it suffices to apply Theorem 2.13 and the fact that $\bar{A}$ and $\bar{B}$ are direct products of simple algebras.

4. N-内 G-代数上的除子

4. Divisors on N-interior G-algebras

4.1. 设 G 是一个有限群; 我们的主要意图是研究 G 的群代数 $\mathcal{O}G$, 和任意 $\mathcal{O}G$-模 M, 也就是说, 任意群同态 $G \to \mathrm{End}_{\mathcal{O}}(M)^*$, 其中 M 是一个自由 $\mathcal{O}$-模. 更一般来说, 能考虑任意群的同态 $G \to A^*$, 其中 A 是 $\mathcal{O}$-代数; 例如, 如果 $\varphi\colon G \to \mathrm{Aut}(B)$ 是在一个 $\mathcal{O}$-代数 B 上的一个 G-作用, 那么 B 上的 G 的群代数 BG 是基 G 上的自由 B-模 $\oplus_{x\in G} Bx$, 跟下面的乘法

4.1. Let G be a finite group; our main purpose is the study of the group algebra $\mathcal{O}G$ and of any $\mathcal{O}G$-module M or, equivalently, any group homomorphism $G \to \mathrm{End}_{\mathcal{O}}(M)^*$, where M is an $\mathcal{O}$-free $\mathcal{O}$-module. More generally, we may consider any group homomorphism $G \to A^*$, where A is an $\mathcal{O}$-algebra; for instance, whenever $\varphi\colon G \to \mathrm{Aut}(B)$ is an action of G on an $\mathcal{O}$-algebra B, the *group algebra BG of G over B* is the free B-module $\oplus_{x\in G} Bx$ over the set G, endowed with the following product

4.1.1 $$\Big(\sum_{x\in G} a_x x\Big)\Big(\sum_{y\in G} b_y y\Big) = \sum_{x,y\in G} a_x \varphi(x)(b_y)xy\,.$$

但是, 对 G 的任意子群 H, 这个同态一定决定另一个群同态 $C_G(H) \to (A^H)^*$ 还决定在 A^H 上的一个 $N_G(H)$-作用, 其中 A^H 是 H 在 A 中的中心化子.

But, for any subgroup H of G, such a group homomorphism determines both, another group homomorphism $C_G(H) \to (A^H)^*$ and an action of $N_G(H)$ on A^H, where A^H denotes the subalgebra of all the elements fixed by H.

4.2. 这样, 最好是从开始已经考虑一个 G 的正规子群 N 和下述情况. 称 A 是一个 N-内 G-代数如果在 $\mathcal{O}$-代数 A 上固定两个群同态

4.2. Thus, the best solution is to start with a normal subgroup N of G and with the following data. An *N-interior G-algebra* is an $\mathcal{O}$-algebra A endowed with two group homomorphisms

4.2.1 $$\varphi : G \longrightarrow \mathrm{Aut}(A)\,, \quad \psi : N \longrightarrow A^*$$

使得对任意 $x \in G$ 与 $y \in N$ 与 $a \in A$ 有

such that, for any $x \in G$, $y \in N$ and $a \in A$, we have

4.2.2 $$(y{\cdot}a)^x = y^x{\cdot}a^x\,, \quad a^y = y^{-1}{\cdot}a{\cdot}y\,,$$

其中 $y{\cdot}a$ 与 $a{\cdot}y$ 分别记 $\psi(y)a$ 与 $a\psi(a)$, 而 a^x 记 $\varphi(x)^{-1}(a)$; 请注意, 有 $\psi(y) = y{\cdot}1 = 1{\cdot}y$, 和 A 也是 $\mathcal{O}(N\times N)G$-模, 其中 $\mathcal{O}(N\times N)G$ 是 $\mathcal{O}(N\times N)$

where $y{\cdot}a$ and $a{\cdot}y$ respectively denote $\psi(y)a$ and $a\psi(y)$, and a^x denote $\varphi(x)^{-1}(a)$; note that we have $\psi(y) = y{\cdot}1 = 1{\cdot}y$ and that A still has an structure of $\mathcal{O}(N\times N)G$-module, where $\mathcal{O}(N\times N)G$ is the group

上的 G 的群代数. 这样, 对 G 的任意子群 H, A^H 成为一个 $C_N(H)$-内 $N_G(H)$-代数.

algebra of G over $\mathcal{O}(N \times N)$. In this way, for any subgroup H of G, A^H becomes a $C_N(H)$-interior $N_G(H)$-algebra.

4.3. 设 B 是另一个 N-内 G-代数, 张量积 $A \otimes_{\mathcal{O}} B$ 显然有一个自然的 N-内 G-代数结构. 另一方面, 一个 N-内 G-代数同态 $f: A \to B$ 是指一个 $\mathcal{O}$-代数同态使得对任意 $x \in G$ 与 $y \in N$ 与 $a \in A$ 有 $f(a^x) = f(a)^x$, 以及 $f(y{\cdot}a) = y{\cdot}f(a)$.

4.3. Let B be another N-interior G-algebra; it is quite clear that the tensor product $A \otimes_{\mathcal{O}} B$ has a canonical N-interior G-algebra structure. On the other hand, an N-interior G-algebra homomorphism $f: A \to B$ is an $\mathcal{O}$-algebra homomorphism such that, for any $x \in G$ and any $y \in N$, we have $f(a^x) = f(a)^x$ and $f(y{\cdot}a) = y{\cdot}f(a)$.

4.4. 今后, 总设 N 是 G 的正规子群, 而 A 是一个 N-内 G-代数. 对 G 的任意子群 H, 令 $\mathrm{Res}^G_H(A)$ 记相应的 $(N \cap H)$-内 H-代数. 另一方面, 考虑上术 A 的子 $\mathcal{O}$-代数 A^H 和考虑其它的点, 称为 A 上的 H-点, 也它的除子, 称为 A 上的 H-除子; 令 $\mathcal{P}_A(H)$ 与 $\mathcal{D}_A(H)$ 记对应的集合; 特别是, 对任意 A 上的 H-点 β 我们称偶对 (H,β) 是一个 A 上的点群, 而简写为 H_β, 我们还简写

4.4. Moreover, always considering a normal subgroup N of G and an N-interior G-algebra A, for any subgroup H of G, we denote by $\mathrm{Res}^G_H(A)$ the corresponding $(N \cap H)$-interior H-algebra. On the other hand, as above, consider the $\mathcal{O}$-subalgebra A^H together with all its points, called *points of H on A*, and all its divisors, called *divisors of H on A*; denote by $\mathcal{P}_A(H)$ and $\mathcal{D}_A(H)$ the corresponding sets; in particular, for any point β of H on A, we call the pair (H,β) a *pointed group on A*, always writing H_β instead of (H,β), and also writing

4.4.1 $$A^H(\beta) = A(H_\beta)\,;$$

别忘了, $s_\beta: A^H \to A(H_\beta)$ 记对应的自然同态 (见 3.3). 这样, $A(H_\beta)$ 是一个 $C_N(H)$-内 $N_G(H_\beta)$-代数, 其中 $N_G(H_\beta)$ 在 $N_G(H)$ 里是 β 的稳定化子.

recall that $s_\beta: A^H \to A(H_\beta)$ denotes the corresponding canonical homomorphism (see 3.3). Hence, $A(H_\beta)$ has an structure of $C_N(H)$-interior $N_G(H_\beta)$-algebra, where $N_G(H_\beta)$ is the stabilizer of β in $N_G(H)$.

4.5. 另一方面, 已经知道, A 上的一个 H-除子 $\omega \in \mathcal{D}_A(H)$ 决定下面的 $\mathcal{O}$-代数

4.5. On the other hand, we have already seen that any divisor $\omega \in \mathcal{D}_A(H)$ of H on A determines the following $\mathcal{O}$-algebra

4.5.1 $$(A^H)_\omega = \mathrm{End}_{A^H}\Big(\bigoplus_{\beta \in \mathcal{P}_A(H)} (A^H i_\beta)^{\omega(\beta)}\Big)^\circ,$$

其中 i_β 是一个 β 的元素; 现在, 我们也能造下面的 $(N\cap H)$-内 H-代数

where i_β is an element of β ; this time, we also can construct the following $(N\cap H)$-interior H-algebra

4.5.2 $$A_\omega = \mathrm{End}_A\Big(\bigoplus_{\beta\in\mathcal{P}_A(H)} (Ai_\beta)^{\omega(\beta)}\Big)^\circ .$$

这是因为在A里的左边乘以A与在 A 里的右边乘以N与在A上的G-作用, 三者一起决定一个 A 上的 $(A\otimes_\mathcal{O}\mathcal{O}N)G$-模结构, 其中 $(A\otimes_\mathcal{O}\mathcal{O}N)G$ 是$A\otimes_\mathcal{O}\mathcal{O}N$上的 G 的群代数; 所以, 直和如下

Indeed, in A, the multiplication on the left by A, the multiplication on the right by N and the action of G all together determine an $(A\otimes_\mathcal{O}\mathcal{O}N)G$-module structure on A, where $(A\otimes_\mathcal{O}\mathcal{O}N)G$ is the group algebra of G over $A\otimes_\mathcal{O}\mathcal{O}N$; hence, the direct sum

4.5.3 $$M_\omega = \bigoplus_{\beta\in\mathcal{P}_A(H)} (Ai_\beta)^{\omega(\beta)}$$

仍然保持一个 $(A\otimes_\mathcal{O}\mathcal{O}N_H)H$-模结构, 其中 $N_H = N\cap H$; 也就是说, 我们得到如下一个 $\mathcal{O}$-代数同态

always holds an $(A\otimes_\mathcal{O}\mathcal{O}N_H)H$-module structure, where $N_H = N\cap H$; in other terms, we obtain an $\mathcal{O}$-algebra homomorphism as follows

4.5.4 $$(A\otimes_\mathcal{O}\mathcal{O}N_H)H \longrightarrow \mathrm{End}_\mathcal{O}(M_\omega) ,$$

从而 A_ω 成为 N_H-内 H-代数.

and A_ω becomes an N_H-interior H-algebra.

4.6. 虽然 A_ω 依赖于 $\{i_\beta\}_{\beta\in\mathcal{P}_A(H)}$ 的选择, 可是如果 $\{i'_\beta\}_{\beta\in\mathcal{P}_A(H)}$ 是另一个选择, 那么相应的$(A\otimes_\mathcal{O}\mathcal{O}N_H)H$-模

4.6. Although A_ω depends on the choice of $\{i_\beta\}_{\beta\in\mathcal{P}_A(H)}$, if $\{i'_\beta\}_{\beta\in\mathcal{P}_A(H)}$ is another choice then the corresponding $(A\otimes_\mathcal{O}\mathcal{O}N_H)H$-modules

4.6.1 $$\bigoplus_{\beta\in\mathcal{P}_A(H)} (Ai_\beta)^{\omega(\beta)} , \qquad \bigoplus_{\beta\in\mathcal{P}_A(H)} (Ai'_\beta)^{\omega(\beta)}$$

是同构的, 并且每一个这样的 $(A\otimes_\mathcal{O}\mathcal{O}N_H)H$-模同构决定一个$N_H$-内 H-代数同构如下, 称它为可嵌入同构

are isomorphic; moreover, any such an isomorphism of $(A\otimes_\mathcal{O}\mathcal{O}N_H)H$-modules induces an N_H-interior H-algebra isomorphism — called *embeddable isomorphism* — as follows

4.6.2 $$\mathrm{End}_A\Big(\bigoplus_{\beta\in\mathcal{P}_A(H)} (Ai_\beta)^{\omega(\beta)}\Big) \cong \mathrm{End}_A\Big(\bigoplus_{\beta\in\mathcal{P}_A(H)} (Ai'_\beta)^{\omega(\beta)}\Big) ;$$

特别是, A_ω 的可嵌入的 N_H-内 H-代数自同构都是被 $(A_\omega^H)^*$ 决定的. 仍然, 如果 ℓ 是 A^H 的幂等元而 $\omega = \mu_{A^H}^\ell$, 那么 $\ell A\ell$ 就

in particular, all the embeddable N_H-interior H-algebra automorphisms of A_ω are determined by $(A_\omega^H)^*$. Furthermore, if ℓ is an idempotent of A^H and we have $\omega = \mu_{A^H}^\ell$,

有一个由其 $\mathcal{O}(N_H \times N_H)H$-模结构决定的 N_H-内 H-代数结构, 并且不难证明

then the $\mathcal{O}(N_H \times N_H)H$-module structure of $\ell A \ell$ induces an N_H-interior H-algebra structure on it, and it is easy to check that

4.6.3 $$A_{\mu^\ell_{AH}} \cong \ell A \ell\,.$$

4.7. 如同在 3.15 里一样, 既然在上述情况里 $\mathcal{O}$-模同构 3.16.2 成为 $\mathcal{O}(N_H \times N_H)H$-模同构, 只要考虑基数为 $\omega(\beta)$ 的有限集合 Ω_β, 以及从 $\Omega_\beta \times \Omega_{\beta'}$ 到 $i_\beta A i_{\beta'}$ 的所有映射的 $\mathcal{O}(N_H \times N_H)H$-模, 令 $(i_\beta A i_{\beta'})^{\Omega_\beta \times \Omega_{\beta'}}$ 其中 β, β' 属于 $\mathcal{P}_A(H)$, A_ω 就成为

4.7. Analogously as in 3.15, since in the situation above the $\mathcal{O}$-module isomorphism 3.16.2 becomes an $\mathcal{O}(N_H \times N_H)H$-module isomorphism, it suffices to consider finite sets Ω_β of cardinal $\omega(\beta)$, and the $\mathcal{O}(N_H \times N_H)H$-module, noted $(i_\beta A i_{\beta'})^{\Omega_\beta \times \Omega_{\beta'}}$, of all the maps from $\Omega_\beta \times \Omega_{\beta'}$ to $i_\beta A i_{\beta'}$ where $\beta, \beta' \in \mathcal{P}_A(H)$, to get the following form of A_ω

4.7.1 $$A_\omega = \bigoplus_{\beta, \beta' \in \mathcal{P}_A(H)} (i_\beta A i_{\beta'})^{\Omega_\beta \times \Omega_{\beta'}}\,,$$

这里乘法定义如下: 对任意 $(i_\beta A i_{\beta'})^{\Omega_\beta \times \Omega_{\beta'}}$, $(i_{\gamma'} A i_\gamma)^{\Omega_{\gamma'} \times \Omega_\gamma}$ 的分别元素 $a_{\beta,\beta'}$, $b_{\gamma',\gamma}$, 如果 $\beta' \neq \gamma'$ 那么规定 $0 = a_{\beta,\beta'}\, b_{\gamma',\gamma}$ 并如果 $\beta' = \gamma'$ 那么 $a_{\beta,\beta'}\, b_{\beta',\gamma}$ 在 $(i_\beta A i_\gamma)^{\Omega_\beta \times \Omega_\gamma}$ 中对任意 $u \in \Omega_\beta$ 与 $w \in \Omega_\gamma$ 的取值是

where the product is defined as follows: for any $a_{\beta,\beta'}$ and any $b_{\gamma',\gamma}$ respectively belonging to $(i_\beta A i_{\beta'})^{\Omega_\beta \times \Omega_{\beta'}}$ and $(i_{\gamma'} A i_\gamma)^{\Omega_{\gamma'} \times \Omega_\gamma}$, if $\beta' \neq \gamma'$ then we just set $a_{\beta,\beta'}\, b_{\gamma',\gamma} = 0$, whereas if $\beta' = \gamma'$ then $a_{\beta,\beta'}\, b_{\beta',\gamma}$ is the element of $(i_\beta A i_\gamma)^{\Omega_\beta \times \Omega_\gamma}$ which, for any $u \in \Omega_\beta$ and any $w \in \Omega_\gamma$, takes the value

4.7.2 $$(a_{\beta,\beta'}\, b_{\beta',\gamma})(u, w) = \sum_{v \in \Omega_{\beta'}} a_{\beta,\beta'}(u, v)\, b_{\beta',\gamma}(v, w)$$

其中 $\beta, \beta', \gamma, \gamma' \in \mathcal{P}_A(H)$; 请注意, M_ω 的 $\mathcal{O}(N_H \times N_H)H$-模结构决定 A_ω 的 N_H-内 H-代数结构.

where $\beta, \beta', \gamma, \gamma' \in \mathcal{P}_A(H)$; note that the structure of $\mathcal{O}(N_H \times N_H)H$-module of M_ω induces the N_H-interior H-algebra structure of A_ω.

4.8. 最后, 由于上面的公式 4.7.1, 显然 $(A_\omega)^H = (A^H)_\omega$, 从而显然有

4.8. Finally, according to the equality 4.7.1 above, clearly $(A_\omega)^H = (A^H)_\omega$, and therefore we have

4.8.1 $$\mathcal{P}_{A_\omega}(H) \subset \mathcal{P}_A(H)\,, \quad \mathcal{D}_{A_\omega}(H) \subset \mathcal{D}_A(H)\,;$$

特别是, A 上的点群 H_β 也是 A_ω 上的点群当且仅当 $\beta \subset \omega$ (见 3.15); 而且, 在上述情况里同构 3.16.2 显然成为一个

in particular, a pointed group H_β on A also is a pointed group on A_ω if and only if $\beta \subset \omega$ (see 3.15); moreover, in the situation above, the isomorphism 3.16.2 clearly becomes an

$\mathcal{O}(N_H\times N_H)H$-模同构. 这样如同在3.17 里, 如果$\theta\in\mathcal{D}_{A_\omega}(H)$ 那么同构 3.17.3 也决定一个 N_H-内 H-代数同构如下, 也称可嵌入同构

isomorphism of $\mathcal{O}(N_H \times N_H)H$-modules. Thus, analogously as in 3.17, if $\theta\in\mathcal{D}_{A_\omega}(H)$ then the isomorphism 3.17.3 also determines an N_H-interior H-algebra isomorphism, still called *embeddable isomorphism*, as follows

4.8.2 $$(A_\omega)_\theta \cong A_\theta\,.$$

注意 4.9. 如同在注意 3.18 里一样, 选择满足$\omega\subset n\mu^1_{A^H}$ 的 $n\in\mathbb{N}$并且设$\mathcal{O}^n$是n个单$\mathcal{O}G$-模的直和; 令 $E=\mathrm{End}_{\mathcal{O}}(\mathcal{O}^n)$. 那么$A\otimes_{\mathcal{O}}E$成为$N$-内$G$-代数; 而且, 如果$e$是$E$的本原幂等元, $1\otimes e$就是一个$(A\otimes_{\mathcal{O}}E)^G$的幂等元; 这样, $A\otimes_{\mathcal{O}}Ee$ 也是一个 $(A\otimes_{\mathcal{O}}\mathcal{O}N)G$-模 (见 4.6), 而 $A\otimes_{\mathcal{O}}eEe$ 有一个被其 $\mathcal{O}(N\times N)G$-模结构决定的 N-内 G-代数结构; 现在, 同构 3.14.2 成为下面的 N-内 G-代数同构

Remark 4.9. As in Remark 3.18, choose $n\in\mathbb{N}$ such that $\omega\subset n\mu^1_{A^H}$ and let $\mathcal{O}^n$ be the direct sum of n copies of the trivial $\mathcal{O}G$-module; consider $E=\mathrm{End}_{\mathcal{O}}(\mathcal{O}^n)$. Thus, $A\otimes_{\mathcal{O}}E$ becomes an N-interior G-algebra; moreover, if e is a primitive idempotent of E, $1\otimes e$ obviously is an idempotent of $(A\otimes_{\mathcal{O}}E)^G$; consequently, $A\otimes_{\mathcal{O}}Ee$ still is an $(A\otimes_{\mathcal{O}}\mathcal{O}N)G$-module (see 4.6), and moreover $A\otimes_{\mathcal{O}}eEe$ has an N-interior G-algebra structure induced by its structure of $\mathcal{O}(N\times N)G$-module; now, isomorphism 3.14.2 becomes the following N-interior G-algebra isomorphism

4.9.1 $$(1\otimes e)(A\otimes_{\mathcal{O}}E)(1\otimes e)\cong \mathrm{End}_{\mathcal{O}}\big((A\otimes_{\mathcal{O}}E)(1\otimes e)\big)^\circ\,.$$

4.10. 特别是, 我们也得到下面的 N_H-内H-代数同构 (见 3.15.4 和 4.5.2)

4.10. In particular, we also get the following N_H-interior H-algebra isomorphism (see 3.15.4 and 4.5.2)

4.10.1 $$\mathrm{Res}^G_H(A)\cong A\otimes_{\mathcal{O}}eEe\cong (A\otimes_{\mathcal{O}}E)_{\mu^{1\otimes e}_{(A\otimes_{\mathcal{O}}E)^H}}\;;$$

这样, ω 与 $\mu^1_{A^H}$ 成为两个 $A^H\otimes_{\mathcal{O}}E$ $(=(A\otimes_{\mathcal{O}}E)^H)$ 的除子 (见4.8.1), 并显然$n\mu^1_{A^H}=\mu^{1\otimes 1}_{(A\otimes_{\mathcal{O}}E)^H}$; 所以有 $(A\otimes_{\mathcal{O}}E)^H$的幂等元$\ell$使得

thus, ω and $\mu^1_{A^H}$ become two divisors of $A^H\otimes_{\mathcal{O}}E$ $(=(A\otimes_{\mathcal{O}}E)^H)$ (see 4.8.1), and moreover, clearly $n\mu^1_{A^H}=\mu^{1\otimes 1}_{(A\otimes_{\mathcal{O}}E)^H}$; consequently, there exists an idempotent ℓ in $(A\otimes_{\mathcal{O}}E)^H$ such that

4.10.2 $$\omega=\mu^\ell_{(A\otimes_{\mathcal{O}}E)^H}\,.$$

仍然, $\ell(A\otimes_{\mathcal{O}}E)\ell$ 有一个被其 $\mathcal{O}(N_H\times N_H)H$-模结构决定的 N_H-内 H-代数结构, 从而

Furthermore, $\ell(A\otimes_{\mathcal{O}}E)\ell$ has an N-interior G-algebra structure induced by its $\mathcal{O}(N\times N)G$-module structure, and therefore

同构 4.6.3 和 4.8.2 决定下面的可嵌入同构

isomorphisms 4.6.3 and 4.8.2 determine the following *embeddable* isomorphism

4.10.3 $$A_\omega \cong \ell(A \otimes_{\mathcal{O}} E)\ell\,.$$

特别是, 考虑 A 上的 H-除子 ω 的时候, 只要用造当的 $A \otimes_{\mathcal{O}} E$ 代替 A, 就能假定 $\omega = \mu^i_{A^H}$, 其中 i 是一个 A^H 的幂等元.

in particular, when considering a divisor ω of H on A, it suffices to replace A by this construction $A \otimes_{\mathcal{O}} E$ and we may assume that $\omega = \mu^i_{A^H}$, where i is an idempotent of A^H.

4.11. 设 B 是另一个 N-内 G-代数; 因为 $(A \otimes_{\mathcal{O}} B)^H$ 包含 $A^H \otimes_{\mathcal{O}} B^H$ 所以映射 3.20.2 和 3.21.3 决定一个双线性映射

4.11. Let B be another N-interior G-algebra; since $(A \otimes_{\mathcal{O}} B)^H$ contains $A^H \otimes_{\mathcal{O}} B^H$, the maps 3.20.2 and 3.21.3 determine a bilinear map

4.11.1 $$t^H_{A,B} : \mathcal{D}_A(H) \times \mathcal{D}_B(H) \longrightarrow \mathcal{D}_{A \otimes_{\mathcal{O}} B}(H)$$

而我们记 $t^H_{A,B}(\omega,\theta) = \omega \otimes \theta$, 其中 $(\omega,\theta) \in \mathcal{D}_A(H) \times \mathcal{D}_B(H)$; 如同在3.20 里一样, 式子3.20.1 和注意 4.9 决定下面的 N_H-内 H-代数同构

and, for any $(\omega,\theta) \in \mathcal{D}_A(H) \times \mathcal{D}_B(H)$, we set $t^H_{A,B}(\omega,\theta) = \omega \otimes \theta$; as in 3.20, the formula 3.20.1 and Remark 4.9 determine the following N_H-interior H-algebra isomorphism

4.11.2 $$(A \otimes_{\mathcal{O}} B)_{\omega \otimes \theta} \cong A_\omega \otimes_{\mathcal{O}} B_\theta\,.$$

4.12. 另一方面, 设 $f: A \to B$ 是一个 N-内 G-代数同态; 因为 $f(A^H) \subset B^H$ 所以映射 3.21.3 决定一个线性映射

4.12. On the other hand, let $f: A \to B$ be an N-interior G-algebra homomorphism; since $f(A^H) \subset B^H$ then the map 3.21.3 determines a linear map

4.12.1 $$\mathrm{res}_{f^H} : \mathcal{D}_A(H) \longrightarrow \mathcal{D}_B(H)$$

并, 由注意4.6, 如果 $\omega \in \mathcal{D}_A(H)$ 那么 f 决定一个 N_H-内 H-代数幺同态

and, by Remark 4.9, if $\omega \in \mathcal{D}_A(H)$ then f induces a N_H-interior H-algebra homomorphism

4.12.2 $$f_\omega : A_\omega \longrightarrow B_{\mathrm{res}_{f^H}(\omega)}\,.$$

特别是, 如果 K_γ 是 A 上的点群使得 (见下面的 5.1),

in particular, if K_γ is a pointed group on A such that (see 5.1 below),

4.12.3 $$K \subset H\,, \quad \gamma \subset \mathrm{res}^H_K(\omega)$$

那么只要使 $N_G(K_\gamma)$, $C_N(K)$, A^K, $A(K_\gamma)$, s_γ 分别代替 G, N, A, B, f, 这个同态就显然决定下面的 $C_N(K)$-内 $N_G(K_\gamma)$-

then, replacing G, N, A, B and f by $N_G(K_\gamma)$, $C_N(K)$, A^K, $A(K_\gamma)$ and s_γ respectively, the corresponding homomorphism determines the following $C_N(K)$-interior

代数同构 (见下面的 5.2)

$N_G(K_\gamma)$-algebra isomorphism (see 5.2 below)

4.12.4 $$A_\omega(K_\gamma) \cong A(K_\gamma)_{\omega_\gamma}$$

其中 (见下面的 5.1)

where we set (see 5.1 below)

4.12.5 $$\omega_\gamma = \mathrm{res}_{(s_\gamma)^K}\big(\mathrm{res}^H_{N_H(K_\gamma)}(\omega)\big)\,.$$

最后, 如果 C 是另一个 N-内 G-代数而 $g\colon B \to C$ 是另一个 N-内 G-代数同态, 显然得到

Finally, if C is another N-interior G-algebra and $g\colon B \to C$ is another N-interior G-algebra homomorphism, we clearly get

4.12.6 $$\mathrm{res}_{g^H} \circ \mathrm{res}_{f^H} = \mathrm{res}_{(g\circ f)^H}\,.$$

4.13. 讨论 N-内 G-代数同态的时候, 满射的条件没有意思; 我们用如下所谓覆盖的条件如下代替满射性. 设 $f\colon A \to B$ 是一个N-内 G-代数同态; 如果对 G 的任意子群 H, res_{f^H} 是双射, 而对任意 $\alpha \in \mathcal{P}_A(H)$, f 决定一个同构

4.13. When discussing on N-interior G-algebra homomorphisms, the simple surjectivity has no interest; below, to replace surjectivity, we employ the so-called *covering condition.* Let $f\colon A \to B$ be a N-interior G-algebra; if res_{f^H} is bijective for any subgroup H of G and, for any $\alpha \in \mathcal{P}_A(H)$, f determines an isomorphism

4.13.1 $$f^\alpha : A(H_\alpha) \cong B(H_{\mathrm{res}_{f^H}(\alpha)})$$

使得 $f^\alpha \circ s_\alpha = s_{\mathrm{res}_{f^H}(\alpha)} \circ f$, 那么 f 称为严格覆盖同态; 也就是说, 如果 f 是一个严格覆盖同态有

such that $f^\alpha \circ s_\alpha = s_{\mathrm{res}_{f^H}(\alpha)} \circ f$, then we call f a *strict covering homomorphism*; thus, if f is a strict covering homomorphism then we have

4.13.2 $$A^H/J(A^H) \cong \prod_{\alpha\in\mathcal{P}_A(H)} A(H_\alpha) \cong \prod_{\beta\in\mathcal{P}_B(H)} B(H_\beta) \cong B^H/J(B^H)\,,$$

从而还有

and therefore we still have

4.13.3 $$B^H = f(A^H) + J(B^H)\,;$$

特别是, 从 $A/J(A) \cong B/J(B)$ 可推出 $J(A)$ 包含 $\mathrm{Ker}(f)$. 反过来, 如果 $\mathrm{Ker}(f) \subset J(A)$ 和任意子群 $H \subset G$ 满足等式 4.13.3 那么 f 是严格覆盖的; 第二个条件就是定义覆盖的条件, 可是我们将只用到严格覆盖同态.

in particular, from $A/J(A) \cong B/J(B)$ it is easy to check that $J(A)$ contains $\mathrm{Ker}(f)$. Conversely, if $\mathrm{Ker}(f) \subset J(A)$ and any subgroup $H \subset G$ fulfills equality 4.13.3, then f is a strict covering homomorphism; the second condition is just the *covering condition,* but we will only employ strict covering homomorphisms.

4.14. 最后, 假定 f 是严格覆盖的. 对 G 的任意子群 H 与 $\omega \in \mathcal{D}_A(H)$, 上述的 H_N-内 H-代数幺同态

4.14. Finally, assume that f is a strict covering homomorphism. For any subgroup H of G and any $\omega \in \mathcal{D}_A(H)$, the N_H-interior H-algebra homomorphism

4.14.1 $$f_\omega: A_\omega \longrightarrow B_{\mathrm{res}_{fH}(\omega)}$$

也是严格覆盖的; 确实, 由注意 4.9, 对某个 $n \in \mathbb{N}$ 与 $(A\otimes_{\mathcal{O}} E)^H$ 的幂等元 ℓ 有 $\omega = \mu^{\ell}_{(A\otimes_{\mathcal{O}} E)^H}$, 其中 $E = \mathrm{End}_{\mathcal{O}}(\mathcal{O}^n)$; 此时, 不难验证 $f \otimes \mathrm{id}_{\mathcal{O}^n}$ 是严格覆盖的; 因为 (见 4.6.3)

is a strict covering homomorphism too; indeed, by Remark 4.9, for suitable $n \in \mathbb{N}$ and $\ell \in (A\otimes_{\mathcal{O}} E)^H$, we have $\omega = \mu^{\ell}_{(A\otimes_{\mathcal{O}} E)^H}$, where $E = \mathrm{End}_{\mathcal{O}}(\mathcal{O}^n)$; moreover, it is easy to check that $f \otimes \mathrm{id}_{\mathcal{O}^n}$ is a strict covering homomorphism; since (see 4.6.3)

4.14.2 $$A_\omega \cong \ell(A\otimes_{\mathcal{O}} E)\ell\,, \quad B_{\mathrm{res}_{fH}(\omega)} \cong f(\ell)(B\otimes_{\mathcal{O}} E)f(\ell)\,,$$

所以 f_ω 也是严格覆盖的. 而且, 不太难验证一个 N-内 G-代数同态 $h: C \to A$ 是严格覆盖的当且仅当 $f \circ h$ 是严格覆盖的.

f_ω is also a strict covering homomorphism. Further, an N-interior G-algebra homomorphism $h: C \to A$ is a strict covering homomorphism if and only if $f \circ h$ is so.

例子 4.15. 设 M 是 $\mathcal{O}G$-模而令 $E = \mathrm{End}_{\mathcal{O}}(M)$; 那么 E 是一个 G-内代数并对任意 G 的子群 H, E^H 就是 $\mathrm{Res}^G_H(M)$ 的自同态的 $\mathcal{O}$-代数. 这样, 任意幂等元 $j \in E^H$ 决定一个 $\mathrm{Res}^G_H(M)$ 的直和项 $j(M)$.

Example 4.15. Let M be an $\mathcal{O}G$-module and set $E = \mathrm{End}_{\mathcal{O}}(M)$; then E is a G-interior algebra and, for any subgroup H of G, E^H is just the endomorphism $\mathcal{O}$-algebra of $\mathrm{Res}^G_H(M)$. Hence, any idempotent $j \in E^H$ determines a direct summand $j(M)$ of $\mathrm{Res}^G_H(M)$.

而且, 如果 $j' \in E^H$ 是另一个幂等元, 那么 $j'(M) \cong j(M)$ 当且仅当 j' 与 j 在 E^H 中是相互共轭的; 确实, 这个同构还推出 $(\mathrm{id}_M - j')(M)$ 与 $(\mathrm{id}_M - j)(M)$ 是同构的 (见推论 3.7), 从而这两个同构决定 $f \in (E^H)^*$ 使得

Moreover, if $j' \in E^H$ is another idempotent, then we have $j'(M) \cong j(M)$ if and only if j' and j are conjugate to each other; indeed, such an isomorphism implies that $(\mathrm{id}_M - j')(M)$ and $(\mathrm{id}_M - j)(M)$ are also isomorphic (see Corollary 3.7), and then such pair of isomorphisms determines an element $f \in (E^H)^*$ such that

4.15.1 $$f\big(j'(M)\big) = j(M)\,, \quad f\big((\mathrm{id}_M - j')(M)\big) = (\mathrm{id}_M - j)(M)\,;$$

也就是说, j^f 是一个 E^H 的幂等元, 并且它的像与核和 j' 的像与核是一样的, 所以得到 $j^f = j'$.

in other words, j^f is an idempotent of E^H and moreover, its image and its kernel coincide with the image and the kernel of j', so that we get $j^f = j'$.

这样, E 上的 H-点 β 的集合与 $\mathrm{Res}_H^G(M)$ 的不可分解直和项的同构类的集合是一一对应的; 而且, $E_\beta = \mathrm{End}_{\mathcal{O}}\big(j(M)\big)$, 这里 $j \in \beta$.

Thus, we have a bijective correspondence between the set of points β of H on E and the set of isomorphism classes of indecomposable direct summands of $\mathrm{Res}_H^G(M)$; moreover, $E_\beta = \mathrm{End}_{\mathcal{O}}\big(j(M)\big)$, where $j \in \beta$.

5. 除子的限制和除子的诱导

5. Restriction and Induction of Divisors

5.1. 仍设 G 是一个有限群, N 是 G 的一个正规子群, A 是一个 N-内 G-代数. 再设 H 与 K 是 G 的两个子群使得 $K \subset H$; 现在我们把 $\mathcal{O}K$-模与 $\mathcal{O}H$-模之间的限制和诱导要扩张到 A 上的 H-除子和 K-除子之间的限制和诱导. 首先, 显然 $A^H \subset A^K$, 从而存在唯一线性映射

5.1. Once again, let G be a finite group, N a normal subgroup of G and A an N-interior G-algebra. Let H and K be two subgroups of G such that $K \subset H$; presently, we want to extend the ordinary restriction and the ordinary induction between the $\mathcal{O}H$- and the $\mathcal{O}K$-modules, to a restriction and an induction between the divisors of H and K on A. First of all, we clearly have $A^H \subset A^K$ and therefore we have a unique linear map

5.1.1 $$\mathrm{res}_K^H : \mathcal{D}_A(H) \longrightarrow \mathcal{D}_A(K)$$

使得对任意 A^H 的幂等元 j 有 $\mathrm{res}_K^H(\mu_{A^H}^j) = \mu_{A^K}^j$ (见 3.21), 称为限制映射.

such that, for any idempotent j in A^H, fulfills $\mathrm{res}_K^H(\mu_{A^H}^j) = \mu_{A^K}^j$ (see 3.21), which we simply call *restriction map*.

5.2. 那么, 如果 H_β 与 K_γ 是 A上的点群使得 $\gamma \subset \mathrm{res}_K^H(\beta)$ 我们就说 H_β 包含 K_γ, 写作 $K_\gamma \subset H_\beta$. 别忘了 $\gamma \subset \mathrm{res}_K^H(\beta)$ 和 $\gamma \in \mathcal{P}_{A_\beta}(K)$ 是相互等价的 (见 3.16); 也就是说, 由同构式 4.12.4, 知有 $K_\gamma \subset H_\beta$ 当且仅当有 $\beta_\gamma \neq 0$. 当然, 如果 L_ε 是一个 A 上的点群使得 $L_\varepsilon \subset K_\gamma$ 那么也得到 $L_\varepsilon \subset H_\beta$. 请注意, 对任意 $\beta \in \mathcal{P}_A(H)$ 存在 $\gamma \in \mathcal{P}_A(K)$ 使得 $K_\gamma \subset H_\beta$; 反过来, 如果 $\gamma \in \mathcal{P}_A(K)$ 那么存在 $\beta \in \mathcal{P}_A(H)$ 使得 $K_\gamma \subset H_\beta$.

5.2. Then, if H_β and K_γ are pointed groups over A such that $\gamma \subset \mathrm{res}_K^H(\beta)$, we just say that H_β contains K_γ, and simply write $K_\gamma \subset H_\beta$. Recall that $\gamma \subset \mathrm{res}_K^H(\beta)$ is equivalent to $\gamma \in \mathcal{P}_{A_\beta}(K)$ (see 3.16); in other terms, by isomorphism 4.12.4 and with the notation there, we have $K_\gamma \subset H_\beta$ if and only if we have $\beta_\gamma \neq 0$. Obviously, if L_ε is a pointed group on A such that $L_\varepsilon \subset K_\gamma$ then we still have $L_\varepsilon \subset H_\beta$. Note that, for any $\beta \in \mathcal{P}_A(H)$ there exists $\gamma \in \mathcal{P}_A(K)$ such that $K_\gamma \subset H_\beta$; conversely, for any $\gamma \in \mathcal{P}_A(K)$ we also have $\beta \in \mathcal{P}_A(H)$ such that $K_\gamma \subset H_\beta$.

5.3. 另一方面, 对任意 $\omega \in \mathcal{D}_A(H)$, 就有到下面的可嵌入同构

5.3. On the other hand, we claim that for any $\omega \in \mathcal{D}_A(H)$, we can construct the following *embeddable* isomorphism

5.3.1 $$A_{\mathrm{res}_K^H(\omega)} \cong \mathrm{Res}_K^H(A_\omega);$$

确实, 由注意 4.9, 能假定 $\omega = \mu_{A^H}^j$, 其中 j 是一个 A^H 的幂等元; 那么有 $\mathrm{res}_K^H(\omega) = \mu_{A^K}^j$,

indeed, by Remark 4.9, we may assume that $\omega = \mu_{A^H}^j$, where j is an idempotent of A^H; then we get $\mathrm{res}_K^H(\omega) = \mu_{A^K}^j$ and there-

从而同构式 5.3.1 的两边都等于 jAj. 特别是, A_ω 上的 K-除子 θ 也是 A 上的 K-除子 (见 4.8.1), 并且得到一个可嵌入同构

fore both members in isomorphism 5.3.1 are equal to jAj. In particular, any divisor θ of K on A_ω is also a divisor of K on A (see 4.8.1), and we get the *embeddable* isomorphism

5.3.2 $$A_\theta \cong (A_\omega)_\theta \,.$$

5.4. 现在, 我们考虑所谓相对迹映射 $\mathrm{Tr}_K^H\colon A^K \to A^H$; 选择一个 $K\backslash H$ 的代表集合 T; 那个映射的定义是 $\mathrm{Tr}_K^H(a) = \sum_{x\in T} a^x$, 其中 $a \in A^K$; 不难验证, 这个定义不依赖 T 的选择, 并映射的像 $A_K^H = \mathrm{Tr}_K^H(A^K)$ 是 A^H 的理想, 这是因为对任意 $b, c \in A^H$ 我们有

5.4. Now, we consider the so-called *relative trace map* $\mathrm{Tr}_K^H\colon A^K \to A^H$; choosing a set of representatives T for $K\backslash H$, we define $\mathrm{Tr}_K^H(a) = \sum_{x\in T} a^x$ for any $a \in A$; it is quite easy to check that this definition does not depend on the choice of T and, moreover, the image of this map $A_K^H = \mathrm{Tr}_K^H(A^K)$ is an ideal of A^H since, for any $b, c \in A^H$, we have

5.4.1 $$b\mathrm{Tr}_K^H(a)c = \mathrm{Tr}_K^H(bac)\,.$$

一般来说, 即使 i 是 A^K 的幂等元, $\mathrm{Tr}_K^H(i)$ 也不必是幂等元的.

In general, even if i is an idempotent of A^K, $\mathrm{Tr}_K^H(i)$ need not be an idempotent.

5.5. 但是, 只要假定对任意 $x \in H-K$ 有 $i\,i^x = 0$ 就得到 $\mathrm{Tr}_K^H(i)$ 是幂等元的, 这是因为

5.5. However, it suffices to assume that for any $x \in H-K$ we have $i\,i^x = 0$, to obtain that $\mathrm{Tr}_K^H(i)$ is an idempotent since

5.5.1 $$\mathrm{Tr}_K^H(i)\mathrm{Tr}_K^H(i) = \sum_{x,y\in T} i^x i^y = \mathrm{Tr}_K^H(i)$$

其中 T 是一个 $K\backslash H$ 的代表集合. 我们称这种幂等元 i 有正交 H/K-迹; 请注意, 此时任意 A^K 的幂等元 j 使得 $ji = j = ij$ 也有正交 H/K-迹. 例如, 只要在例子里 4.15 存在一个 $\mathcal{O}H$-模 N 使得

where T is a set of representatives for $K\backslash H$. To consider this kind of idempotents, we say that i *has an orthogonal* H/K*-trace*; note that any idempotent j in A^K such that $ji = j = ij$ also has an orthogonal H/K-trace. For instance, it suffices that in the example 4.15 we had an $\mathcal{O}H$-module N such that

5.5.2 $$M = \mathrm{Ind}_H^G(N) = \bigoplus_{x\in S} x \otimes N\,,$$

其中 S 是 G/H 的代表集合, E^H 就包含一个满足 $\mathrm{id}_M = \mathrm{Tr}_H^G(j)$ 的有正交 G/H-迹的幂等元 j; 确实, $\mathrm{Res}_H^G(M)$ 的直和项 $1\otimes N$ 决定此幂等元.

where S is a set of representatives for G/H, to exhibit in E^H an idempotent j having an orthogonal G/H-trace and fulfilling $\mathrm{id}_M = \mathrm{Tr}_H^G(j)$; indeed, the direct summand $1\otimes N$ of $\mathrm{Res}_H^G(M)$ determines such an idempotent.

命题 5.6. 假定对任意 A 上的点群 H_β 存在有正交 G/H-迹的幂等元 $j \in \beta$. 那么, 对任意 G 的子群 H 与 K 使得 $K \subset H$, 存在唯一线性映射

Proposition 5.6. *Assume that, for any pointed group H_β on A, some $j \in \beta$ has an orthogonal G/H-trace. Then, for any subgroups H and K of G such that $K \subset H$ there is a unique linear map*

5.6.1 $$\mathrm{ind}_K^H : \mathcal{D}_A(K) \longrightarrow \mathcal{D}_A(H)$$

使得对任意 A^K 的有正交 H/K-迹的幂等元 i 有

such that, for any idempotent j of A^K which has an orthogonal H/K-trace, we have

5.6.2 $$\mathrm{ind}_K^H(\mu^i_{A^K}) = \mu^{\mathrm{Tr}_K^H(i)}_{A^H}\,.$$

而且, 对满足 $\omega' \subset \omega$ 的任意 $\omega, \omega' \in \mathcal{D}_A(K)$ 有

Moreover, for any $\omega, \omega' \in \mathcal{D}_A(K)$ fulfilling $\omega' \subset \omega$, we have

5.6.3 $$\mathrm{ind}_K^H(\omega') \subset \mathrm{ind}_K^H(\omega)\,.$$

证明: 对任意 $\gamma \in \mathcal{P}_A(K)$, 设 $j \in \gamma$ 是一个有正交 G/K-迹的幂等元; 显然, j 也有正交 H/K-迹的, 从而我们能定义

Proof: For any $\gamma \in \mathcal{P}_A(K)$, choose an idempotent $j \in \gamma$ having an orthogonal G/K-trace; clearly, j still has an orthogonal H/K-trace and thus we can define

5.6.4 $$\mathrm{ind}_K^H(\gamma) = \mu^{\mathrm{Tr}_K^H(j)}_{A^H}\,.$$

这个除子不依赖 j 的选择; 确实, 如果 $i \in \gamma$ 是另一个有正交 H/K-迹的幂等元, 那么存在 $a \in (A^K)^*$ 使得 $j = i^a$, 从而得到

This divisor does not depend on the choice of j; indeed, if $i \in \gamma$ is another idempotent having an orthogonal H/K-trace, then there exists $a \in (A^K)^*$ such that $j = i^a$ and therefore we get

5.6.5 $$\mathrm{Tr}_K^H(a^{-1}i)\mathrm{Tr}_K^H(ia) = \sum_{x,y\in T} (a^{-1}i)^x(ia)^y = \mathrm{Tr}_K^H(j)\,;$$

类似地, 因为有 $a^{-1}i = ja^{-1}$ 与 $ia = aj$ 所以我们得到

similarly, since we have $a^{-1}i = ja^{-1}$ and $ia = aj$, we also get

5.6.6 $$\mathrm{Tr}_K^H(i) = \mathrm{Tr}_K^H(ia)\mathrm{Tr}_K^H(a^{-1}i)\,.$$

这样, 由推论 3.8, $\mathrm{Tr}_K^H(i)$ 与 $\mathrm{Tr}_K^H(j)$ 在 A^H 里是相互共轭, 这是因为 $\mathrm{Tr}_K^H(ia)$, $\mathrm{Tr}_K^H(a^{-1}i)$ 两者都属于 A^H. 现在, 对任意 $\omega \in \mathcal{D}_A(K)$, 就定义

Thus, according to Corollary 3.8, $\mathrm{Tr}_K^H(i)$ and $\mathrm{Tr}_K^H(j)$ are conjugate to each other in A^H since $\mathrm{Tr}_K^H(ia)$ and $\mathrm{Tr}_K^H(a^{-1}i)$ both belong to A^H. Now, for any divisor $\omega \in \mathcal{D}_A(K)$, we define

5.6.7 $$\mathrm{ind}_K^H(\omega) = \sum_{\gamma\in\mathcal{P}_A(K)} \omega(\gamma)\,\mathrm{ind}_K^H(\gamma)\,.$$

最后, 设 i, i', i'' 是三个 A^K 的幂等元使得 $i'i'' = 0 = i''i'$ 与 $i = i' + i''$; 如果 i 是有正交 H/K-迹的, 显然 i' 与 i'' 也是有正交 H/K-迹, 并且得到

Finally, let i, i' and i'' be three idempotents in A^K fulfilling $i'i'' = 0 = i''i'$ and $i = i' + i''$; if i has an orthogonal H/K-trace, it is quite clear that i' and i'' still have an orthogonal H/K-trace, moreover we get

5.6.8 $$\mathrm{Tr}_K^H(i')\mathrm{Tr}_K^H(i'') = 0 = \mathrm{Tr}_K^H(i'')\mathrm{Tr}_K^H(i');$$

所以, 由命题 3.19 我们也得到

hence, by Proposition 3.19, we also get

5.6.9 $$\mu^{i}_{A^K} = \mu^{i'}_{A^K} + \mu^{i''}_{A^K}, \quad \mu^{\mathrm{Tr}_K^H(i)}_{A^H} = \mu^{\mathrm{Tr}_K^H(i')}_{A^H} + \mu^{\mathrm{Tr}_K^H(i'')}_{A^H}.$$

那么, 只要对

Then, arguing by induction on

5.6.10 $$\ell = \sum_{\gamma \in \mathcal{P}_A(K)} \mu^{i}_{A^K}(\gamma),$$

使用归纳法, 从等式 5.6.4 和 5.6.9 可推出等式 5.6.2 .

it is not difficult to prove equality 5.6.2 from equalities 5.6.4 and 5.6.9.

推论 5.7. **同上的假定之下. 设 H, K, L 是 G 的三个子群使得 H 包含 K 与 L. 对任意 $\omega \in \mathcal{D}_A(K)$ 有 (Mackey 公式)**

Corollary 5.7. *With the hypothesis of the proposition, let H, K and L be subgroups of G such that H contains K and L. For any $\omega \in \mathcal{D}_A(K)$ we have (Mackey formula)*

5.7.1 $$\mathrm{res}_L^H\big(\mathrm{ind}_K^H(\omega)\big) = \sum_{x \in T} \mathrm{ind}_{L \cap K^x}^L\big(\mathrm{res}_{L \cap K^x}^{K^x}(\omega^x)\big)$$

其中 T 是一个 $K\backslash H/L$ 的代表集合. 特别是, 有

where T is a set of representatives for $K\backslash H/L$. In particular, we have

5.7.2 $$\omega \subset \mathrm{res}_K^H\big(\mathrm{ind}_K^H(\omega)\big).$$

而且, 如果 $K \subset L$ 那么有

Moreover, if $K \subset L$ then we have

5.7.3 $$\mathrm{ind}_L^H\big(\mathrm{ind}_K^L(\omega)\big) = \mathrm{ind}_K^H(\omega).$$

证明: 由于 ind_K^H 是线性的, 能假定 ω 是 A 上的 K-点; 那么, 选择一个有正交 H/K-迹的 $i \in \omega$; 这样, 有 (见等于 5.6.2)

Proof: Since ind_K^H is linear, we may assume that ω is a point of K on A; then, choosing an idempotent $i \in \omega$ having an orthogonal H/K-trace, we have (see equality 5.6.2)

5.7.4 $$\mathrm{ind}_K^H(\omega) = \mu^{\mathrm{Tr}_K^H(i)}_{A^H};$$

可是, 如果 T 是 $K\backslash H/L$ 的代表集合, 并 U_x 是 $(L \cap K^x)\backslash L$

but, if T is a set of representatives for $K\backslash H/L$ and, for any $x \in T$, U_x is a set of

的代表集合，其中 $x \in T$，不难验证

representatives for $(L \cap K^x)\backslash L$, then it is easily checked that

5.7.5 $$\mathrm{Tr}_K^H(i) = \sum_{x\in T}\sum_{y\in U_x} i^{xy} = \sum_{x\in T}\mathrm{Tr}_{L\cap K^x}^L(i^x),$$

也验证 i^x 是有正交 $L/(L\cap K^x)$-迹的，其中 $x \in T$；所以，对 $\mathcal{O}$-代数 A^L 用命题 3.19，我们就得到

and that i^x has an orthogonal $L/(L \cap K^x)$-trace for any $x \in T$; hence, by applying Proposition 3.19 to the $\mathcal{O}$-algebra A^L, we get

5.7.6 $$\mu_{A^L}^{\mathrm{Tr}_K^H(i)} = \sum_{x\in T}\mu_{A^L}^{\mathrm{Tr}_{L\cap K^x}^L(i^x)} = \sum_{x\in T}\mathrm{ind}_{L\cap K^x}^L(\mu_{A^{L\cap K^x}}^{i^x}),$$

从而我们也得到 Mackey 式子. 进一步，如果 $K \subset L$ 那么有 $\mathrm{Tr}_K^H(i) = \mathrm{Tr}_L^H(\mathrm{Tr}_K^L(i))$，并 i 有正交 L/K-迹，而 $\mathrm{Tr}_K^L(i)$ 有正交 H/L-迹；所以得到

and therefore we get the Mackey formula too. Furthermore, if $K \subset L$ then we have $\mathrm{Tr}_K^H(i) = \mathrm{Tr}_L^H(\mathrm{Tr}_K^L(i))$, i has an orthogonal L/K-trace, and $\mathrm{Tr}_K^L(i)$ has an orthogonal H/L-trace; consequently, we get

5.7.7 $$\mu_{A^H}^{\mathrm{Tr}_K^H(i)} = \mathrm{ind}_L^H(\mu_{A^L}^{\mathrm{Tr}_K^L(i)}) = \mathrm{ind}_L^H(\mathrm{ind}_K^L(\mu_{A^K}^i)).$$

注意 5.8. 同上的假定之下. 设 H 与 K 是两个 G 的子群使得 H 包含 K 并设 θ 是一个 A 上的 K-除子. 如果 $N = G$，那么我们只要有 A_θ 的结构，就能决定 $A_{\mathrm{ind}_K^H(\theta)}$ 的结构；确实，由注意 4.9 和同构 5.3.2，能假定 A^K 有一个有正交 H/K-迹的幂等元 j 使得 $\theta = \mu_A^j$；那么得到下面的 H-内代数同构 (见同构 3.15.4)

Remark 5.8. With the hypothesis of the proposition, let H and K be subgroups of G such that H contains K and let θ be a divisor of K on A. If $N = G$, it is possible to determine the structure of $A_{\mathrm{ind}_K^H(\theta)}$ from the structure of A_θ; indeed, by Remark 4.9 and isomorphism 5.3.2, we may assume that A^K contains an idempotent j which has an orthogonal H/K-trace and fulfills $\theta = \mu_A^j$; then, we get the following H-interior algebra isomorphism (see isomorphism 3.15.4)

5.8.1 $$A_{\mathrm{ind}_K^H(\theta)} \cong \mathrm{Tr}_K^H(j)A\mathrm{Tr}_K^H(j) \cong \bigoplus_{x,y\in X} j^xAj^y = \bigoplus_{x,y\in X} x^{-1}{\cdot}jAj{\cdot}y,$$

其中 $X \subset H$ 是一个 $K\backslash H$ 的代表集合；而且，因为 j 是有正交 H/K-迹的，所以对任意 $x, x', y, y' \in H$ 与 $a, a' \in jAj$，如果 $yx' \in H - K$ 那么我们得到

where $X \subset H$ is a set of representatives for $K\backslash H$; moreover, since j has an orthogonal H/K-trace, for any $x, x', y, y' \in H$ and any $a, a' \in jAj$, whenever $yx' \in H - K$, we have

5.8.2 $$(x{\cdot}a{\cdot}y)(x'{\cdot}a'{\cdot}y') = x{\cdot}a{\cdot}y{\cdot}j^y j^{x'}{\cdot}x'{\cdot}a'{\cdot}y' = 0$$

并如果 $yx' \in K$ 那么就得到

whereas if $yx' \in K$ then we get

5.8.3 $$(x{\cdot}a{\cdot}y)(x'{\cdot}a'{\cdot}y') = x{\cdot}(a{\cdot}yx'{\cdot}a'){\cdot}y'$$

其中 $a{\cdot}yx'{\cdot}a'$ 属于 jAj.

where $a{\cdot}yx'{\cdot}a'$ still belongs to jAj.

这样, 对任意 K-内代数 B 我们令

Thus, for any K-interior algebra B, we set

5.8.4 $$\mathrm{Ind}_K^H(B) = \mathcal{O}H \otimes_{\mathcal{O}K} B \otimes_{\mathcal{O}K} \mathcal{O}H = \bigoplus_{x,y\in X} x^{-1} \otimes B \otimes y\,,$$

并在 $\mathrm{Ind}_K^H(B)$ 中定义如下的运算

and, in $\mathrm{Ind}_K^H(B)$, we define the following operations

5.8.5 $$\begin{gathered}(x\otimes b\otimes y)(x'\otimes b'\otimes y') = (x\otimes \iota_K^H(yx'){\cdot}bb'\otimes y')\,,\\ x'{\cdot}(x\otimes b\otimes y) = x'x\otimes b\otimes y\,,\quad (x\otimes b\otimes y){\cdot}y' = x\otimes b\otimes yy'\,,\end{gathered}$$

其中 $x, x', y, y' \in H$ 与 $b, b' \in B$ 而, $\iota_K^H(y) = y$ 或 0 当 $y \in K$ 或 $y \in H - K$. 那么不难验证 $\mathrm{Ind}_K^H(B)$ 就是 H-内代数. 此时, 因为 $A_\theta \cong jAj$, 所以得到

where $x, x', y, y' \in H$ and $b, b' \in B$, and moreover, $\iota_K^H(y) = y$ or 0 according to $y \in K$ or $y \in H - K$. Then, it is not difficult to check that $\mathrm{Ind}_K^H(B)$ is an H-interior algebra. At that point, since $A_\theta \cong jAj$, we get

5.8.6 $$A_{\mathrm{ind}_K^H(\theta)} \cong \mathrm{Ind}_K^H(A_\theta)\,.$$

5.9. 如果对任意 A 上的点群 H_β 存在一个有正交 G/H-迹的幂等元 $j \in \beta$, 我们就说 A 是一个完备诱导的 N-内 G-代数. 不难构造一个完备诱导的 N-内 G-代数: 令 子(G) 记 G 的子群的集合, 并考虑下面的 N-内 G-代数

5.9. If, for any pointed group H_β on A, there exists $j \in \beta$ which has an orthogonal G/H-trace, we say that A is *inductively complete.* It is not difficult to construct inductively complete N-interior G-algebras: denote by 子(G) † the set of all the subgroups of G, and consider the following N-interior G-algebra

5.9.1 $$E = \mathrm{End}_{\mathcal{O}}\Big(\bigoplus_{H\in \text{子}(G)} \mathrm{Ind}_H^G(\mathcal{O})\Big)\,;$$

请注意, 任意 $H \in$ 子(G) 决定一个 E^H 的有正交 G/H-迹的幂等元 j_H (见例子4.15 和 5.4);

note that any $H \in$ 子(G) determines in E^H an idempotent j_H which has an orthogonal G/H-trace (see example 4.15 and 5.4);

† The Chinese character 子 is pronounced "zi" as in "zest" and, in this context, means "sub-".

而且，如果 H 跑遍 子(G)，那么幂等元 $\mathrm{Tr}_H^G(j_H)$ 都是相互正交的，还满足

moreover, the idempotents $\mathrm{Tr}_H^G(j_H)$, when H runs over 子(G), are pairwise orthogonal and fulfill

5.9.2 $$1_E = \sum_{H\in 子(G)} \mathrm{Tr}_H^G(j_H)\,.$$

命题 5.10. 上述 N-内 G-代数 E 是完备诱导的. 而且，如果 B 是完备诱导的 N-内 G-代数，那么张量积 $A\otimes_{\mathcal{O}} B$ 也是完备诱导的.

Proposition 5.10. *The N-interior G-algebra E above is inductively complete. Moreover, if B is an inductively complete N-interior G-algebra, the tensor product $A\otimes_{\mathcal{O}} B$ still is inductively complete.*

证明: 设 H_β 是一个 E 上的点群；已经知道存在 $K\in$子(G) 使得 β 决定一个 $\mathrm{Res}_H^G(\mathrm{Ind}_K^G(\mathcal{O}))$ 的不可分解直和项的同构类 (见例子 4.15)；但是显然有

Proof: Let H_β be a pointed group on E; we already know that β determines an isomorphism class of indecomposable direct summands of $\mathrm{Res}_H^G(\mathrm{Ind}_K^G(\mathcal{O}))$ for some $K\in$子(G) (see example 4.15); but, we clearly have

5.10.1 $$\mathrm{Res}_H^G\big(\mathrm{Ind}_K^G(\mathcal{O})\big) \cong \bigoplus_{x\in T} \mathrm{Ind}_{H\cap H^x}^H(\mathcal{O})\,,$$

其中 T 是一个 $H\backslash G/K$ 的代表集合；所以，存在 $L\in$子(H) 使得 β 决定一个 $\mathrm{Ind}_L^H(\mathcal{O})$ 的不可分解直和项的同构类；这样，因为存在一个 E^L 的有正交 G/L-迹的本原幂等元 j_L 使得

where T is a set of representatives for $H\backslash G/K$; hence, there is $L\in$子(H) such that β determines an isomorphism class of indecomposable direct summands of $\mathrm{Ind}_L^H(\mathcal{O})$; thus, since E^L contains a primitive idempotent j_L which has an orthogonal G/L-trace and fulfills

5.10.2 $$\mathrm{Ind}_L^H(\mathcal{O}) \cong \mathrm{Tr}_L^H(j_L)\big(\mathrm{Ind}_L^G(\mathcal{O})\big)\,,$$

所以 $\mathrm{Tr}_L^H(j_L)$ 是有正交 G/H-迹的，并且存在 $j\in\beta$ 使得 (见命题 3.16)

then $\mathrm{Tr}_L^H(j_L)$ has an orthogonal G/H-trace, and there is $j\in\beta$ such that (see Proposition 3.16)

5.10.3 $$j\mathrm{Tr}_L^H(j_L) = j = \mathrm{Tr}_L^H(j_L)j\,,$$

从而 j 也是有正交 G/H-迹的.

and thus j still has an orthogonal G/H-trace.

而且，令 $f\colon B\to A\otimes_{\mathcal{O}}B$ 记自然的映射，即 $f(b) = 1\otimes b$，其中 $b\in B$；请注意 f 只是一个 1-内 G-代数同态. 如果 H_ρ 是

Moreover, denote by $f\colon B \to A\otimes_{\mathcal{O}} B$ the canonical map, i.e. $f(b) = 1\otimes b$ for any $b\in B$; note that f simply is an 1-interior G-algebra homomorphism. For any pointed

$A \otimes_{\mathcal{O}} B$ 上的点群, 因为 $f(1) = 1 \otimes 1$ 所以显然存在 $\beta \in \mathcal{P}_B(H)$ 使得 (见 3.21)

group H_ρ on $A \otimes_{\mathcal{O}} B$, since we have $f(1) = 1 \otimes 1$, it is quite clear that there exists $\beta \in \mathcal{P}_B(H)$ fulfilling (see 3.21)

5.10.4 $$\big(\mathrm{res}_{f^H}(\beta)\big)(\rho) \neq 0$$

其中

where

5.10.5 $$f^H : B^H \longrightarrow (A \otimes_{\mathcal{O}} B)^H$$

是 f 的限制; 这样, 只要选择一个有正交 G/H-迹的幂等元 $j \in \beta$ 就存在 $\ell \in \rho$ 使得 $\ell f(j) = \ell = f(j)\ell$ (见 3.14.1 和命题 3.19); 那么 ℓ 也是有正交 G/H-迹的.

denotes the corresponding restriction of f; consequently, choosing an idempotent $j \in \beta$ which has an orthogonal G/H-trace, there exists $\ell \in \rho$ such that $\ell f(j) = \ell = f(j)\ell$ (see equality 3.14.1 and Proposition 3.19); then ℓ still has an orthogonal G/H-trace.

定理 5.11. 存在一个完备诱导的 N-内 G-代数 B 并一个除子 $\omega \in \mathcal{D}_B(G)$ 使得

5.11.1. 有 $A \cong B_\omega$.

5.11.2. 对任意完备诱导的 N-内 G-代数 B' 并 $\omega' \in \mathcal{D}_{B'}(G)$ 使得 $A \cong B'_{\omega'}$, 存在 $\theta' \in \mathcal{D}'_B(G)$ 与 $g: B \cong B'_{\theta'}$ 使得 $\mathrm{res}_g(\omega') = \omega$ 并 $\omega' \subset n\theta'$, 其中 $n \in \mathbb{N}$.

Theorem 5.11. *There exist an inductively complete N-interior G-algebra B and a divisor $\omega \in \mathcal{D}_B(G)$ such that*

5.11.1. *We have $A \cong B_\omega$.*

5.11.2. *For any inductively complete N-interior G-algebra B' and any $\omega' \in \mathcal{D}_{B'}(G)$ such that $A \cong B'_{\omega'}$, there are $\theta' \in \mathcal{D}'_B(G)$ and $g: B \cong B'_{\theta'}$ fulfilling $\mathrm{res}_g(\omega') = \omega$ and $\omega' \subset n\theta'$ for some $n \in \mathbb{N}$.*

证明: 由命题 5.10, $A \otimes_{\mathcal{O}} E$ 是完备诱导的; 而且, 因为有 N-内 G-代数同构 $j_G E j_G \cong \mathcal{O}$, 所以还有一个 N-内 G-代数同构 $(A \otimes_{\mathcal{O}} E)_\omega \cong A$, 其中 $\omega = \mu^{1\otimes j_G}_{(A\otimes_{\mathcal{O}} E)^G}$ (见 3.15 和 3.20). 特别是, A 上的点群也是 $A \otimes_{\mathcal{O}} E$ 上的点群, 而令

Proof: By Proposition 5.10, $A \otimes_{\mathcal{O}} E$ is inductively complete; moreover, since we have an N-interior G-algebra isomorphism $j_G E j_G \cong \mathcal{O}$, we also have an N-interior G-algebra isomorphism $(A \otimes_{\mathcal{O}} E)_\omega \cong A$, where $\omega = \mu^{1\otimes j_G}_{(A\otimes_{\mathcal{O}} E)^G}$ (see 3.15 and 3.20). In particular, any pointed group on A still is a pointed group on $A \otimes_{\mathcal{O}} E$, and we set

5.11.3 $$\theta = \bigcup_{H_\beta} \mathrm{ind}_H^G(\beta)\,, \quad B = \big(A \otimes_{\mathcal{O}} E\big)_\theta$$

其中 H_β 跑遍 A 上的点群的集合; 请注意, 任意 $\alpha \in \mathcal{P}_A(G)$ 包含在 θ 中, 从而存在 $n \in \mathbb{N}$ 使得 $\omega \subset n\theta$.

where H_β runs over the set of pointed groups on A; note that any $\alpha \in \mathcal{P}_A(G)$ is contained in θ and therefore there exists $n \in \mathbb{N}$ fulfilling $\omega \subset n\theta$.

再次, B 上的点群 H_ρ 也是一个 $A\otimes_{\mathcal{O}} E$ 上的点群使得 $\rho\subset \mathrm{res}^G_H(\theta)$; 所以存在一个 A 上的点群 K_γ 使得 $\rho\subset \mathrm{res}^G_H\big(\mathrm{ind}^G_K(\gamma)\big)$; 这样, 由 Mackey 式子 (见推论 5.7), 存在 $x\in G$ 使得

Once again, any pointed group H_ρ on B still is a pointed group on $A\otimes_{\mathcal{O}} E$ which fulfills $\rho\subset \mathrm{res}^G_H(\theta)$; consequently, there is a pointed group K_γ on A such that $\rho\subset \mathrm{res}^G_H\big(\mathrm{ind}^G_K(\gamma)\big)$; then, it follows from the Mackey formula (see Corollary 5.7) that there exists $x\in G$ fulfilling

5.11.4 $$\rho\subset \mathrm{ind}^H_{H\cap K^x}\big(\mathrm{res}^{K^x}_{H\cap K^x}(\gamma^x)\big)\,.$$

也就是说, 事实上存在一个 A 上的点群 L_δ 使得 $\rho\subset \mathrm{ind}^H_L(\delta)$; 所以得到

That is to say, in fact there exists a pointed group L_ρ on A such that $\rho\subset \mathrm{ind}^H_L(\delta)$; hence, we get

5.11.5 $$\mathrm{ind}^G_H(\rho)\subset \mathrm{ind}^G_L(\delta)\subset\theta\,,$$

从而 $\mathrm{ind}^G_H(\rho)$ 也是 B 上的 G-除子 (见 3.16); 这样, B 也是完备诱导的. 类似地, ω 也是 B 上的 G-除子, 从而还得到 $A\cong B_\omega$ (见 4.8.2).

and therefore $\mathrm{ind}^G_H(\rho)$ is still a divisor of G on B (see 3.16); consequently, B is also inductively complete. Similarly, ω is also a divisor of G on B, and therefore we still get $A\cong B_\omega$ (see 4.8.2).

以后, 设 B' 是一个完备诱导的 N-内 G-代数, 并设 ω' 是一个 B' 的 G-除子使得 $A\cong B'_{\omega'}$; 同上推理, 有下面的 N-内 G-代数的同构, 还有下面的式子

Further, let B' be an inductively complete N-interior G-algebra and ω' a divisor of G on B' such that $A\cong B'_{\omega'}$; following the argument in the begining, we get the following N-interior G-algebra isomorphism and equality

5.11.6 $$h'\colon B'\cong (B'\otimes_{\mathcal{O}} E)_{\mu^{1\otimes j_G}_{(B'\otimes_{\mathcal{O}} E)^G}}\,,\qquad \mathrm{res}_{h'}(\omega')=\omega'\otimes\mu^{j_G}_E\,;$$

而且, 由同构 3.20.3, 我们得到另一个 N-内 G-代数同构

moreover, from isomorphism 3.20.3, we get another N-interior G-algebra isomorphism

5.11.7 $$h: A\otimes_{\mathcal{O}} E\cong (B'\otimes_{\mathcal{O}} E)_{\omega'\otimes\mu^1_E}$$

使得 $\mathrm{res}_h(\omega)=\omega'\otimes\mu^{j_G}_E$.

such that we have $\mathrm{res}_h(\omega)=\omega'\otimes\mu^{j_G}_E$.

特别是, 对任意 A 上的点群 H_β, $\mathrm{ind}^G_H(\beta)$ 也是 $B'\otimes_{\mathcal{O}} E$ 上的 G-除子; 另一方面, H_β 也是 B' 上的点群, 而 B' 是完备诱导的; 所以, 得到 $\mathrm{ind}^G_H(\beta)\subset\theta'$. 也就是说, 在 $B'\otimes_{\mathcal{O}} E$ 上有 $\theta\subset\theta'$ (见定义 5.11.3); 最后, 由同构 5.3.2, 得到一下 N-内

In particular, for any pointed group H_β on A, $\mathrm{ind}^G_H(\beta)$ is still a divisor of G on $B'\otimes_{\mathcal{O}} E$; on the other hand, H_β is also a pointed group on B', and B' is inductively complete; hence, we get $\mathrm{ind}^G_H(\beta)\subset\theta'$. In other words, in the divisors of $B'\otimes_{\mathcal{O}} E$ we obtain $\theta\subset\theta'$ (see definition 5.11.3); finally, from isomorphism 5.3.2 above, we get

G-代数同构使得 $\mathrm{res}_g(\omega') = \mathrm{res}_{h^{-1}}(\omega' \otimes \mu_E^{j_G}) = \omega$

the N-interior G-algebra isomorphism below such that $\mathrm{res}_g(\omega') = \mathrm{res}_{h^{-1}}(\omega' \otimes \mu_E^{j_G}) = \omega$

5.11.8
$$\begin{aligned} g : B'_\theta &\cong \big((B' \otimes_{\mathcal{O}} E)_{\theta'}\big)_\theta \cong (B' \otimes_{\mathcal{O}} E)_\theta \\ &\cong \big((B' \otimes_{\mathcal{O}} E)_{\omega' \otimes \mu_E^1}\big)_\theta \cong B\,. \end{aligned}$$

定理 5.12. 假定 A 是完备诱导的. 设 $H\,,K$ 是 G 的子群使得 $K \subset H$ 并设 i,j 分别是 A^H, A^K 的幂等元. 那么, 有 $\mu^i_{A^H} \subset \mathrm{ind}_K^H(\mu^j_{A^K})$ 当且仅当存在 $a,b \in A^K$ 使得 $i = \mathrm{Tr}_K^H(ajb)$.

Theorem 5.12. *Assume that A is inductively complete. Let $H\,,K$ be subgroups of G such that $K \subset H$ and $i\,,j$ respective idempotents of A^H and A^K. Then, we have $\mu^i_{A^H} \subset \mathrm{ind}_K^H(\mu^j_{A^K})$ if and only if there are $a,b \in A^K$ such that $i = \mathrm{Tr}_K^H(ajb)$.*

证明: 由注意4.9, 存在一个 N-内 G-代数 B, 还存在一个 B^G 的幂等元 e 使得 $A = eBe$ 并且有一个 B^H 的幂等元 ℓ 使得 $\mathrm{ind}_K^H(\mu^j_{B^K}) = \mu^\ell_{B^H}$; 那么不难证明存在 $b \in (B^K)^*$ 使得 j^b 是有正交 H/K-迹的; 进一步, 由命题 3.19, 存在 $a \in (B^H)^*$ 使得 i^a 与 $\mathrm{Tr}_K^H(j^b)$ 交换; 这样, 从 $\mu^i_{B^H} \subset \mu^\ell_{B^H}$ 就推出 $i^a = i^a\mathrm{Tr}_K^H(j^b)$; 所以, 如果我们有 $\mu^i_{A^H} \subset \mathrm{ind}_K^H(\mu^j_{A^K})$ 就得到

Proof: According to Remark 4.9, there are an N-interior G-algebra B, an idempotent e in B^G and an idempotent ℓ in B^H such that we have $A = eBe$ and $\mathrm{ind}_K^H(\mu^j_{B^K}) = \mu^\ell_{B^H}$; then, it is not difficult to prove the existence of $b \in (B^K)^*$ such that j^b has an orthogonal H/K-trace; moreover, it follows from Proposition 3.19 that there is $a \in (B^H)^*$ such that i^a centralizes $\mathrm{Tr}_K^H(j^b)$; consequently, the inclusion $\mu^i_{B^H} \subset \mu^\ell_{B^H}$ implies that $i^a\mathrm{Tr}_K^H(j^b) = i^a$; hence, if $\mu^i_{A^H} \subset \mathrm{ind}_K^H(\mu^j_{A^K})$ then we get

5.12.1
$$i = ia\,\mathrm{Tr}_K^H(b^{-1}jb)\,a^{-1}i = \mathrm{Tr}_K^H\big((iab^{-1}j)\,j\,(jba^{-1}i)\big)$$

并且, 因为 $ei = i = ie$ 与 $ej = j = je$, 所以 $iab^{-1}j$, $jba^{-1}i$ 两者都属于 $eB^Ke = A^K$.

and moreover, since $ei = i = ie$ and $ej = j = je$, the elements $iab^{-1}j$ and $jba^{-1}i$ both belong to $eB^Ke = A^K$.

反过来,假定 $i = \mathrm{Tr}_K^H(ajb)$, 其中 $a,b \in A^K$; 在 B 中能假定 j 有正交 H/K-迹; 那么, 只要选择一个 H/K 的代表的集合 T, 就得到

Conversely, assume that $i = \mathrm{Tr}_K^H(ajb)$, where $a,b \in A^K$; in B, we may assume that j has an orthogonal H/K-trace; then, choosing a set of representatives T for H/K, we get

5.12.2
$$\begin{aligned} \mathrm{Tr}_K^H(aj)\,\mathrm{Tr}_K^H(jb) &= \Big(\sum_{x\in T} a^x j^x\Big)\Big(\sum_{y\in T} j^y b^y\Big) \\ &= \sum_{x\in T} (ajb)^x = i\,; \end{aligned}$$

这样，$\ell = \mathrm{Tr}_K^H(jbi)\,\mathrm{Tr}_K^H(iaj)$ 是 B^H 的幂等元，这是因为有

consequently, $\ell = \mathrm{Tr}_K^H(jbi)\,\mathrm{Tr}_K^H(iaj)$ is an idempotent of B^H since we have

5.12.3
$$\begin{aligned}\ell^2 &= \mathrm{Tr}_K^H(jbi)\,i\,\mathrm{Tr}_K^H(aj)\,\mathrm{Tr}_K^H(jb)\,i\,\mathrm{Tr}_K^H(iaj)\\ &= \mathrm{Tr}_K^H(jbi)\,i\,\mathrm{Tr}_K^H(iaj) = \ell\,;\end{aligned}$$

而且，由推论 3.8，ℓ 与 i 在 B^H 中是相互共轭的. 另一方面，$\mathrm{Tr}_K^H(j)$ 是 B^H 的幂等元，并有

moreover, by Corollary 3.8, ℓ and i are conjugate to each other in B^H. On the other hand, $\mathrm{Tr}_K^H(j)$ is an idempotent of B^H, we get

5.12.4
$$\begin{aligned}\ell\,\mathrm{Tr}_K^H(j) &= \mathrm{Tr}_K^H(jbi)\,\mathrm{Tr}_K^H(iaj)\,\mathrm{Tr}_K^H(j)\\ &= \mathrm{Tr}_K^H(jbi)\Big(\sum_{x\in T} i^x a^x j^x\Big)\Big(\sum_{y\in T} j^y\Big) = \ell\end{aligned}$$

以及类似地有 $\mathrm{Tr}_K^H(j)\ell = \ell$；也就是说，由命题 3.19，我们得到

and, similarly, we still get $\mathrm{Tr}_K^H(j)\ell = \ell$; in other terms, by Proposition 3.19, we obtain

5.12.5
$$\mu_{B^H}^i = \mu_{B^H}^\ell \subset \mu_{B^H}^{\mathrm{Tr}_K^H(j)} = \mathrm{ind}_K^H(\mu_{B^K}^j)$$

从而也得到

and therefore we still obtain

5.12.6
$$\mu_{A^K}^i \subset \mathrm{ind}_K^H(\mu_{A^K}^j)\,.$$

注意 5.13. 假定 A 是完备诱导的. 设 H 与 K 是 G 的子群使得 $K \subset H$ 并设 ω 与 θ 分别是 A 上的 H- 与 K-除子使得

Remark 5.13. Assume that A is inductively complete. Let H and K be subgroups of G such that $K \subset H$, and let ω and θ respectively be divisors of H and K on A such that

5.13.1
$$\theta \subset n\cdot\mathrm{res}_K^H(\omega)\,, \quad \omega \subset m\cdot\mathrm{ind}_K^H(\theta)$$

其中 $n, m \in \mathbb{N}-\{0\}$. 那么 A_ω 与 A_θ 是相互 Morita 等价 $\mathcal{O}$-代数；确实，由同构 4.8.2，我们得到 $A_\theta \cong (A_{n\cdot\omega})_\theta$；另一方面，由注意 4.9 和式子 5.6.2，不难证明 $\mathcal{P}(A_{\mathrm{ind}_K^H(\theta)}) = \mathcal{P}(A_\theta)$（见 5.7.2)；所以我们得到

for some $n, m \in \mathbb{N}-\{0\}$. Then A_ω and A_θ are *Morita equivalent* $\mathcal{O}$*-algebras*; indeed, from isomorphism 4.8.2 we get $A_\theta \cong (A_{n\cdot\omega})_\theta$; on the other hand, from Remark 4.9 and equality 5.6.2 it is not difficult to prove that $\mathcal{P}(A_{\mathrm{ind}_K^H(\theta)}) = \mathcal{P}(A_\theta)$ (see 5.7.2); consequently, we get

5.13.2
$$\mathcal{P}(A_\theta) \subset \mathcal{P}(A_{n\cdot\omega}) = \mathcal{P}(A_\omega) \subset \mathcal{P}(A_{\mathrm{ind}_K^H(m\cdot\theta)}) = \mathcal{P}(A_\theta)\,,$$

从而，A_θ 和 A_ω 两者都是 Morita 等价于 $A_{n\cdot\omega}$（见 3.15).

and therefore A_θ and A_ω are both Morita equivalent to $A_{n\cdot\omega}$ (see 3.15).

6. N-内 G-代数上的局部点群

6.1. 如同在 §5 里一样, G 是一个有限群, N 是一个 G 的正规子群, A 是一个 N-内 G-代数. 请注意, 如果 A 不是完备诱导的, 那么由定理 5.11 我们能取一个完备诱导的 N-内 G-代数 B 和 B 的 G-除子 ω 使得 $A \cong B_\omega$, 这样关于除子的诱导和限制总可在 B 中实现. 所以, 不失一般性, 以下我们设 A 是完备诱导的.

6.2. 在 §5 里, 我们在 A 上的点群里研究了那些从其真子群的除子的诱导可以得到的点群; 现在, 我们研究 A 上的其它点群, 即所谓局部点群: 我们称 A 上的点群 P_γ 是局部的 (也称其点 γ 是局部的) 如果对任意 A 上的点群 Q_δ 使得 P 真包含 Q 有 $\gamma \not\subset \mathrm{ind}_Q^P(\delta)$; 请注意, A 上的 1-点都是局部的. 令 $\mathcal{LP}_A(P)$ 记 A 上的局部 P-点的集合. 由定理 5.12, 有下面的等价定义.

命题 6.3. 一个 A 上的点群 P_γ 是局部的当且仅当对任意 P 的真子群 Q 有 $\gamma \not\subset A_Q^P$. 特别是, 对任意局部点群 P_γ, 其群 P 是 p-群.

证明: 如果 P 有真子群 Q 使得 γ 包含在 A_Q^P 里, 那么对任意 $i \in \gamma$ 存在 $a \in A^Q$ 使得 $i = \mathrm{Tr}_Q^P(a) = \mathrm{Tr}_Q^P(iai)$; 这样, 只要

6. Local Pointed Groups on N-interior G-algebras

6.1. As in Section 5, G is a finite group, N a normal subgroup of G and A a N-interior G-algebra. Note that whenever A is not inductively complete, it follows from Theorem 5.11 that we can find an inductively complete N-interior G-algebra B, together with a divisor ω of G on B such that $A \cong B_\omega$, so that all the questions concerning induction and restriction of divisors can be discussed in B. Hence, without loss of generality we may assume that A is inductively complete.

6.2. In §5 we have discussed on the pointed groups on A which can be obtained from the induction of divisors of their proper subgrups; now, we discuss on the others, i.e. on the so-called *local pointed groups:* we say that a pointed group P_γ on A is *local* (or that the point γ is *local*) whenever, for any pointed group Q_δ on A such that P properly contains Q, we have $\gamma \not\subset \mathrm{ind}_Q^P(\delta)$; note that all the points of 1 on A are local. Denote by $\mathcal{LP}_A(P)$ the set of local points of P on A. By Theorem 5.12, we have the following equivalent definition.

Proposition 6.3. *A pointed group P_γ on A is local if and only if, for any proper subgroup Q of P, we have $\gamma \not\subset A_Q^P$. In particular, for any local pointed group P_γ, the group P is a p-group.*

Proof: If P has a proper subgroup Q such that γ is contained in A_Q^P then, for any $i \in \gamma$ there is $a \in A^Q$ such that $i = \mathrm{Tr}_Q^P(a) = \mathrm{Tr}_Q^P(iai)$; in particular, choosing a set J of

选择一个 iA^Qi 的相互正交本原幂等元的集合 J 使得 $\sum_{j\in J} j = i$, 就得到

pairwise orthogonal primitive idempotents of iA^Qi in such a way that $\sum_{j\in J} j = i$, we get

6.3.1 $$i = \sum_{j\in J} \mathrm{Tr}_Q^P(iaj)$$

从而存在 $j \in J$ 使得 $\mathrm{Tr}_Q^P(iaj)$ 不属于 $J(iA^Pi)$. 另一方面, 商 $iA^Pi/J(iA^Pi)$ 显然是可除代数; 所以元素 $\mathrm{Tr}_Q^P(iaj)$ 在 A^P 中是可逆的 (见 2.9); 也就是说, 存在 $b \in A^P$ 使得

and therefore there exists $j \in J$ such that $\mathrm{Tr}_Q^P(iaj)$ does not belong to $J(iA^Pi)$. On the other hand, it is clear that $iA^Pi/J(iA^Pi)$ is a division algebra; hence, $\mathrm{Tr}_Q^P(iaj)$ is an inversible element on A^P (see 2.9); that is to say, there is $b \in A^P$ such that

6.3.2 $$i = b\mathrm{Tr}_Q^P(iaj) = \mathrm{Tr}_Q^P(biaj)$$

从而, 由定理 5.12, 我们得到 $\gamma \subset \mathrm{ind}_Q^P(\delta)$, 其中 $\delta \in \mathcal{P}_A(Q)$ 是 j 的共轭类. 反过来, 如果 P_γ 不是局部的, 那么存在一个 A 上的点群 Q_δ 使得 P 真包含 Q 与有 $\gamma \subset \mathrm{ind}_Q^P(\delta)$, 从而, 仍引用定理5.12, 得到 $i = \mathrm{Tr}_Q^P(ajb)$, 其中 $j \in \delta$ 与 $a, b \in A^Q$; 也就是说, 有 $\gamma \subset A_Q^P$. 最后, 只要 Q 是一个 P 的 Sylow p-子群, 就有 $A_Q^P = A^P$, 这是因为对任意 $a \in A^P$ 有

and therefore, by Theorem 5.12, we get $\gamma \subset \mathrm{ind}_Q^P(\delta)$, where $\delta \in \mathcal{P}_A(Q)$ is the conjugacy class of j. Conversely, if P_γ is not local, then there exists a pointed group Q_δ on A such that P properly contains Q and we have $\gamma \subset \mathrm{ind}_Q^P(\delta)$, and therefore, according to Theorem 5.12 again, we still get $i = \mathrm{Tr}_Q^P(ajb)$, where $j \in \delta$ and $a, b \in A^Q$; in other terms, we get $\gamma \subset A_Q^P$. Finally, if Q is a Sylow p-subgroup of P then we get $A_Q^P = A^P$ since, for any $a \in A^P$, we have

6.3.3 $$a = \frac{1}{|P\colon Q|}\mathrm{Tr}_Q^P(a)\,.$$

6.4. 设 P 是一个 G 的 p-子群; 上述命题引导考虑下面的 $C_N(P)$-内 $N_G(P)$-代数, 称为 Brauer 商

6.4. Let P be a p-subgroup of G; the proposition above leads to consider the following $C_N(P)$-interior $N_G(P)$-algebra, called *Brauer quotient,*

6.4.1 $$A(P) = k \otimes_{\mathcal{O}} \left(A^P \Big/ \sum_Q A_Q^P\right)$$

其中 Q 跑遍 P 的真子群的集合 (请注意, $|P\colon Q|A^P \subset A_Q^P$,

where Q runs over the set of proper subgroups of P (note that $|P\colon Q|A^P \subset A_Q^P$,

张量积乘以 k 的原因就是这个事实); 而且, 令

which is the reason for tensoring by k); moreover, denote by

6.4.2 $$\mathrm{Br}_P : A^P \longrightarrow A(P)$$

(或者 Br_P^A) 记自然的映射. 设H是G的子群使得 $P \subset H$; 因为A^P包含A^H与$A(P)^{N_H(P)}$包含$\mathrm{Br}_P(A^H)$, 所以任意A的H-除子 ω 决定下面的$A(P)$的 $N_H(P)$-除子

(or by Br_P^A) the canonical map. Let H be a subgroup of G such that $P \subset H$; since A^P contains A^H and $A(P)^{N_H(P)}$ contains $\mathrm{Br}_P(A^H)$, any divisor ω of H on A determines the following divisor of $N_H(P)$ on $A(P)$

6.4.3 $$\omega_P = \mathrm{res}_{(\mathrm{Br}_P)^H}(\omega)$$

其中

where

6.4.4 $$(\mathrm{Br}_P)^H : A^H \longrightarrow A(P)^{N_H(P)}$$

是相应的由 Br_P 决定的同态 (见 3.21.3). 而且, 由同构 5.3.1 和同态 4.12.2, Br_P 决定下面同态

is the corresponding homomorphism induced by Br_P (see 3.21.3). Moreover, from isomorphism 5.3.1 and homomorphism 4.12.2, Br_P determines the homomorphism

6.4.5 $$\big(\mathrm{Res}^H_{N_H(P)}(A_\omega)\big)^P \cong (A^P)_{\mathrm{res}^H_{N_H(P)}(\omega)} \longrightarrow A(P)_{\omega_P}\,.$$

6.5. 最后, 这个同态也决定下面的 $C_N(P)$-内 $N_H(P)$-代数同构

6.5. Finally, this homomorphism induces the $C_N(P)$-interior $N_H(P)$-algebra isomorphism

6.5.1 $$A_\omega(P) \cong A(P)_{\omega_P}\,;$$

确实, 由注意 4.9, 能假定 $A_\omega = iAi$, 其中i是一个A^H的幂等元; 那么, 一方面显然

indeed, by Remark 4.9, we may assume that $A_\omega = iAi$ where i is an idempotent in A^H; then, on the one hand, we clearly have

6.5.2 $$A(P)_{\omega_P} = \mathrm{Br}_P(i)A(P)\mathrm{Br}_P(i)\,;$$

另一方面对任意 P 的子群 Q 有 $(iAi)^Q = iA^Qi$, 从而还有 $(iAi)^P_Q = iA^P_Qi$; 设$\bar\imath$是i的像; 所以我们得到

on the other hand, for any subgroup Q of P, we have $(iAi)^Q = iA^Qi$ and therefore we still have $(iAi)^P_Q = iA^P_Qi$; hence, denoting by $\bar\imath$ the image of i, we get

6.5.3 $$(iAi)^P \Big/ \sum_Q (iAi)^P_Q \cong iA^Pi \Big/ \sum_Q iA^P_Qi \cong \bar\imath\Big(A^P \Big/ \sum_Q A^P_Q\Big)\bar\imath\,.$$

推论 6.6. 一个 A 上的点群 P_γ 是局部的当且仅当 $\mathrm{Br}_P(\gamma)$ 不是零. 特别是, Br_P 决定一个从 $\mathcal{LP}_A(P)$ 到 $\mathcal{P}\big(A(P)\big)$ 的双射, 而且对任意 $\gamma \in \mathcal{LP}_A(P)$ 它也决定下面的 $C_N(P)$-内 $N_G(P_\gamma)$-代数同构

Corollary 6.6. *A pointed group P_γ on A is local if and only if $\mathrm{Br}_P(\gamma)$ is not zero. In particular, Br_P determines a bijective map from $\mathcal{LP}_A(P)$ to $\mathcal{P}\big(A(P)\big)$, and moreover, for any $\gamma \in \mathcal{LP}_A(P)$, it determines the following $C_N(P)$-interior $N_G(P_\gamma)$-algebra isomorphism*

6.6.1 $$A(P_\gamma) \cong \big(A(P)\big)\big(\mathrm{Br}_P(\gamma)\big)\,.$$

证明: 如果 $\mathrm{Br}_P(\gamma) \neq 0$ 那么对任意 P 的真子群 Q 有 $\gamma \not\subset A_Q^P$, 从而 γ 是局部的 (见命题 6.3). 反过来, 假定 $\mathrm{Br}_P(\gamma)=0$; 只要在同构 6.5.1 里使 P,γ 代替 H,ω 就得到 $A_\gamma(P) = \{0\}$; 也就是说, 有

Proof: If $\mathrm{Br}_P(\gamma) \neq 0$ then we have $\gamma \not\subset A_Q^P$ for any proper subgroup Q of P and therefore γ is local (see Proposition 6.3). Conversely, assume that we have $\mathrm{Br}_P(\gamma) = 0$; then, by replacing H and ω by P and γ in isomorphism 6.5.1, we get $A_\gamma(P) = \{0\}$; that is to say, we have

6.6.2 $$iA^Pi = \sum_Q iA_Q^Pi + J(\mathcal{O})\cdot iA^Pi\,,$$

其中 $i \in \gamma$ 并 Q 跑遍 P 的真子群的集合; 可是, 因为商 $iA^Pi/J(iA^Pi)$ 是可除代数, 并且 iA_Q^Pi 的像在这个商里是一个理想 (见 5.4.1), 所以存在一个 P 的真子群 Q 使得 $iA^Pi = iA_Q^Pi$, 从而 $\gamma \subset A_Q^P$; 再次引用命题 6.3, γ 不是局部的. 最后, 只要把命题 3.23 用到 $\mathrm{Br}_P\colon A^P \to A(P)$ 就证明完毕.

where $i \in \gamma$ and Q runs over the set of proper subgroups of P; but, since the quotient $iA^Pi/J(iA^Pi)$ is a division algebra and the image of iA_Q^Pi in this quotient is an ideal (see 5.4.1), there exists a proper subgroup Q of P such that $iA^Pi = iA_Q^Pi$ and therefore $\gamma \subset A_Q^P$; quoting Proposition 6.3 again, γ is not local. Finally, it suffices to apply Proposition 3.23 to the homomorphism $\mathrm{Br}_P\colon A^P \to A(P)$, to complete the proof.

6.7. 下面的定理是关于局部点群的基本事实; 如同注意 6.9 指出, 这个定理推广通常的 Sylow 定理, 可是别忘了我们在命题 6.3 的证明中已经使用了 Sylow 定理.

6.7. Next theorem states a fundamental fact on local pointed groups; as Remark 6.9 below explains, this result generalizes the ordinary Sylow Theorems, but note that we have already employed the Sylow Theorems in the proof of Proposition 6.3.

定理 6.8. 设 H_β 是 A 上的点群. 假定 P_γ 是满足 $P \subset H$ 与 $\beta \subset \mathrm{ind}_P^H(\gamma)$ 的 A 上的极小点群. 那么 P_γ 是局部的, 并且

Theorem 6.8. *Let H_β be a pointed group on A. Assume that P_γ is a minimal pointed group on A which fulfills $P \subset H$ and $\beta \subset \mathrm{ind}_P^H(\gamma)$. Then P_γ is local and we have*

$P_\gamma \subset H_\beta$. **而且, 对 A 上的任意在 H_β 里包含的局部点群 Q_δ 存在 $x \in H$ 使得 $(Q_\delta)^x \subset P_\gamma$.**

$P_\gamma \subset H_\beta$. Moreover, for any local pointed group Q_δ on A which is contained in H_β, there exists $x \in H$ such that $(Q_\delta)^x \subset P_\gamma$.

证明: 设 R_ε 是一个 A 上的点群使得 $R_\varepsilon \subset P_\gamma$ 与 $\gamma \subset \mathrm{ind}_R^P(\varepsilon)$; 由命题 5.6 和推论 5.7, 得到

Proof: Let R_ε be a pointed group on A fulfilling $R_\varepsilon \subset P_\gamma$ and $\gamma \subset \mathrm{ind}_R^P(\varepsilon)$; by Proposition 5.6 and Corollary 5.7, we get

6.8.1 $$\beta \subset \mathrm{ind}_P^H(\gamma) \subset \mathrm{ind}_P^H\big(\mathrm{ind}_R^P(\varepsilon)\big) = \mathrm{ind}_R^H(\varepsilon)\,;$$

那么, 极小的条件推出 $R_\varepsilon = P_\gamma$, 从而 P_γ 是局部的. 另一方面, 别忘了有 $Q_\delta \subset H_\beta$ 当且仅当有 $Q \subset H$ 与 $\mathrm{res}_Q^H(\beta)_\delta \neq 0$ (见 5.2); 因为 $\beta \subset \mathrm{ind}_P^H(\gamma)$ 所以我们还有 $\mathrm{res}_Q^H\big(\mathrm{ind}_P^H(\gamma)\big)_\delta \neq 0$. 可是, 由推论 5.7, 得到

then, the minimality condition forces $R_\varepsilon = P_\gamma$ and therefore P_γ is local. On the other hand, recall that we have $Q_\delta \subset H_\beta$ if and only if we have $Q \subset H$ and $\mathrm{res}_Q^H(\beta)_\delta \neq 0$ (see 5.2); moreover, since $\beta \subset \mathrm{ind}_P^H(\gamma)$, we still have $\mathrm{res}_Q^H\big(\mathrm{ind}_P^H(\gamma)\big)_\delta \neq 0$. But it follows from Corollary 5.7 that we have

6.8.2 $$\mathrm{res}_Q^H\big(\mathrm{ind}_P^H(\gamma)\big) = \sum_{x\in X} \mathrm{ind}_{Q\cap P^x}^Q\big(\mathrm{res}_{Q\cap P^x}^{P^x}(\gamma^x)\big)\,,$$

其中 X 是一个 $P\backslash H/Q$ 的代表集合. 这样, 因为 δ 是局部的, 所以得到

where X is a set of representatives for $P\backslash H/Q$. Consequently, since δ is local, we get

6.8.3 $$\mathrm{res}_Q^H\big(\mathrm{ind}_P^H(\gamma)\big)_\delta = \sum_{x\in Y} \mathrm{res}_Q^{P^x}(\gamma^x)_\delta\,,$$

其中 Y 是满足 $Q \subset P^x$ 的 $x \in X$ 的集合 (见等式 4.12.5); 因此存在 $y \in Y$ 使得 $0 \neq \mathrm{res}_Q^{P^y}(\gamma^y)_\delta$, 从而有 $Q_\delta \subset (P_\gamma)^y$ (见 5.2).

where Y is the set of elements $x \in X$ which fulfill $Q \subset P^x$ (for the notation, see equality 4.12.5); hence, there exists $y \in Y$ such that $\mathrm{res}_Q^{P^y}(\gamma^y)_\delta \neq 0$ and therefore we have $Q_\delta \subset (P_\gamma)^y$ (see 5.2).

特别是, 只要把上面的证明用于 A_β 并且只要选择满足 $\beta \subset \mathrm{ind}_Q^H(\delta)$ 的 A_β 上的极小点群 Q_δ, 就已经知道 δ 是局部的并且有 $Q_\delta \subset H_\beta$; 那么, 因为 P_γ 是满足 $P \subset H$ 并 $\beta \subset \mathrm{ind}_P^H(\gamma)$ 的 A 上的极小点群, 所以, 由于

In particular, by applying the argument above to A_β, and by choosing a pointed group Q_δ on A_β which is minimal fulfilling $\beta \subset \mathrm{ind}_Q^H(\delta)$, we already know that Q_δ is local and that $Q_\delta \subset H_\beta$; then, since P_γ is a minimal pointed group on A fulfilling $P \subset H$ and $\beta \subset \mathrm{ind}_P^H(\gamma)$, and since we have

6.8.4 $$(Q_\delta)^{y^{-1}} \subset P_\gamma\,, \quad \beta \subset \mathrm{ind}_{Q^{y^{-1}}}^H(\delta^{y^{-1}})\,,$$

我们得到 $P_\gamma = (Q_\delta)^{y^{-1}}$ 从而也得到 $P_\gamma \subset H_\beta$.

we obtain $P_\gamma = (Q_\delta)^{y^{-1}}$ and therefore we also obtain $P_\gamma \subset H_\beta$.

注意 6.9. 如果 $A = \mathcal{O}$ 那么 A 上的点群的点都等于 $\{1\}$; 这样, 点群的包含关系就是其群的包含关系; 所以, 只要用对应的完备诱导的 N-内 G-代数代替 $\mathcal{O}$, 上述的条件 $\beta \subset \mathrm{ind}_P^H(\gamma)$ 就等价于 p 不整除 $|H\colon P|$, 也就是说, 就等价于 p-群 P 是一个 H 的 Sylow p-子群; 而且, Q_δ 是局部的当且仅当 Q 是一个 p-群.

Remark 6.9. If $A = \mathcal{O}$ then the points of all the pointed groups on A are equal to $\{1\}$; thus, the inclusion between pointed groups coincides with the inclusion between their groups; hence, replacing $\mathcal{O}$ by the corresponding inductively complete N-interior G-algebra, the condition $\beta \subset \mathrm{ind}_P^H(\gamma)$ above is equivalent to the fact that p does not divide $|H\colon P|$; that is to say, to the fact that P is a Sylow p-subgroup of H; moreover Q_δ is local if and only if Q is a p-group.

6.10. 上述定理中的 P_γ 称为 H_β 的亏点群 (也称 P 是 H_β 的亏群), 而 A_γ 称为 A_β 的源代数; 别忘了, 上述源代数是一个 $(N\cap P)$-内 P-代数; 而且, 由 3.16, A_γ 是 Morita 等价于 A_β 的. 由上面的定理, 一个 H_β 的亏点群也是一个在 H_β 里包含的极大局部点群. 下面的推论给出亏点群的第三等价定义.

6.10. We call the pointed group P_γ in the theorem a *defect pointed group of* H_β (we still call P a *defect group of* H_β), and A_γ a *source algebra of* A_β; recall that this source algebra actually is an $(N\cap P)$-interior P-algebra; moreover, by 3.16, A_γ is Morita equivalent to A_β. Note that, by the theorem above, a defect pointed group of H_β is also a maximal local pointed group contained in H_β. Next corollary gives a third equivalent definition.

6.11. 如果 $A = \mathrm{End}_{\mathcal{O}}(M)$, 其中 M 是一个 $\mathcal{O}G$-模, 那么对任意 $i \in \beta$ 与 $j \in \gamma$ 已经知道 $i(M)$ 与 $j(M)$ 分别是不可分解 $\mathcal{O}H$-模与 $\mathcal{O}P$-模 (见例子 4.15); 此时, P 并 $j(M)$ 分别是所谓 $i(M)$ 的顶点并源.

6.11. If $A = \mathrm{End}_{\mathcal{O}}(M)$, where M is an $\mathcal{O}G$-module, then, for any $i \in \beta$ and any $j \in \gamma$, we already know that $i(M)$ and $j(M)$ respectively are indecomposable $\mathcal{O}H$- and $\mathcal{O}P$-modules (see example 4.15); in this case, P and $j(M)$ respectively are the so-called *vertex* and *source* of $i(M)$.

推论 6.12. 设 H_β 是 A 上的点群. 一个 A 上的局部点群 P_γ 是 H_β 的亏点群当且仅当 $P_\gamma \subset H_\beta$ 并且 $\beta \subset A_P^H$.

Corollary 6.12. *Let H_β a pointed group on A. A local pointed group P_γ on A is a defect pointed group of H_β if and only if $P_\gamma \subset H_\beta$ and $\beta \subset A_P^H$.*

证明: 如果 P_γ 是 H_β 的亏点群那么, 由上面的定理, 有 $P_\gamma \subset H_\beta$ 与 $\beta \subset A_P^H$. 反过来, 显然存在一个 H_β 的亏点群 Q_δ 使得 $P_\gamma \subset Q_\delta$; 因为 $Q_\delta \subset H_\beta$

Proof: If P_γ is a defect pointed group of H_β then, according to the theorem above, we have $P_\gamma \subset H_\beta$ and $\beta \subset A_P^H$. Conversely, it is quite clear that there exists a defect pointed group Q_δ of H_β such that $P_\gamma \subset Q_\delta$, and

所以 $\delta \subset \mathrm{res}_Q^H(\beta)$；可是不难证明 (见 Mackey 公式 5.7.1)

we have $\delta \subset \mathrm{res}_Q^H(\beta)$ since $Q_\delta \subset H_\beta$; but, it is easily checked (see Mackey Formula 5.7.1)

6.12.1 $$A_P^H \subset \sum_{x\in T} \mathrm{Tr}_{Q\cap P^x}^Q (A_{Q\cap P^x}^{P^x}),$$

其中T是一个$P/H\backslash Q$的代表集合; 这样, 因为δ是局部的, 所以存在$x\in T$使得$Q=P^x$ (见命题 6.3), 从而$Q_\delta = P_\gamma$.

where T is a set of representatives for $P/H\backslash Q$; consequently, since δ is local, there is $x \in H$ such that $Q = P^x$ (see Proposition 6.3) and therefore we get $Q_\delta = P_\gamma$.

6.13. 这样, 任意A上的点群决定唯一一个A上的局部点群的G-共轭类. 反过来, 当固定A上的局部点群P_γ的时候, 下面的定理指出我们能用$C_N(P)$-内$N_G(P_\gamma)$-代数$A(P_\gamma)$来决定A上的所有这种点群H_β使得P_γ是H_β的亏点群; 令$\mathcal{P}_A(H)_\gamma$记满足P_γ是H_β的亏点群的点$\beta\in\mathcal{P}_A(H)$的集合. 再令

6.13. Thus, any pointed group on A determines a unique G-conjugacy class of local pointed groups on A. Conversely, once we fix a local pointed group P_γ on A, the theorem below allows us to employ the $C_N(P)$-interior $N_G(P)$-algebra $A(P)$ to determine all the pointed groups H_β on A such that P_γ is a defect pointed group of H_β ; denote by $\mathcal{P}_A(H)_\gamma$ the set of points $\beta \in \mathcal{P}_A(H)$ such that P_γ is a defect pointed group of H_β. Set

6.13.1 $$\bar{N}_G(P_\gamma) = N_G(P_\gamma)/P\,,\quad \bar{C}_N(P) = C_N(P)/\big(N\cap Z(P)\big)\,.$$

请注意, $A(P_\gamma)$ 也是$\bar{C}_N(P)$-内$\bar{N}_G(P_\gamma)$-代数.

Note that $A(P_\gamma)$ is also a $\bar{C}_N(P)$-interior $\bar{N}_G(P_\gamma)$-algebra.

定理 6.14. **设P_γ是A上的局部点群并设H是G的子群使得$P\subset H$. 对任意$\beta\in\mathcal{P}_A(H)$, P_γ是H_β的亏点群当且仅当有**

Theorem 6.14. *Let P_γ be a local pointed group on A and H a subgroup of G such that $P \subset H$. For any $\beta \in \mathcal{P}_A(H)$, P_γ is a defect pointed group of H_β if and only if we have*

6.14.1 $$s_\gamma(\beta) \subset A(P_\gamma)_1^{\bar{N}_H(P_\gamma)} - \{0\}\,.$$

此时, $s_\gamma(\beta)$是$A(P_\gamma)^{\bar{N}_H(P_\gamma)}$的点. 而且同态$s_\gamma\colon A^P \to A(P_\gamma)$给出下面的双射

In this case, $s_\gamma(\beta)$ is a point of $A(P_\gamma)^{\bar{N}_H(P_\gamma)}$. Further, the homomorphism $s_\gamma\colon A^P \to A(P_\gamma)$ induces a bijective map

6.14.2 $$\mathcal{P}_A(H)_\gamma \longrightarrow \{\bar\beta \in \mathcal{P}_{A(P_\gamma)}\big(\bar{N}_H(P_\gamma)\big) \mid \bar\beta \subset A(P_\gamma)_1^{\bar{N}_H(P_\gamma)}\}$$

并, 对任意相应的偶对$(\beta,\bar\beta)$, 它决定下面的k-代数同构

and, for any pair $(\beta,\bar\beta)$ in the correspondence, it determines the following k-algebra isomorphism

6.14.3 $$A(H_\beta) \cong A(P_\gamma)\big(\bar{N}_H(P_\gamma)_{\bar\beta}\big)\,.$$

证明: 由命题 3.23 和 5.2, s_γ 决定一个从满足 $P_\gamma \subset H_\beta$ 的点 $\beta \in \mathcal{P}_A(H)$ 的集合到 $\mathcal{P}(s_\gamma(A^H))$ 的双射, 并且对任意相应的偶对 $(\beta, s_\gamma(\beta))$ 它也决定一个 k-代数同构

Proof: It follows from Proposition 3.23 and 5.2 that s_γ determines a bijective map from the set of $\beta \in \mathcal{P}_A(H)$ fulfilling $P_\gamma \subset H_\beta$ onto $\mathcal{P}(s_\gamma(A^H))$, and moreover, for any pair $(\beta, s_\gamma(\beta))$, it still determines a k-algebra isomorphism

6.14.4 $$A(H_\beta) \cong \big(s_\gamma(A^H)\big)\big(s_\gamma(\beta)\big);$$

而且, 有 $s_\gamma(\beta) \subset s_\gamma(A_P^H)$ 当且仅当有 $\beta \subset A_P^H$; 也就是说, 由推论 6.12, 同态 s_γ 决定一个双射如下

furthermore, we have $s_\gamma(\beta) \subset s_\gamma(A_P^H)$ if and only if we have $\beta \subset A_P^H$; in other terms, by Corollary 6.12, the homomorphism s_γ induces a bijective map

6.14.5 $$\mathcal{P}_A(H)_\gamma \longrightarrow \{\bar\beta \in \mathcal{P}\big(s_\gamma(A^H)\big) \mid \bar\beta \subset s_\gamma(A_P^H)\}.$$

这样, 只要使用下面的引理并且注意 3.9 就完证明.

Thus, to complete the proof, it suffices to apply the next lemma and Remark 3.9.

引理 6.15. 设 P_γ 是 A 上的局部点群并设 H 是 G 的子群使得 $P \subset H$. 如果 $a \in \bigcap_\delta \mathrm{Ker}(s_\delta)$ 其中 δ 跑遍 $\mathcal{LP}_A(P) - \{\gamma\}$ 那么有

Lemma 6.15. *Let P_γ be a local pointed group on A and H a subgroup of G such that $P \subset H$. If $a \in \bigcap_\delta \mathrm{Ker}(s_\delta)$, where δ runs over $\mathcal{LP}_A(P) - \{\gamma\}$ then we have*

6.15.1 $$s_\gamma\big(\mathrm{Tr}_P^H(a)\big) = \mathrm{Tr}_1^{\bar N_H(P_\gamma)}\big(s_\gamma(a)\big).$$

特别是有

In particular, we have

6.15.2 $$s_\gamma(A_P^H) = A(P_\gamma)_1^{\bar N_H(P_\gamma)}.$$

证明: 设 T 是一个 $P/H\backslash P$ 的代表集合; 因为显然 $P\backslash H = \sqcup_{x\in T} P\backslash PxP$ 并对任意 $x \in T$ 左边乘 x 决定一个从 $(P \cap P^x)\backslash P$ 到 $P\backslash PxP$ 的双射, 所以对任意 $a \in A^P$ 得到

Proof: Let T be a set of representatives for $P/H\backslash P$; since we have $P\backslash H = \sqcup_{x\in T} P\backslash PxP$ and, for any $x \in T$, the multiplication on the left by x defines a bijective map from $(P \cap P^x)\backslash P$ onto $P\backslash PxP$, for any $a \in A^P$ we have

6.15.3 $$\mathrm{Tr}_P^H(a) = \sum_{x\in T} \mathrm{Tr}_{P\cap P^x}^P(a^x),$$

从而, 由同构 6.6.1, 也得到

and thus, from isomorphism 6.6.1, we get

6.15.4 $$s_\gamma\big(\mathrm{Tr}_P^H(a)\big) = \sum_{x\in N_T(P)} s_\gamma(a^x) = \sum_{y\in S} \mathrm{Tr}_1^{\bar N_H(P_\gamma)}\big(s_\gamma(a^y)\big),$$

其中S是$N_H(P)/N_H(P_\gamma)$的代表集合; 这样, $A(P_\gamma)_1^{\bar{N}_H(P_\gamma)}$包含$s_\gamma(A_P^H)$; 最后, 如果

where S denotes a set of representatives for $N_H(P)/N_H(P_\gamma)$; thus, $A(P_\gamma)_1^{\bar{N}_H(P_\gamma)}$ contains $s_\gamma(A_P^H)$; finally, if

6.15.5 $$a \in \bigcap_{\delta \in \mathcal{LP}_A(P) - \{\gamma\}} \mathrm{Ker}(s_\delta)\,, \quad y \in N_T(P) - N_T(P_\gamma)\,,$$

那么a属于$\mathrm{Ker}(s_{\gamma^{y^{-1}}})$, 从而有$s_\gamma(a^y) = 0$.

a belongs to $\mathrm{Ker}(s_{\gamma^{y^{-1}}})$ and therefore we get $s_\gamma(a^y) = 0$.

推论 6.16. **设P_γ是A上的局部点群并设H是G的子群使得$P \subset H$. 如果β与ν分别是A上的H-点与$N_H(P_\gamma)$-点使得**

Corollary 6.16. *Let P_γ be a local pointed group on A and H a subgroup of G such that $P \subset H$. If β and ν respectively are points of H and $N_H(P_\gamma)$ on A such that*

6.16.1 $$P_\gamma \subset N_H(P_\gamma)_\nu \subset H_\beta\,,$$

那么P_γ是H_β的亏点群当且仅当P_γ是$N_H(P_\gamma)_\nu$的亏点群. 此时β与ν彼此决定.

P_γ is a defect pointed group of H_β if and only if P_γ is a defect pointed group of $N_H(P_\gamma)_\nu$. Then, β and ν determine each other.

证明: 如果P_γ是一个在H_β中包含的极大局部点群, 那么在$N_H(P_\gamma)_\nu$中也是极大的. 反过来, 如果$s_\gamma(\nu) \subset A(P_\gamma)_1^{\bar{N}_H(P_\gamma)}$那么$s_\gamma(\nu)$是一个$A(P_\gamma)^{\bar{N}_H(P_\gamma)}$的点, 从而存在$\beta' \in \mathcal{P}_A(H)$使得$P_\gamma \subset H_{\beta'}$与$s_\gamma(\beta') = s_\gamma(\nu)$ (见定理 6.14). 另一方面, 由命题 3.23 并 5.2, $s_\gamma(\beta)$与$s_\gamma(\beta')$是两个$s_\gamma(A^H)$的点; 而且, 因为H_β包含$N_H(P_\gamma)_\nu$所以存在$i \in \beta$与$j \in \nu$使得$ij = j = ji$, 从而得到

Proof: If P_γ is a local pointed group which is maximal contained in H_β, then it is also maximal contained in $N_H(P_\gamma)_\nu$. Conversely, if $s_\gamma(\nu) \subset A(P_\gamma)_1^{\bar{N}_H(P_\gamma)}$ then $s_\gamma(\nu)$ is a point of $A(P_\gamma)^{\bar{N}_H(P_\gamma)}$ and therefore there exists $\beta' \in \mathcal{P}_A(H)$ such that we have $P_\gamma \subset H_{\beta'}$ and $s_\gamma(\beta') = s_\gamma(\nu)$ (see Theorem 6.14). On the other hand, according to Proposition 3.23 and to 5.2, $s_\gamma(\beta)$ and $s_\gamma(\beta')$ are two points of $s_\gamma(A^H)$; moreover, since H_β contains $N_H(P_\gamma)_\nu$, there exist $i \in \beta$ and $j \in \nu$ fulfilling $ij = j = ji$, and therefore we get the equality

6.16.2 $$s_\gamma(i)s_\gamma(j) = s_\gamma(j) = s_\gamma(j)s_\gamma(i)\,.$$

最后, 因为在$s_\gamma(A^H)$里$s_\gamma(i)$是本原的, 所以得到$s_\gamma(\beta) = s_\gamma(\beta')$, 从而也得到$\beta = \beta'$ (见命题 3.23).

Finally, since the idempotent $s_\gamma(i)$ is primitive in $s_\gamma(A^H)$, we get $s_\gamma(\beta) = s_\gamma(\beta')$ and therefore we still get $\beta = \beta'$ (see Proposition 3.23).

推论 6.17. 如果 P_γ 与 Q_δ 是 A 上的局部点群使得 P_γ 真包含 Q_δ 那么存在一个 P_γ 的局部子点群 R_ε 使得 Q_δ 就是 R_ε 的正规真子点群.

Corollary 6.17. *If P_γ and Q_δ are local pointed groups on A such that P_γ properly contains Q_δ, then there exists a local pointed group R_ε contained in P_γ such that Q_δ is a proper normal subgroup of R_ε.*

证明: 别忘了, 因为 $Q_\delta \subset P_\gamma$ 所以 Q_δ 也是 A_γ 上的点群, 并且有 A_γ 上的 $N_P(Q_\delta)$-点 ν 使得 $Q_\delta \subset N_P(Q_\delta)_\nu$ (见 5.2); 又, ν 也是 A 上的 $N_P(Q_\delta)$-点使得

Proof: Recall that, since $Q_\delta \subset P_\gamma$, Q_δ is a pointed group on A_γ too, and then there exists a point ν of $N_P(Q_\delta)$ on A_γ such that $Q_\delta \subset N_P(Q_\delta)_\nu$ (see 5.2); moreover, ν still is a point of $N_P(Q_\delta)$ on A such that

6.17.1 $$Q_\delta \subset N_P(Q_\delta)_\nu \subset P_\gamma$$

并且, 因为 Q_δ 不是 P_γ 的亏点群, 所以 Q_δ 也不是 $N_P(Q_\delta)_\nu$ 的亏点群 (见推论 6.16); 这样, Q_δ 在 $N_P(Q_\delta)_\nu$ 的局部子点群之间不是极大的.

and, since Q_δ is not a defect pointed group of P_γ, Q_δ is not a defect pointed group of $N_P(Q_\delta)_\nu$ too (see Corollary 6.16); thus, Q_δ is not maximal in the set of local pointed groups contained in $N_P(Q_\delta)_\nu$.

6.18. 一般来说, 关于点群的很多问题的解答只是依赖于局部点群的集合. 设 B 是一个 N-内 G-代数, 并设 $f: A \to B$ 是幺 N-内 G-代数同态; 下面的结果表示 f 是严格覆盖的条件 (见 4.13) 就是这种问题; 事实上, 我们在超聚焦子代数的存在性的证明里要用这个结果. 请注意, 如果 f 是严格覆盖的, 那么对任意 A 上的点群 H_β 与 K_γ 有 $K_\gamma \subset H_\beta$ 当且仅当 $K_{\mathrm{res}_{fK}(\gamma)} \subset H_{\mathrm{res}_{fH}(\beta)}$; 而且, 不难验证 B 也是完备诱导的, 并且如果 $K \subset H$ 那么有

6.18. Generally speaking, the answer of many questions about pointed groups only depends on the set of local pointed groups. Let B be an N-interior G-algebra and $f: A \to B$ a unitary N-interior G-algebra homomorphism; the next result shows that the question whether or not f is a strict covering (see 4.13) is one of this kind; as a matter of fact, we need this result in the proof of the existence of the hyperfocal subalgebra. Note that, if f is a strict covering, for any pointed groups H_β and K_γ on A, we have $K_\gamma \subset H_\beta$ if and only if we have $K_{\mathrm{res}_{fK}(\gamma)} \subset H_{\mathrm{res}_{fH}(\beta)}$; further, B is inductively complete, and if $K \subset H$ then

6.18.1 $$\mathrm{res}_{fH} \circ \mathrm{ind}_K^H = \mathrm{ind}_K^H \circ \mathrm{res}_{fK} ;$$

特别是, H_β 是局部的当且仅当 $H_{\mathrm{res}_{fH}(\beta)}$ 是局部的.

in particular, H_β is local if and only if $H_{\mathrm{res}_{fH}(\beta)}$ is local.

推论 6.19. 设B是N-内G-代数并设$f\colon A \to B$是N-内G-代数幺同态. 假定对任意G的p-子群P, 映射

Corollary 6.19. *Let B be an N-interior G-algebra and $f\colon A \to B$ a unitary N-interior G-algebra homomorphism. Assume that, for any p-subgroup P of G, the map*

6.19.1 $$\mathrm{res}_{f^P}; \mathcal{LP}_A(P) \longrightarrow \mathcal{LP}_B(P)$$

是双射并对任意$\gamma \in \mathcal{LP}_A(P)$, f^P 诱导下面的同构

is bijective, and that, for any $\gamma \in \mathcal{LP}_A(P)$, f^P induces the following isomorphism

6.19.2 $$A(P_\gamma) \cong B(P_{\mathrm{res}_{f^P}(\gamma)}) .$$

那么f是严格覆盖的.

Then, f is a strict covering.

证明: 设H_β是A上的点群, 并设P_γ是H_β的亏点群; 由定理 6.14, $s_\gamma(\beta)$是$A(P_\gamma)_1^{\bar{N}_H(P_\gamma)}$的点, 并且有

Proof: Let H_β be a pointed group on A and P_γ a defect pointed group of H_β; by Theorem 6.14, $s_\gamma(\beta)$ is a point of $A(P_\gamma)_1^{\bar{N}_H(P_\gamma)}$ and we have

6.19.3 $$A(H_\beta) \cong A(P_\gamma)\big(\bar{N}_H(P_\gamma)_{s_\gamma(\beta)}\big) ;$$

记$\gamma^f = \mathrm{res}_{f^P}(\gamma)$; 因为显然有$\bar{N}_H(P_\gamma) = \bar{N}_H(P_{\gamma^f})$ 并且有下面的同构 (见 6.19.2)

set $\gamma^f = \mathrm{res}_{f^P}(\gamma)$; since we clearly have the equality $\bar{N}_H(P_\gamma) = \bar{N}_H(P_{\gamma^f})$ and have the following isomorphism (see 6.19.2)

6.19.4 $$f^\gamma\colon A(P_\gamma) \cong B(P_{\gamma^f})$$

使得$f^\gamma \circ s_\gamma = s_{\gamma^f} \circ f$, 所以$s_{\gamma^f}(f(\beta))$是$B(P_{\gamma^f})_1^{\bar{N}_H(P_{\gamma^f})}$的点; 这样, 只要再次应用定理 6.14, 就存在$\beta^f \in \mathcal{P}_B(H)$使得$P_{\gamma^f}$是$H_{\beta^f}$的亏点群, 并且有

fulfilling $f^\gamma \circ s_\gamma = s_{\gamma^f} \circ f$, the image $s_{\gamma^f}\big(f(\beta)\big)$ is a point of $B(P_{\gamma^f})_1^{\bar{N}_H(P_{\gamma^f})}$; hence, by Theorem 6.14 again, there exists a point $\beta^f \in \mathcal{P}_B(H)$ such that P_{γ^f} is a defect pointed group of H_{β^f}, and moreover we have

6.19.5 $$B(H_{\beta^f}) \cong B(P_{\gamma^f})\big(\bar{N}_H(P_{\gamma^f})_{s_{\gamma^f}(f(\beta))}\big) .$$

也就是说, 对G的任意子群H存在一个被$\iota^H(\beta) = \beta^f$定义的映射 $\iota^H\colon \mathcal{P}_A(H) \to \mathcal{P}_B(H)$.

that is to say, for any subgroup H of G, there exists a map $\iota^H\colon \mathcal{P}_A(H) \to \mathcal{P}_B(H)$ defined by $\iota^H(\beta) = \beta^f$.

此时, 对任意$\beta \in \mathcal{P}_A(H)$有$A(H_\beta) \cong B(H_{\beta^f})$, 并且$H_\beta$与$H_{\beta^f}$的亏群一样; 进一步, 如果$P_\gamma$是$H_\beta$的亏点群, 那么$P_{\gamma^f}$是$H_{\beta^f}$的亏点群; 从而, 由定理 6.14, ι^H是单射的, 并且因为P_{γ^f}跑遍B上的局部点

At the same time, for any $\beta \in \mathcal{P}_A(H)$ we have $A(H_\beta) \cong B(H_{\beta^f})$ and the defect groups of H_β and H_{β^f} coincide; further, if P_γ is a defect pointed group of H_β then P_{γ^f} is a defect pointed group of H_{β^f}; hence, it follows from the Theorem 6.14 that ι^H is injective, and moreover, since P_{γ^f} runs over

群的集合, 所以 H_{β^f} 跑遍 B 上的所有点群的集合 (见定理 6.8). 最后, 我们得到 f 决定一个满同态

the set of all the local pointed groups on B, H_{β^f} runs over the set of all the pointed groups on B (see Theorem 6.8). Finally, f induces a surjective homomorphism

6.19.6 $$A^H \longrightarrow \prod_{\beta\in\mathcal{P}_A(H)} B(H_{\beta^f}) \cong B^H/J(B^H)\,,$$

从而 $B^H = f(A^H) + J(B^H)$; 而且, 由上术的假定, 如果 $P=1$ 那么得到 $A/J(A) \cong B/J(B)$, 从而 $\mathrm{Ker}(f) \subset J(A)$. 这样, f 是严格覆盖的 (见 4.13).

and thus we have $B^H = f(A^H) + J(B^H)$; moreover, according to our hypothesis above, if $P = 1$ then we get $A/J(A) \cong B/J(B)$, and therefore we still get $\mathrm{Ker}(f) \subset J(A)$. Thus, f is a strict covering (see 4.13).

7. 论 Green 不可分解性定理

7.1. 如同在§6里一样, G 是一个有限群, N 是 G 的正规子群, A 是一个 N-内 G-代数, 可是这里我们特别假定 G 是 p-群并假定 k 是代数闭的; 也与 6.1 所述, 能假定 A 是完备诱导的. 在这个情况里我们将看到: 任意点群 H_β 的点被 H_β 的亏点群所决定. 如果 $A = \mathrm{End}_{\mathcal{O}}(M)$, 其中 M 是一个 $\mathcal{O}G$-模, 那么这个事实就是所谓 Green 不可分解性定理. 为了证明这个事实我们需要下面的引理; 事实上, 我们在§10里将仍使用这个引理.

定理 7.2. **假定 k 是代数闭的, 还假定 G 是一个 p-群. 对任意 A 上的点群 H_β, 除子 $\mathrm{ind}_H^G(\beta)$ 是 A 上的 G-点.**

证明: 对阶 $|G|$ 用归纳法. 显然能假定 $H \neq G$; 设 K 是 G 的极大子群使得 $H \subset K$; 因为 G 是 p-群, 所以 K 是正规的, 而 $|G/K| = p$. 那么只要考虑 $\mathrm{Res}_K^G(A)$ 就得到 $\mathrm{ind}_H^K(\beta)$ 是一个 A 上的 K-点.

也就是说, 能假定 $H \lhd G$ 与 $|G/H| = p$ (见 5.7.3); 事实上, 因为 $\mathrm{ind}_H^G(\beta)$ 是 A^G 的除子, 并且 $A^G = (A^H)^{G/H}$, 所以还能假定有 $|G| = p$ 与 $H = 1$. 那么, 选择有正交 $G/1$-迹的 $j \in \beta$, 而记 $i = \mathrm{Tr}_1^G(j)$; 因为 j 在 A 里是本原的, 所以下面的引理推出 j 在 AG 里也是本原的; 也就是说, AGj 是一个不可分解投射 AG-模 (见

7. On Green's Indecomposability Theorem

7.1. As in §6, G is a finite group, N is a normal subgroup of G and A is an N-interior G-algebra, but this time we assume further that G is a p-group and that k is algebraically closed; as explained in 6.1 above, we may assume that A is inductively complete. In this particular situation we will see that the point of any pointed group H_β is determined by the defect pointed group of H_β. Whenever $A = \mathrm{End}_{\mathcal{O}}(M)$, where M is an $\mathcal{O}G$-module, this fact is just the so-called *Green Indecomposability Theorem.* In order to prove that statement we need the lemma below; actually, in §10 we will employ this lemma again.

Theorem 7.2. *Assume that k is algebraically closed and that G is a p-group. For any pointed group H_β on A, the divisor $\mathrm{ind}_H^G(\beta)$ is a point of G on A.*

Proof: We argue by induction on the order $|G|$ of G and may assume that $H \neq G$; let K be a maximal subgroup of G such that $H \subset K$; since G is a p-group, K is normal in G and we have $|G/K| = p$. Then, it suffices to consider $\mathrm{Res}_K^G(A)$, to get that $\mathrm{ind}_H^K(\beta)$ is a point of K on A.

That is to say, we may assume that $H \lhd G$ and that $|G/H| = p$ (see 5.7.3); actually, since $\mathrm{ind}_H^G(\beta)$ is a divisor of A^G and we have $A^G = (A^H)^{G/H}$, we still may assume that $|G| = p$ and $H = 1$. In this case, choose $j \in \beta$ having an orthogonal $G/1$-trace and set $i = \mathrm{Tr}_1^G(j)$; since j is primitive in A, the next lemma implies that j is also primitive in AG; consequently, AGj is an indecomposable projective AG-module (see 3.14.2); moreover, it is not difficult to prove

3.14.2); 而且, 不难证明有下面的 AG-模同构

that we have the following AG-module isomorphisms

7.2.1 $$\mathcal{O}G \otimes_{\mathcal{O}} Aj \cong AG \otimes_A Aj \cong AGj\,.$$

另一方面,在 A 里左边的乘以 A 以及 A 上的 G-作用在 Ai 上决定一个 AG-模结构; 那么, 因为有下面的 G-稳定直和分解

On the other hand, the multiplication on the left by A and the action of G, both on A, induce an AG-module structure on Ai; then, since we have the following G-stable direct sum decomposition

7.2.2 $$Ai \cong \bigoplus_{x\in G} Aj^x\,,$$

所以有 AG-模同构

we get the AG-module isomorphism

7.2.3 $$Ai \cong \mathcal{O}G \otimes_{\mathcal{O}} Aj$$

从而 Ai 是不可分解的 AG-模; 可是, 因为 $\mathrm{End}_A(Ai)^\circ \cong iAi$ (见 3.14.2), 所以还有

and therefore Ai is an indecomposable AG-module; but, since $\mathrm{End}_A(Ai)^\circ \cong iAi$ (see 3.14.2), we still get

7.2.4 $$\mathrm{End}_{AG}(Ai)^\circ \cong iA^G i\,;$$

因此, i 是 A^G 的本原幂等元.

hence, i is a primitive idempotent in A^G.

引理 7.3. 假定 k 是代数闭的, 并设 B 是一个 $\mathcal{O}$-代数. 假定存在一个有限 p-群 P 跟 B 的子 $\mathcal{O}$-模的一个集合 $\{B_u\}_{u\in P}$ 使得有

Lemma 7.3. *Assume that k is algebraically closed and let B be an $\mathcal{O}$-algebra. Assume that there exist a finite p-group P together with a family $\{B_u\}_{u\in P}$ of $\mathcal{O}$-submodules of B, which fulfill*

7.3.1 $$1 \in B_1\,, \quad B_u{\cdot}B_v \subset B_{uv}\,, \quad \sum_{u\in P} B_u = B$$

其中 $u,v \in P$. 那么有

for any $u,v\in P$. Then, we have

7.3.2 $$B_1 \cap J(B) = J(B_1)$$

并且 B_1 的任意本原幂等元在 B 中也是本原的.

and any primitive idempotent of B_1 remains primitive in B.

证明: 由推论 2.18, 能假定 $\mathcal{O} = k$; 当然还能假定 $P \neq 1$. 我们对阶 $|P|$ 使用归纳法; 设 Z 是 $Z(P)$ 的子群使得 $|Z| = p$; 令 $\bar{P} = P/Z$ 并对任意 $\bar{u} \in \bar{P}$ 再令 $B_{\bar{u}} = \sum_{u\in\bar{u}} B_u$; 那么不

Proof: According to Corollary 2.18, we may assume that $\mathcal{O} = k$; obviously, we may assume that $P \neq 1$. We argue by induction on $|P|$; let Z be a subgroup of $Z(P)$ such that $|Z| = p$; set $\bar{P} = P/Z$ and, for any $\bar{u} \in \bar{P}$, set $B_{\bar{u}} = \sum_{u\in\bar{u}} B_u$; then, it is easily checked

难验证 $\bar{P}$ 与 $\{B_{\bar{u}}\}_{\bar{u}\in\bar{P}}$ 也满足上述的条件. 因此我们得到 $B_Z \cap J(B) = J(B_Z)$ 并且 B_Z 的任意本原幂等元在B中也是本原的.

that $\bar{P}$ and $\{B_{\bar{u}}\}_{\bar{u}\in\bar{P}}$ still fulfill the condition above. Hence, we get $B_Z \cap J(B) = J(B_Z)$ and any primitive idempotent of B_Z remains primitive in B.

也就是说, 能假定 $|P| = p$. 令 $B{\cdot}J(B_1)^n{\cdot}B$ 记 B 的由 $J(B_1)^n$ 生成的理想,其中$n\geq 1$; 首先, 我们要证明 $B{\cdot}J(B_1){\cdot}B$ 是幂零的; 更精解地, 我们证明

That is to say, we may assume that $|P| = p$. Denote by $B{\cdot}J(B_1)^n{\cdot}B$ the ideal of B generated by $J(B_1)^n$, where $n \geq 1$; first of all, we have to prove that $B{\cdot}J(B_1){\cdot}B$ is nilpotent; more precisely, we prove that

7.3.3 $$(B{\cdot}J(B_1)^n{\cdot}B)^{p+1} \subset B{\cdot}J(B_1)^{2n}{\cdot}B\,,$$

其中$n \geq 1$; 确实, 任意左边项 $(B{\cdot}J(B_1)^n{\cdot}B)^{p+1}$ 的元素都是一些如下形式的项之和

where $n \geq 1$; indeed, all the elements in the left member $(B{\cdot}J(B_1)^n{\cdot}B)^{p+1}$ are sums of products having the following pattern

7.3.4 $$b_{u_0}\Big(\prod_{i=1}^{p+1} c_i\, b_{u_i}\Big)$$

其中 $u_i \in P$ 与 $b_{u_i} \in B_{u_i}$ 与 $c_i \in J(B_1)^n$; 可是, 只要考虑 $v_m = \prod_{i=1}^m u_i$, 其中 $1\leq m\leq p$, 与 $v_\circ = 1$, 就存在 j,k 使得 $0\leq j<k\leq p$ 与 $v_j = v_k$; 所以还有 $\prod_{i=j+1}^k u_i = 1$. 那么, 因为

where $u_i \in P$, $b_{u_i} \in B_{u_i}$ and $c_i \in J(B_1)^n$; but, considering the sequence $v_\circ = 1$ and $v_m = \prod_{i=1}^m u_i$, where $1 \leq m \leq p$, there are j and k such that $0 \leq j < k \leq p$ and $v_J = v_k$; consequently, we get $\prod_{i=j+1}^k u_i = 1$. Then, since we have

7.3.5 $$\Big(\prod_{i=j+1}^{k} c_i\, b_{u_i}\Big)c_{k+1} = c_{j+1}\Big(\prod_{i=j+1}^{k-1} b_{u_i}c_{i+1}\Big)\, b_{u_k}c_{k+1}\,,$$

所以这个积属于 $c_{j+1}B_1c_{k+1}$; 可是 $c_{j+1}B_1c_{k+1}$ 显然包含在 $J(B_1)^{2n}$ 中. 特别是, $J(B)$ 包含$J(B_1)$, 从而有 $B_1\cap J(B) = J(B_1)$.

this product belongs to $c_{j+1}B_1c_{k+1}$; but, $c_{j+1}B_1c_{k+1}$ is clearly contained in $J(B_1)^{2n}$. In particular, $J(B)$ contains $J(B_1)$ and therefore we have $B_1 \cap J(B) = J(B_1)$.

这样能假定 $J(B) = \{0\}$; 别忘了, 还要证明 B_1 的本原幂等元 i 在B里也是本原的; 我们对维数 $\dim_k(B)$ 使用归纳法. 不太难验证 P 与 $\{iB_ui\}_{u\in P}$ 也满足上述的条件; 而且, 因为 $J(B_1) = \{0\}$ 并且 i 是本原的, 所以 iB_1i 是一个可除 k-代

Thus, we may assume that $J(B) = \{0\}$; it remains to prove that any primitive idempotent i of B_1 is alsothis kind primitive in B; we argue by induction on $\dim_k(B)$. It is quite easy to check that P and $\{iB_ui\}_{u\in P}$ also fulfill the condition above; moreover, since $J(B_1) = \{0\}$ and i is primitive, iB_1i is a division k-algebra; thus, since k is alge-

数; 这样, 因为 k 是代数闭的, 所以得到 $iB_1i = k{\cdot}i$.

braically closed, we get $iB_1i = k{\cdot}i$.

最后, 假定 $B_1 = k$; 只要得到 $B = k + J(B)$ 就完成证明. 分两种情况; 首先也假定对任意 $u \in P - \{1\}$ 有 $B_u{\cdot}B_{u^{-1}} = \{0\}$; 那么有更精确的结论如下

Finally, assume that $B_1 = k$; in order to complete the proof, it suffices to prove that $B = k+J(B)$. We consider two cases; firstly, we also assume that, for any $u \in P - \{1\}$, we have $B_u{\cdot}B_{u^{-1}} = \{0\}$; then, we get the following more precise result

7.3.6
$$\Big(\sum_{u\in P-\{1\}} B_u\Big)^p = \{0\}\,;$$

确实, 任意左边之展开式的项之和都形如 $\prod_{i=1}^{p} b_{u_i}$, 其中 $u_i \in P - \{0\}$ 而 $b_{u_i} \in B_{u_i}$; 仍使用上一段的推理可得到 $1 \le j < k \le p$ 使得 $\prod_{i=j}^{k} u_i = 1$, 从而有 $\prod_{i=j}^{k} b_{u_i} = 0$.

indeed, any term in the displayed sum of the left member looks like $\prod_{i=1}^{p} b_{u_i}$, where $u_i \in P - \{0\}$ and $b_{u_i} \in B_{u_i}$; applying the argument above, this time we obtain $1 \le j < k \le p$ such that $\prod_{i=j}^{k} u_i = 1$ and therefore we obtain $\prod_{i=j}^{k} b_{u_i} = 0$.

不然, 就存在 $u \in P - \{0\}$ 使得 $B_u{\cdot}B_{u^{-1}} \neq \{0\}$; 因为 $B_u{\cdot}B_{u^{-1}} \subset B_1$ 与 $B_1 = k$, 所以 B_1 有一个可逆元素 b_u; 特别是, 得到

Otherwise, there exists $u \in P - \{0\}$ such that $B_u{\cdot}B_{u^{-1}} \neq \{0\}$; since $B_u{\cdot}B_{u^{-1}} \subset B_1$ and $B_1 = k$, B_1 contains an inversible element b_u; in particular, we get

7.3.7
$$B_{u^n} = B_1(b_u)^n = k{\cdot}(b_u)^n\,;$$

而且 $(b_u)^p = \lambda \in k^*$, 从而存在 $\mu \in k^*$ 使得 $(\mu b_u)^p = 1$, 这是因为 k 是代数闭的. 这样, 还存在一个满 k-代数同态 $\rho\colon kP \to B$ 使得 $\rho(u) = \mu b_u$; 那么, 因为有 $kP = k + J(kP)$, 所以还有 $B = k + J(B)$,

moreover, we have $(b_u)^p = \lambda \in k^*$ and therefore, since k is algebraically closed, there is $\mu \in k^*$ such that $(\mu b_u)^p = 1$. Consequently, there exists a surjective k-algebra homomorphism $\rho(u) = \mu b_u$; then, since we have $kP = k + J(kP)$, we still have $B = k + J(B)$,

推论 7.4. **假定 k 是代数闭的, 还假定 G 是 p-群. 在 A 里存在一个 G-稳定的相互正交幂等元的集合 I 使得任意 $i \in I$ 在 A^{G_i} 中是本原的, 并且满足 $\mathrm{Br}_{G_i}(i) \neq 0$, 其中 G_i 是 i 的**

Corollary 7.4. *Assume that k is algebraically closed and that G is a p-group. In A there exists a G-stable set I of pairwise orthogonal idempotents such that any $i \in I$ is primitive in A^{G_i} and fulfills $\mathrm{Br}_{G_i}(i) \neq 0$, where G_i is the stabilizer of i. Moreover,*

稳定化子. 而且, 如果 J 是 A 的 G-稳定的相互正交幂等元的集合, 那么存在 $a\in(A^G)^*$ 使得

if J is a G-stable set of pairwise orthogonal idempotents in A then there exists $a \in (A^G)^*$ fulfilling

7.4.1 $$\Big(\sum_{j\in J}\mathcal{O}{\cdot}j\Big)^a \subset \sum_{i\in I}\mathcal{O}{\cdot}i\,.$$

证明: 设 α 是 A 上的 G-点并设 P_γ 是 G_α 的亏点群; 这样, 如果 $i\in\gamma$ 是有正交 G/P-迹, 那么从定理 7.2 可推出 $\ell=\mathrm{Tr}_P^G(i)$ 属于 α; 也就是说, 对任意 $\ell\in\alpha$ 存在一个 G-稳定的相互正交幂等元的集合 I_ℓ 使得 P 就是 I_ℓ 的某一元素 i 的稳定化子, 并且 i 在 A^P 中是本原的, 以及 $\mathrm{Br}_P(i)\neq 0$. 设 L 是一个 A^G 的相互正交本原幂等元的集合使得 $\sum_{\ell\in L}\ell=1$; 现在, 只要令 $I=\bigcup_{\ell\in L}I_\ell$, 就证明了第一个断言.

设 J 是一个 G-稳定的相互正交幂等元的集合; 记 $e=\sum_{j\in J}j$; 只要使 $J\cup\{1-e\}$ 代替 J 就能假定 $e=1$. 而且, 对任意 $j\in J$ 在 jAj 里存在一个 G_j-稳定的相互正交幂等元的集合 I_j 使得任意 $i\in I_j$ 在 $A^{(G_j)_i}$ 中是本原的, 以及 $\mathrm{Br}_{(G_j)_i}(i)\neq 0$, 其中 G_j 是 j 的稳定化子; 还能选择这个集合 $\{I_j\}_{j\in J}$ 使得对任意 $x\in G$ 与 $j\in J$ 有 $(I_j)^x=I_{j^x}$; 这样, 只要使 $\bigcup_{j\in J}I_j$ 代替 J 就能假定 J 也满足上面的条件.

那么, 由定理 7.2, 对任意 $j\in J$, $\mathrm{Tr}_{G_j}^G(j)$ 是 A^G 的本原幂等元; 更精确的, 选择一个 J 的 G-轨道集合的代表集合 $M\subset J$; 此时 $\{\mathrm{Tr}_{G_j}^G(j)\}_{j\in M}$ 是 A^G 的相互正交本原幂等元的

Proof: Let α be a point of G on A and P_γ a defect pointed group of G_α; thus, if $i\in\gamma$ has an orthogonal G/P-trace then it follows from Theorem 7.2 that $\ell=\mathrm{Tr}_P^G(i)$ belongs to α; that is to say, for any $\ell\in\alpha$, there is a G-stable set I_ℓ of pairwise orthogonal idempotents such that P is the stabilizer of some element i of I_ℓ, i is a primitive idempotent of A^P and we have $\mathrm{Br}_P(i)\neq 0$. Let L be a set of pairwise orthogonal primitive idempotents in A^G such that $\sum_{\ell\in L}\ell=1$; now, in order to prove the first statement, it suffices to consider $I=\bigcup_{\ell\in L}I_\ell$.

Let J be a G-stable set of pairwise orthogonal idempotents; set $e=\sum_{j\in J}j$; up to the replacement of J by $J\cup\{1-e\}$, we may assume that $e=1$. Moreover, for any $j\in J$, in jAj there exists a G_j-stable set I_j of pairwise orthogonal idempotents such that any $i\in I_j$ is primitive in $A^{(G_j)_i}$ and fulfills $\mathrm{Br}_{(G_j)_i}(i)\neq 0$, where G_j denotes the stabilizer of j; actually, we may choose this family $\{I_j\}_{j\in J}$ in such a way that we have $(I_j)^x=I_{j^x}$ for any $x\in G$ and any $j\in J$; consequently, replacing J by $\bigcup_{j\in J}I_j$, we still may assume that J fulfills the conditions above.

Then, it follows from Theorem 7.2 that, for any $j\in J$, $\mathrm{Tr}_{G_j}^G(j)$ is primitive in A^G; more precisely, choosing a set $M\subset J$ of representatives for the orbits of G in J, $\{\mathrm{Tr}_{G_j}^G(j)\}_{j\in M}$ is a set of pairwise orthogonal primitive idempotents of A^G such that

集合使得 $\sum_{j\in M}\mathrm{Tr}_{G_j}^G(j)=1$. 这样由命题3.4存在 $a\in(A^G)^*$ 使得 $L^a=\{\mathrm{Tr}_{G_j}^G(j)\}_{j\in M}$; 特别是, 如果 $i\in I$ 与 $j\in J$ 满足 $\mathrm{Tr}_{G_i}^G(i)^a=\mathrm{Tr}_{G_j}^G(j)$ 那么 A 上的被 $\mathrm{Tr}_{G_i}^G(i)$ 决定的 G-点和被 $\mathrm{Tr}_{G_j}^G(j)$ 决定的 G-点是同一个 G-点 α, 并且 G_i 与 G_j 都是 G_α 的亏群 (见推论 6.12).

$\sum_{j\in M}\mathrm{Tr}_{G_j}^G(j)=1$. Thus, according to Proposition 3.4, there exists $a\in(A^G)^*$ such that $L^a=\{\mathrm{Tr}_{G_j}^G(j)\}_{j\in M}$; in particular, if $i\in I$ and $j\in J$ fulfill $\mathrm{Tr}_{G_i}^G(i)^a=\mathrm{Tr}_{G_j}^G(j)$ then the points of A^G determined by $\mathrm{Tr}_{G_i}^G(i)$ and $\mathrm{Tr}_{G_j}^G(j)$ are equal to the same point α of G on A, and moreover, G_i and G_j are both defect groups of G_α (see Corollary 6.12).

所以, 由定理 6.8, 只要使某个 i 的 G-共轭代替 i 就能假定 $G_i=G_j$ 并假定 A 上的被 i 与 j 分别决定的G_i-点 (或者 G_j-点) 是同样的; 此时, 存在 $b_j\in(A^{G_j})^*$ 使得 $i^{b_j}=j$.

Consequently, it follows from Theorem 6.8 that, replacing i by some G-conjugate, we may assume that $G_i=G_j$ and that the points of $G_j=G_i$ on A determined by i and j coincide; in this case, there exists $b_j\in(A^{G_j})^*$ such that $i^{b_j}=j$.

这样, 对任意$j\in M$存在 $(A^{G_j})^*$ 的元素 b_j 使得 j^{b_j} 属于I; 最后, 令

Thus, for any $j\in M$ there exists an element b_j of $(A^{G_j})^*$ such that j^{b_j} belongs to I; finally, set

7.4.2
$$b=\sum_{j\in M}\mathrm{Tr}_{G_j}^G\big((b_j)^{-1}j\big);$$

不太难验证

it is not difficult to check that

7.4.3
$$b\Big(\sum_{j\in M}\mathrm{Tr}_{G_j}^G(j\,b_j)\Big)=\sum_{j\in M}\mathrm{Tr}_{G_j}^G(j^{b_j})=\sum_{\ell\in L}\ell=1,$$

从而 b 是一个 A^G 的可逆元素; 而且, 对任意 $j\in M$ 得到

and therefore b is inversible in A^G; moreover, for any $j\in M$ we get

7.4.4
$$bj\,b^{-1}=j^{b_j}\in I,$$

从而得到 $J^b\subset I$ (事实上, 对经一系列适当替换后的目前的 J 有 $J^b=I$).

and therefore we have $J^b\subset I$ (actually, after all the replacements, for the present set J we get $J^b=I$).

8. 融合在 N-内 G-代数里

8.1. 如同在§5里一样, G 是一个有限群, N 是 G 的正规子群, A 是一个 N-内 G-代数; 而且, 我们能假定 A 是完备诱导的. 已经知道 G 作用在 A 上的点群的集合上; 并且如果 H_β 是 A 上的点群, 那么任意 $x\in G$ 当然决定群同态 $\kappa_x: H \to H^x$ 使得 $\kappa_x(y) = y^x$, 其中 $y \in H$.

8.2. 可是, 假定我们有一个 A 上的点群 L_ε 使得虽然 $x \notin L$ 然而 $H_\beta \subset L_\varepsilon$, $(H_\beta)^x \subset L_\varepsilon$ 都成立; 这样, H_β, $(H_\beta)^x$ 也是 A_ε 上的点群: 从这个 $(N\cap L)$-内 L-代数我们能不能决定 κ_x? 只要 $x \in N$ 就能决定如下.

8.3. 设 H 与 K 是 G 的子群. 我们要考虑那些被 A^* 决定的群同态 $\varphi: K \to H$; 这样, 我们假定 φ 是单射的, 并且假定对任意 $y \in K$ 有

8. Fusions in N-interior G-algebras

8.1. As in §5, G is a finite group, N is a normal subgroup of G and A is an N-interior G-algebra; moreover, we may assume that A is inductively complete. It is already clear that G acts on the set of all the pointed groups on A; furthermore, if H_β is a pointed group on A then any $x \in G$ naturally determines a group homomorphism $\kappa_x: H \to H^x$ such that $\kappa_x(y) = y^x$ for any $y \in H$.

8.2. But, assume that we have a pointed group L_ε on A such that, although $x \notin L$, however it happens that $H_\beta \subset L_\varepsilon$ and $(H_\beta)^x \subset L_\varepsilon$; thus, H_β and $(H_\beta)^x$ are also pointed groups on A_ε : Is it possible to determine κ_x from this $(N\cap L)$-interior L-algebra? If $x \in N$, this is possible in the following way.

8.3. Let H and K be subgroups of G. We want to consider group homomorphisms $\varphi: K \to H$ induced by elements of A^*; thus, we assume that φ is injective, and that, for any $y \in K$, we have

8.3.1 $$\varphi(y) \in yN$$

(也就是说, 在 $\bar G = G/N$ 中有 $\bar K \subset \bar H$ 并 φ 就引导包含映射).

(in other terms, in $\bar G = G/N$ we have $\bar K \subset \bar H$ and φ just induces the inclusion map).

8.4. 一般来说, 我喜欢使用所谓外同态: $K \times H$ 作用在 $\mathrm{Hom}(K,H)$ 上并精确的定义是

8.4. In general, we like to employ the so-called *exomorphisms*: $H \times K$ acts on $\mathrm{Hom}(K,H)$ and explicitly, we set

8.4.1 $$\varphi^{(y,x)}(z) = \varphi(z^{y^{-1}})^x = \varphi(z)^{\varphi(y)^{-1}x},$$

其中 $\varphi \in \mathrm{Hom}(K,H)$, $x \in H$, $y,z \in K$; 那么, 一个群外同态 $\tilde\varphi: K \to H$ 就是 $\mathrm{Hom}(K,H)$ 上的 $K\times H$-轨道; 令 $\widetilde{\mathrm{Hom}}(K,H)$ 记所有 $K\times H$-轨道的集合;

where $\varphi \in \mathrm{Hom}(K,H)$, $x \in H$ and $y,z \in K$; then, an *exterior group homomorphism* or, simply, a *group exomorphism* $\tilde\varphi: K \to H$ is just an orbit of $K \times H$ on $\mathrm{Hom}(K,H)$; denote by $\widetilde{\mathrm{Hom}}(K,H)$ the set of all the

一般的说, 即使 φ 满足上面的条件 8.3.1, $\tilde{\varphi}$ 的元素并不一定都满足该条件. 请注意, 如果 L 是 G 的子群并且 $\psi \in \mathrm{Hom}(L,K)$, 那么对任意 $\varphi' \in \tilde{\varphi}$ 与 $\psi' \in \tilde{\psi}$ 有

$K \times H$-orbits; generally speaking, even if φ fulfills condition 8.3.1 above, certainly not all the elements of $\tilde{\varphi}$ need to fulfill this condition. Note that, if L is a subgroup of G and $\psi \in \mathrm{Hom}(L,K)$, for any $\varphi' \in \tilde{\varphi}$ and $\psi' \in \tilde{\psi}$ we have

8.4.2 $$\widetilde{\varphi' \circ \psi'} = \widetilde{\varphi \circ \psi},$$

这样, 可定义 $\tilde{\varphi} \circ \tilde{\psi} = \widetilde{\varphi \circ \psi}$.

Thus, we can define $\tilde{\varphi} \circ \tilde{\psi} = \widetilde{\varphi \circ \psi}$.

8.5. 设 ω, θ 分别是 A 上的 H-除子与 K-除子. 我们要考虑的群外同态应该决定一个从 θ 到 ω 的"对应"; 这样, 因为它们是被 A^* 决定的, 所以假定

8.5. Let ω and θ be respective divisors of H and K on A. The group exomorphisms we want to consider should determine a "correspondence" from θ to ω; thus, since they are determined by A^*, we assume that

8.5.1 $$\mathrm{res}_1^K(\theta) \subset \mathrm{res}_1^H(\omega),$$

从而我们得到可嵌入 $\mathcal{O}$-代数同构如下 (见 3.17.3)

and therefore we get the following *embeddable* $\mathcal{O}$-algebra isomorphism (see 3.17.3)

8.5.2 $$\mathrm{Res}_1^K(A_\theta) \cong \mathrm{Res}_1^G(A)_{\mathrm{res}_1^K(\theta)} \cong \big(\mathrm{Res}_1^H(A_\omega)\big)_{\mathrm{res}_1^H(\theta)}.$$

别忘了 (见 4.5.3), 这两个 $\mathcal{O}$-代数的定义是 $A_\omega^\circ = \mathrm{End}_A(M_\omega)$, $A_\theta^\circ = \mathrm{End}_A(M_\theta)$ 其中 M_ω, M_θ 是适当的 $(A\otimes_\mathcal{O}\mathcal{O}(H\cap N))H$- 与 $(A\otimes_\mathcal{O}\mathcal{O}(K\cap N))K$-模的选择; 所以得到这个可嵌入的 $\mathcal{O}$-代数同构就等价于 M_θ 是 M_ω 的直和项; 这样条件 8.5.1 等价于 $\mathrm{Res}_A(M_\theta)$ 是 $\mathrm{Res}_A(M_\omega)$ 的直和项.

Recall (see 4.5.3) that these $\mathcal{O}$-algebras are defined by $A_\omega^\circ = \mathrm{End}_A(M_\omega)$, $A_\theta^\circ = \mathrm{End}_A(M_\theta)$ where M_ω and M_θ are suitable choices of $(A\otimes_\mathcal{O}\mathcal{O}(H\cap N))H$- and $(A\otimes_\mathcal{O}\mathcal{O}(K\cap N))K$-modules; hence, to get this *embeddable* $\mathcal{O}$-algebra isomorphism is equivalent to say that M_θ is a direct summand of M_ω; thus, condition 8.5.1 is equivalent to say that $\mathrm{Res}_A(M_\theta)$ is a direct summand of $\mathrm{Res}_A(M_\omega)$.

8.6. 另一方面, 令 $K \times^{\bar{G}} H$ 记被 $(y,x) \in K\times H$ 使得 $yN = xN$ 构成的 $K\times H$ 的子群, 其中 $\bar{G} = G/N$; 那么, $K \times^{\bar{G}} H$ 在 $\mathrm{Hom}_A(M_\theta, M_\omega)$ 上作用如下

8.6. On the other hand, denote by $K \times^{\bar{G}} H$ the subgroup of $K\times H$ formed by the elements $(x,y) \in K\times H$ fulfilling $yN = xN$, where we set $\bar{G} = G/N$; then, $K \times^{\bar{G}} H$ acts on $\mathrm{Hom}_A(M_\theta, M_\omega)$ as follows

8.6.1 $$f^{(y,x)}(m) = x^{-1}\cdot f\big((xy^{-1}\cdot 1_A)(y\cdot m)\big) = x^{-1}(xy^{-1}\cdot 1_A)\cdot f(y\cdot m),$$

其中$(y,x)\in K\times^{\bar G}H, m\in M_\theta$；确实，对任意 $a\in A$ 我们得到

where $(y,x)\in K\times^{\bar G}H$ and $m\in M_\theta$; indeed, for any $a\in A$ we have

8.6.2
$$\begin{aligned} f^{(y,x)}(a{\cdot}m) &= x^{-1}{\cdot}f\big((xy^{-1}{\cdot}1_A)a^{y^{-1}}{\cdot}(y{\cdot}m)\big)\\ &= x^{-1}{\cdot}f\big(a^{x^{-1}}(xy^{-1}{\cdot}1_A){\cdot}(y{\cdot}m)\big)\\ &= x^{-1}a^{x^{-1}}{\cdot}f\big((xy^{-1}{\cdot}1_A){\cdot}(y{\cdot}m)\big) = a{\cdot}f^{(y,x)}(m)\,. \end{aligned}$$

而且，如果 M'_ω 与 M'_θ 是那个模的另一个选择 (见 4.6)，每两个同构$M'_\omega\cong M_\omega$与$M'_\theta\cong M'_\theta$显然决定 $K\times^{\bar G}H$-稳定同构如下

moreover, if M'_ω and M'_θ are different choices for those modules (see 4.6), any pair of isomorphisms $M'_\omega\cong M_\omega$ and $M'_\theta\cong M'_\theta$ clearly determines a $K\times^{\bar G}H$-stable isomorphism

8.6.3
$$\mathrm{Hom}_A(M'_\theta,M'_\omega)\cong\mathrm{Hom}_A(M_\theta,M_\omega)\,.$$

8.7. 最后，如果 φ 是满足 $\mathrm{Ker}(\varphi)=\{1\}$并且对任意 $y\in K$ 满足 $\varphi(y)\in yN$ 的 $\mathrm{Hom}(K,H)$ 的元素，那么可定义群同态

8.7. Finally, if a group homomorphism $\varphi\in\mathrm{Hom}(K,H)$ fulfills $\mathrm{Ker}(\varphi)=\{1\}$ and $\varphi(y)\in yN$ for any $y\in K$ then we can define a group homomorphism

8.7.1
$$\Delta_\varphi\colon K\longrightarrow K\times^{\bar G}H$$

为 $\Delta_\varphi(y)=\big(y,\varphi(y)\big)$，其中 $y\in K$；请注意，如果 $\Delta_\varphi(K)$ 稳定一个 $f\in\mathrm{Hom}_A(M_\theta,M_\omega)$ 那么$\varphi(K)$ 稳定 $f(M_\theta)$，并且

setting $\Delta_\varphi(y)=\big(y,\varphi(y)\big)$ for any $y\in K$; note that if $\Delta_\varphi(K)$ stabilizes an element $f\in\mathrm{Hom}_A(M_\theta,M_\omega)$ then $\varphi(K)$ stabilizes $f(M_\theta)$, and the map

8.7.2
$$f\colon M_\theta\longrightarrow\mathrm{Res}_{\mathrm{id}_A\otimes\varphi}(M_\omega)$$

是 $A\otimes_{\mathcal O}\mathcal O(N\cap K)$-模同态的 (见 8.6.1).

is an $A\otimes_{\mathcal O}\mathcal O(N\cap K)$-module homomorphism (see 8.6.1).

8.8. 这样，如果我们有 $\mathrm{res}_1^K(\theta)\subset\mathrm{res}_1^H(\omega)$ 并φ 既满足 $\Delta_\varphi(K)$ 在$\mathrm{Hom}_A(M_\theta,M_\omega)$里稳定一个分离的单同态 $f\colon M_\theta\to M_\omega$，也满足 $\varphi(K)$ 在M_ω里稳定一个 $f(M_\theta)$ 的补 A-模 X，我们就说 φ 是一个从 θ 到 ω 的 A-融合，还说$\tilde\varphi$ 是一个从 θ 到ω 的外A-融合；

8.8. Thus, if we have $\mathrm{res}_1^K(\theta)\subset\mathrm{res}_1^H(\omega)$ and φ is a group homomorphism such that $\Delta_\varphi(K)$ fix a direct injection $f\colon M_\theta\to M_\omega$ in $\mathrm{Hom}_A(M_\theta,M_\omega)$ and $\varphi(K)$ stabilizes a complement X of $f(M_\theta)$ as A-module, we say that φ is a *fusion from θ to ω in A*, and that $\tilde\varphi$ is an *exterior fusion from θ to ω in A*; we respectively denote by 融$_A(K_\theta,H_\omega)$† and $F_A(K_\theta,H_\omega)$ the sets of

† The Chinese character 融 is pronounced "rong" as in "jonquil" and means "fuse".

令 融$_A(K_\theta,H_\omega)$，$F_A(K_\theta,H_\omega)$ 分别记从 θ 到 ω 的 A-融合的集合与从 θ 到 ω 的外 A-融合的集合. 别忘了, 处处有 G 作用, 所以对任意 $x\in G$ 有

fusions and of *exterior* fusions from θ to ω in A. Since G acts on the set of its own subgroups, so in the set of group homomorphisms between them, and on the set of pointed groups on A, for any $x\in G$ we have

8.8.1 $$\text{融}_A(K_\theta,H_\omega)^x=\text{融}_A\big((K_\theta)^x,(H_\omega)^x\big)\quad .$$

8.9. 请注意, 如果 L 是 G 的子群而 ζ 是 A 上的 L-除子那么对任意 $f\in \mathrm{Hom}_A(M_\theta,M_\omega)$ 与 $g\in \mathrm{Hom}_A(M_\zeta,M_\theta)$ 与 $(x,y,z)\in H\times K\times L$ 满足 $xN=yN=zN$ 得到

8.9. Note that if L is a subgroup of G and ζ is a divisor of L on A then, for any $f\in\mathrm{Hom}_A(M_\theta,M_\omega)$, any $g\in\mathrm{Hom}_A(M_\zeta,M_\theta)$ and any $(x,y,z)\in H\times K\times L$ fulfilling $xN=yN=zN$ we get

8.9.1 $$f^{(y,x)}\circ g^{(z,y)}=(f\circ g)^{(z,x)}\,,$$

确实, 对任意 $m\in M_\zeta$ 有

Indeed, for any $m\in M_\zeta$, we have

8.9.2 $$\begin{aligned}f^{(y,x)}\big(g^{(z,y)}(m)\big)&=f^{(y,x)}\Big(y^{-1}{\cdot}g\big((yz^{-1}{\cdot}1_A)(z{\cdot}m)\big)\Big)\\&=x^{-1}{\cdot}f\Big((xy^{-1}{\cdot}1_A){\cdot}g\big((yz^{-1}{\cdot}1_A)(z{\cdot}m)\big)\Big)\\&=x^{-1}{\cdot}(f\circ g)\big((xz^{-1}{\cdot}1_A)(z{\cdot}m)\big)\,.\end{aligned}$$

这样, 如果 $\psi\in$ 融$_A(L_\zeta,K_\theta)$ 那么 $\varphi\circ\psi$ 是一个从 ζ 到 ω 的 A-融合; 确实, 因为

Consequently, if $\psi\in$ 融$_A(L_\zeta,K_\theta)$ then $\varphi\circ\psi$ is a fusion from ζ to ω in A; indeed, since we have

8.9.3 $$\mathrm{res}_1^L(\zeta)\subset\mathrm{res}_1^K(\theta)\subset\mathrm{res}_1^H(\omega)\,,$$

所以 $\mathrm{res}_1^L(\zeta)\subset\mathrm{res}_1^H(\omega)$ (见 3.13); 而且, 如果 $\Delta_\psi(L)$ 稳定一个分离的单同态 $g\colon M_\zeta\to M_\theta$ 而 $\psi(L)$ 稳定一个 $g(M_\zeta)$ 的补 A-模 Y，那么只要使用 8.9.1, 就可证明 $\Delta_{\varphi\circ\psi}(L)$ 稳定 $f\circ g$, 并且不难证明 $(\varphi\circ\psi)(L)$ 稳定 $f(Y)\oplus X$.

we still have $\mathrm{res}_1^L(\zeta)\subset\mathrm{res}_1^H(\omega)$ (see 3.13); moreover, if $\Delta_\psi(L)$ fixes a direct injection $g\colon M_\zeta\longrightarrow M_\theta$ of A-modules and $\psi(L)$ stabilizes a complement Y of $g(M_\zeta)$ as A-modules, then it follows from equality 8.9.1 that $\Delta_{\varphi\circ\psi}(L)$ fixes $f\circ g$, and it is quite easy to check that $\Delta_{\varphi\circ\psi}(L)$ stabilizes $f(Y)\oplus X$.

命题 8.10. 设 H 是 G 的子群并设 ω 与 θ 是两个 A 上的 H-除子. 那么有 $\mathrm{id}_H\in$ 融$_A(H_\theta,H_\omega)$ 当且仅当有 $\theta\subset\omega$.

Proposition 8.10. *Let H be a subgroup of G and consider two divisors ω and θ of H on A. Then, we have $\mathrm{id}_H\in$ 融$_A(H_\theta,H_\omega)$ if and only if we have $\theta\subset\omega$.*

证明: 如果 $\mathrm{id}_H \in$ 融$_A(H_\theta, H_\omega)$ 那么有一个A-模的分离的单同态 $f: M_\theta \to M_\omega$ 被 $\Delta_{\mathrm{id}}(H)$ 稳定, 还有$f(M_\theta)$ 的补A-模 X 被H稳定; 也就是说, f 是一个AH-模的分离的单同态; 所以, 由于 M_ω 与 M_θ 的定义 (见 4.5.3), 对任意 $\beta \in \mathcal{P}_A(H)$ 得到 $\theta(\beta) \le \omega(\beta)$. 反过来, 如果 $\theta \subset \omega$ 那么M_θ是AH-模 M_ω 的直和项, 从而有 f 并 X 如上; 所以 $\mathrm{id}_H \in$ 融$_A(H_\theta, H_\omega)$.

Proof: If $\mathrm{id}_H \in$ 融$_A(H_\theta, H_\omega)$ then we have a direct injection $f: M_\theta \to M_\omega$ of A-modules fixed by $\Delta_{\mathrm{id}}(H)$, and an H-stable complement X of $f(M_\theta)$ as A-modules; that is to say, f is a direct injection of AH-modules; consequently, it follows from the definitions of M_ω and M_θ (see 4.5.3) that we have $\theta(\beta) \le \omega(\beta)$ for any $\beta \in \mathcal{P}_A(H)$. Conversely, if $\theta \subset \omega$ then M_θ is a direct summand of M_ω as AH-modules, and therefore we get f and X as above; hence, we have $\mathrm{id}_H \in$ 融$_A(H_\theta, H_\omega)$.

8.11. 例如, 如果 $K \subset H$ 并 $\theta \subset \mathrm{res}^H_K(\omega)$, 那么群包含同态 $\iota: K \to H$ 显然是从 θ 到 ω 的 A-融合, 这是因为 M_θ 是 $\mathrm{Res}^H_K(M_\omega)$ 的直和项 (见 4.5). 而且, 任意 $x \in N$ 决定一个从 ω 到 ω^x 的 A-融合, 这是因为有 $\mathrm{res}^{H^x}_1(\omega^x) = \mathrm{res}^H_1(\omega)$. 更一般来说, 令 $E_N(K_\theta, H_\omega)$ 记从 K 到 H 的这样的群外同态的集合, 它们被满足 $K \subset H^x$ 与 $\theta \subset \mathrm{res}^{H^x}_K(\omega^x)$ 的 $x \in N$ 所决定; 那么我们得到

8.11. For instance, whenever $K \subset H$ and $\theta \subset \mathrm{res}^H_K(\omega)$, then the group homomorphism $\iota: K \to H$ defined by the inclusion is obviously a fusion from θ to ω in A since M_θ is a direct summand of $\mathrm{Res}^H_K(M_\omega)$ (see 4.5). Moreover, any $x \in N$ determines a fusion from ω to ω^x in A since we have $\mathrm{res}^{H^x}_1(\omega^x) = \mathrm{res}^H_1(\omega)$. More generally, denote by $E_N(K_\theta, H_\omega)$ the set of group exomorphisms from K to H determined by the elements $x \in N$ fulfilling $K \subset H^x$ and $\theta \subset \mathrm{res}^{H^x}_K(\omega^x)$; then, we have

8.11.1 $$E_N(K_\theta, H_\omega) \subset F_A(K_\theta, H_\omega).$$

8.12. 另一方面, 如果 $|K| = |H|$ 与 $\mathrm{res}^K_1(\theta) = \mathrm{res}^H_1(\omega)$, 那么任意从 θ 到 ω 的A-融合 $\varphi: K \to H$ 是双射的, 并对应的同态 $f: M_\theta \to M_\omega$ 也是双射的, 这是因为 (见 3.15)

8.12. On the other hand, if we have $|K| = |H|$ and $\mathrm{res}^K_1(\theta) = \mathrm{res}^H_1(\omega)$ then any fusion $\varphi: K \to H$ from θ to ω in A is bijective, and the corresponding homomorphism $f: M_\theta \to M_\omega$ is bijective too since (see 3.15)

8.12.1 $$\mathrm{Res}_A(M_\theta) \cong \mathrm{Res}_A(M_\omega);$$

所以 φ^{-1} 就是一个从 ω 到 θ 的 A-融合. 特别是, 融$_A(H_\omega)$ 是 $\mathrm{Aut}(H)$ 的子群.

consequently, it is quite clear that φ^{-1} is a fusion from ω to θ in A. In particular, the set 融$_A(H_\omega)$ is a subgroup of $\mathrm{Aut}(H)$.

命题 8.13. 设H与K是G的子群并设ω与θ分别是A上的H-除子与K-除子. 我们考虑$\varphi\in$融$_A(K_\theta,H_\omega)$, 令$L=\varphi(K)$再令$\psi: K\to L$记被φ决定的同态. 那么存在唯一A上的L-除子ζ使得

Proposition 8.13. *Let H and K be subgroups of G and consider respective divisors ω and θ of H and K on A. Moreover, consider a fusion $\varphi\in$ 融$_A(K_\theta,H_\omega)$, set $L=\varphi(K)$ and denote by $\psi: K\to L$ the group homomorphism determined by φ. Then, there are a unique divisor ζ of L on A fulfilling*

8.13.1 $$\mathrm{res}_1^K(\theta)=\mathrm{res}_1^L(\zeta),\quad \zeta\subset\mathrm{res}_L^H(\omega),\quad \psi\in\text{融}_A(K_\theta,L_\zeta),$$

还存在一个$(K\cap N)$-内K-代数同构$f_\psi: A_\theta\cong\mathrm{Res}_\psi(A_\zeta)$使得下面的$\mathcal{O}$-代数同构是可嵌如的

and a $(K\cap N)$-interior K-algebra isomorphism $f_\psi: A_\theta\cong\mathrm{Res}_\psi(A_\zeta)$ such that the following $\mathcal{O}$-algebra isomorphism is embeddable

8.13.2 $$\mathrm{Res}_1^G(A)_{\mathrm{res}_1^K(\theta)}\cong\mathrm{Res}_1^K(A_\theta)\overset{f_\psi}{\cong}\mathrm{Res}_1^L(A_\zeta)\cong\mathrm{Res}_1^G(A)_{\mathrm{res}_1^L(\zeta)}.$$

证明: 我们已经知道$\Delta_\varphi(K)$在$\mathrm{Hom}_A(M_\theta,M_\omega)$中稳定一个分离的单同态$h: M_\theta\to M_\omega$, 而$L$在$M_\omega$里稳定一个$h(M_\theta)$的补$A$-模$X$; 特别是, 我们得到一个$L$-稳定$A$-模分解

Proof: By definition, we already know that $\Delta_\varphi(K)$ fixes a direct injection $h: M_\theta\to M_\omega$ in $\mathrm{Hom}_A(M_\theta,M_\omega)$, and that L stabilizes a complement X of $h(M_\theta)$ in M_ω as A-modules; in particular, we get an L-stable A-module direct sum decomposition

8.13.3 $$M_\omega=h(M_\theta)\oplus X;$$

也就是说, 因为A_ω的定义就是$\mathrm{End}_A(M_\omega)$, 所以存在$(A_\omega)^L$的幂等元i使得

in other terms, since $\mathrm{End}_A(M_\omega)$ is just our definition of A_ω, there exists an idempotent i in $(A_\omega)^L$ such that

8.13.4 $$i(M_\omega)=h(M_\theta),\quad (1-i)(M_\omega)=X;$$

那么i决定一个A_ω上的L-除子, 从而决定一个A上的L-除子ζ (见 4.8.1).

then i determines a divisor of L on A_ω and therefore it determines a divisor ζ of L on A (see 4.8.1).

而且, 由同构 4.8.2 与 5.3.2, 得到下面的$(N\cap L)$-内L-代数同构

Moreover, from isomorphisms 4.8.2 and 5.3.2, we get the following $(N\cap L)$-interior L-algebra isomorphisms

8.13.5 $$A_\zeta\cong(A_\omega)_\zeta\cong iA_\omega i\cong\mathrm{End}_A\big(i(M_\omega)\big)^\circ=\mathrm{End}_A\big(h(M_\theta)\big)^\circ;$$

这样,因为$h(M_\theta)$是$\mathrm{Res}_L^H(M_\omega)$的子$(A\otimes_{\mathcal{O}}(L\cap N))L$-模 (见 8.7) 所以能选择$M_\zeta=h(M_\theta)$;

thus, since $h(M_\theta)$ is an $(A\otimes_{\mathcal{O}}(L\cap N))L$-submodule of $\mathrm{Res}_L^H(M_\omega)$ (see 8.7), we can choose $M_\zeta=h(M_\theta)$; in this case, we have

那么有 $A_\zeta = \mathrm{End}_A\big(h(M_\theta)\big)^\circ$ 而 h 决定一个可嵌入同构

$A_\zeta = \mathrm{End}_A\big(h(M_\theta)\big)^\circ$ and h determines an *embeddable* isomorphism

8.13.6 $$f_\psi\colon \mathrm{Res}_1^K(A_\theta) \cong \mathrm{Res}_1^L(A_\zeta)$$

使得对任意 $a \in \mathrm{End}_A(M_\theta)^\circ$ 有 $h \circ a = f_\psi(a) \circ h$.

such that we have $h \circ a = f_\psi(a) \circ h$ for any $a \in A_\theta = \mathrm{End}_A(M_\theta)^\circ$.

特别是, 既然 f_ψ 是可嵌入的, 有 $\mathrm{res}_1^K(\theta) = \mathrm{res}_1^L(\zeta)$; 然后, 因为 Δ_ψ 显然稳定 $h\colon M_\theta \cong M_\zeta$ 所以有 $\psi \in$ 融$_A(K_\theta, L_\zeta)$. 而且, 如果 ζ' 是满足条件 8.13.1 的 A 上的 L-除子, 那么 id_L 属于融$_A(L_\zeta, L_{\zeta'})$ 并 融$_A(L_{\zeta'}, L_\zeta)$, 从而 $\zeta' = \zeta$ (见命题 8.10). 事实上, f_ψ 是一个从 A_θ 到 $\mathrm{Res}_\psi(A_\zeta)$ 的 $(L \cap N)$-内 L-代数同构; 确实, 既然 Δ_ψ 稳定 h, 只要连续使用 M_θ 的 $(A \otimes_{\mathcal{O}} \mathcal{O}(K \cap N))K$-模结构并 M_ζ 的 $(A \otimes_{\mathcal{O}} \mathcal{O}(L \cap N))L$-模结构, 就得到

In particular, since f_ψ is *embeddable*, we obtain $\mathrm{res}_1^K(\theta) = \mathrm{res}_1^L(\zeta)$; furthermore, since Δ_ψ clearly stabilizes $h\colon M_\theta \cong M_\zeta$, we thus get $\psi \in$ 融$_A(K_\theta, L_\zeta)$. Moreover, if ζ' is a divisor of L on A which fulfills condition 8.13.1, then id_L belongs to 融$_A(L_\zeta, L_{\zeta'})$ and to 融$_A(L_{\zeta'}, L_\zeta)$, and therefore we get $\zeta' = \zeta$ (see Proposition 8.10). Actually, f_ψ is an $(L \cap N)$-interior L-algebra isomorphism from A_θ onto $\mathrm{Res}_\psi(A_\zeta)$; indeed, as far as we already know that Δ_ψ fixes h , successively employing the $(A \otimes_{\mathcal{O}} \mathcal{O}(K \cap N))K$-module structure of M_θ and the $(A \otimes_{\mathcal{O}} \mathcal{O}(L \cap N))L$-module structure of M_ζ , we get

8.13.7
$$\begin{aligned} f_\psi(y{\cdot}a^x)\big(h(m)\big) &= h\big(y{\cdot}a^x(m)\big) = \psi(y){\cdot}h\big(x^{-1}{\cdot}a(x{\cdot}m)\big) \\ &= \big(\psi(y){\cdot}(x^{-1}\psi(x){\cdot}1_A)\big){\cdot}\Big(\psi(x)^{-1}{\cdot}h\big(a(x{\cdot}m)\big)\Big) \\ &= \psi(yx^{-1}){\cdot}\Big(\big(\psi(x)x^{-1}{\cdot}1_A\big){\cdot}h\big(a(x{\cdot}m)\big)\Big) \\ &= \psi(yx^{-1}){\cdot}h\Big(a\big((\psi(x)x^{-1}{\cdot}1_A){\cdot}(x{\cdot}m)\big)\Big) \\ &= \psi(yx^{-1}){\cdot}f_\psi(a)\Big(h\big((\psi(x)x^{-1}{\cdot}1_A){\cdot}(x{\cdot}m)\big)\Big) \\ &= \psi(yx^{-1}){\cdot}f_\psi(a)\big(\psi(x){\cdot}h(m)\big) \\ &= \big(\psi(y){\cdot}f_\psi(a)^{\psi(x)}\big)\big(h(m)\big)\,, \end{aligned}$$

其中 $x \in K$, $y \in K \cap N$, $m \in M_\theta$, $a \in A_\theta$; 从而对任意 $a \in A_\theta$ 有

where $x \in K$, $y \in K \cap N$, $m \in M_\theta$ and $a \in A_\theta$; hence, for any $a \in A_\theta$, we have

8.13.8 $$f_\psi(y{\cdot}a^x) = \psi(y){\cdot}f_\psi(a)^{\psi(x)}\,.$$

8.14. 假定 $|K| = |H|$ 与 $\mathrm{res}_1^K(\theta) = \mathrm{res}_1^H(\omega)$；对任意 $\varphi \in$ 融$_A(K_\theta, H_\omega)$，命题 8.13 推出有一个 $(K \cap N)$-内 K-代数同构 $f_\varphi\colon A_\theta \cong \mathrm{Res}_\varphi(A_\omega)$ 使得下面的 $\mathcal{O}$-代数同构是可嵌入的

8.14. Assume that $|K| = |H|$ and that $\mathrm{res}_1^K(\theta) = \mathrm{res}_1^H(\omega)$; for any $\varphi \in$ 融$_A(K_\theta, H_\omega)$, Proposition 8.13 implies the existence of a $(K \cap N)$-interior K-algebra isomorphism $f_\varphi\colon A_\theta \cong \mathrm{Res}_\varphi(A_\omega)$ such that the following $\mathcal{O}$-algebra isomorphism is *embeddable*

8.14.1 $$\mathrm{Res}_1^G(A)_{\mathrm{res}_1^K(\theta)} \cong \mathrm{Res}_1^K(A_\theta) \overset{f_\varphi}{\cong} \mathrm{Res}_1^H(A_\omega) \cong \mathrm{Res}_1^G(A)_{\mathrm{res}_1^H(\omega)} .$$

现在，设 φ 是一个从 K 到 H 的同构使得 $\varphi(y) \in yN$，其中 $y \in K$，并且假定存在 f_φ 如上：φ 是不是从 θ 到 ω 的 A-融合？

Now, let φ be a group isomorphism from K to H such that $\varphi(y) \in yN$ for any $y \in K$, and assume that there exists f_φ as above: is φ a fusion from θ to ω in A?

8.15. 不总是. 因为同构式 8.14.1 是可嵌入的，所以存在一个 A-模同构 $h\colon M_\theta \cong M_\omega$ 使得

8.15. Not always. Since isomorphism 8.14.1 is *embeddable*, there exists an A-module isomorphism $h\colon M_\theta \cong M_\omega$ fulfilling

8.15.1 $$h \circ a = f_\varphi(a) \circ h$$

其中 $a \in A_\theta$；特别是，对任意 $y \in K$ 得到（见 8.9.1）

for any $a \in A_\theta$; in particular, for any $y \in K$, we get (see 8.9.1)

8.15.2 $$\begin{aligned} h^{(y,\varphi(y))} \circ a^y &= (h \circ a)^{(y,\varphi(y))} = (f_\varphi(a) \circ h)^{(y,\varphi(y))} \\ &= f_\varphi(a)^{\varphi(y)} \circ h^{(y,\varphi(y))} = f_\varphi(a^y) \circ h^{(y,\varphi(y))} , \end{aligned}$$

从而还得到

and therefore we still get

8.15.3 $$\begin{aligned} h^{-1} \circ h^{(y,\varphi(y))} \circ a &= h^{-1} \circ f_\varphi(a) \circ h^{(y,\varphi(y))} \\ &= a \circ h^{-1} \circ h^{(y,\varphi(y))} ; \end{aligned}$$

令 $\mathfrak{c}_{\varphi,h}(y) = h^{-1} \circ h^{(y,\varphi(y))}$；所以这个 M_θ 的自同构 $\mathfrak{c}_{\varphi,h}(y)$ 属于 $Z(A_\theta)^*$.

set $\mathfrak{c}_{\varphi,h}(y) = h^{-1} \circ h^{(y,\varphi(y))}$; consequently, this automorphism $\mathfrak{c}_{\varphi,h}(y)$ of M_ω belongs to $Z(A_\theta)^*$.

8.16. 这样，我们得到一个映射 $\mathfrak{c}_{\varphi,h}\colon K \to Z(A_\theta)^*$；而且，对任意 $x, y \in K$ 有

8.16. Thus, we can define a map $\mathfrak{c}_{\varphi,h}\colon K \to Z(A_\theta)^*$; moreover, for any x, y in K, we have

8.16.1 $$\begin{aligned} \mathfrak{c}_{\varphi,h}(xy) &= h^{-1} \circ h^{(xy,\varphi(xy))} \\ &= (h^{-1} \circ h^{(y,\varphi(y))}) \circ (h^{-1} \circ h^{(x,\varphi(x))})^y \\ &= \mathfrak{c}_{\varphi,h}(x)^y \mathfrak{c}_{\varphi,h}(y) ; \end{aligned}$$

此时, 如果 $x \in N \cap K$ 那么有 $\mathfrak{c}_{\varphi,h}(x) = 1$, 这是因为对任意 $m \in M_\theta$ 有

at the same time, if $x \in N \cap K$ then we have $\mathfrak{c}_{\varphi,h}(x) = 1$ since, for any $m \in M_\theta$, we get

8.16.2
$$\begin{aligned} h\big((\varphi(x)x^{-1}{\cdot}1_A){\cdot}(x{\cdot}m)\big) &= (\varphi(x){\cdot}1_A){\cdot}h\big((x{\cdot}1_{A_\theta})(m)\big) \\ &= (\varphi(x){\cdot}1_A){\cdot}f_\varphi(x{\cdot}1_{A_\theta})\big(h(m)\big) \\ &= (\varphi(x){\cdot}1_A){\cdot}(\varphi(x){\cdot}1_{A_\omega})\big(h(m)\big) \\ &= \varphi(x){\cdot}h(m)\,, \end{aligned}$$

从而得到 $h^{(x,\varphi(x))} = h$; 也就是说, $\mathfrak{c}_{\varphi,h}$ 决定一个映射

and therefore we also have $h^{(x,\varphi(x))} = h$; that is to say, $\mathfrak{c}_{\varphi,h}$ determines a map

8.16.3
$$\bar{\mathfrak{c}}_{\varphi,h}\colon \bar{K} = K/(N \cap K) \to Z(A_\theta)^*$$

它是从 $\bar{K}$ 到 $Z(A_\theta)^*$ 的1-循环, 这里我们考虑被 A_θ 上的 K-作用决定的 $Z(A_\theta)^*$ 的 $\mathbb{Z}\bar{K}$-模结构.

which is a 1-cocycle from $\bar{K}$ to $Z(A_\theta)^*$, where we consider the $\mathbb{Z}\bar{K}$-module structure of $Z(A_\theta)^*$ determined by the action of K on A_θ.

8.17. 令 $\mathbf{H}^1\big(\bar{K}, Z(A_\theta)^*\big)$ 记相应的1-同调群, 再令 $\mathfrak{h}_\varphi$ 记被 $\bar{\mathfrak{c}}_{\varphi,h}$ 决定的元素; $\mathfrak{h}_\varphi$ 不依赖于 h 的选择; 确实, 另一个选择 $h'\colon M_\theta \cong M_\omega$ 满足

8.17. Denote by $\mathbf{H}^1\big(\bar{K}, Z(A_\theta)^*\big)$ the corresponding 1-cohomology group and by $\mathfrak{h}_\varphi$ the element determined by $\bar{\mathfrak{c}}_{\varphi,h}$; we claim that $\mathfrak{h}_{\varphi,h}$ does no depend on the choice of h; indeed, another choice $h'\colon M_\theta \cong M_\omega$ fulfills

8.17.1
$$(h^{-1} \circ h') \circ a = h^{-1} \circ f_\varphi(a) \circ h' = a \circ (h^{-1} \circ h')$$

其中 $a \in A_\theta$; 这样 $z = h^{-1} \circ h'$ 属于 $Z(A_\theta)^*$, 并对任意 $y \in K$ 得到 (见 8.9.1)

for any $a \in A_\theta$; thus, the element $z = h^{-1} \circ h'$ belongs to $Z(A_\theta)^*$, and moreover, for any $y \in K$ we get (see 8.9.1)

8.17.2
$$\mathfrak{c}_{\varphi,h'}(y) = z^{-1} \circ h^{-1} \circ h'^{(y,\varphi(y))} = z^{-1} \circ \mathfrak{c}_{\varphi,h}(y) \circ z^y\,.$$

定理 8.18. 设 H, K 是两个 G 的相互同构的子群并设 ω, θ 分别是 A 上的 H-除子与 K-除子使得 $\mathrm{res}^H_1(\omega) = \mathrm{res}^K_1(\theta)$. 那么一个群同构 $\varphi\colon K \cong H$ 是从 θ 到 ω 的 A-融合当且仅当存在一个 $(K \cap N)$-内 K-代数同构

Theorem 8.18. *Let H and K be mutually isomorphic subgroups of G and consider respective divisors ω and θ of H and K on A such that $\mathrm{res}^H_1(\omega) = \mathrm{res}^K_1(\theta)$. Then, a group isomorphism $\varphi\colon K \cong H$ is a fusion from θ to ω in A if and only if there exists a $(K \cap N)$-interior K-algebra isomorphism*

8.18.1
$$f_\varphi : A_\theta \cong \mathrm{Res}_\varphi(A_\omega)$$

使得下面的 $\mathcal{O}$-代数同构是可嵌入的

such that the following $\mathcal{O}$-algebra isomorphism is embeddable

8.18.2 $$\mathrm{Res}_1^G(A)_{\mathrm{res}_1^K(\theta)} \cong \mathrm{Res}_1^K(A_\theta) \overset{f_\varphi}{\cong} \mathrm{Res}_1^H(A_\omega) \cong \mathrm{Res}_1^G(A)_{\mathrm{res}_1^H(\omega)}$$

并且 $\mathbf{H}^1(\bar{K}, Z(A_\theta)^*)$ 的元素 $\mathfrak{h}_\varphi$ 是零.

and the element $\mathfrak{h}_\varphi$ of $\mathbf{H}^1(\bar{K}, Z(A_\theta)^*)$ is equal to zero.

证明: 我们已经知道这个条件是必要的. 反过来, 如果存在 f_φ 如上并且有 $\mathfrak{h}_\varphi = 0$, 那么存在 $z \in Z(A_\theta)^*$ 使得对任意 $y \in K$ 有 $h^{-1} \circ h^{(y,\varphi(y))} = z^{-1} \circ z^y$, 其中 $h: M_\theta \cong M_\omega$ 是一个 A-模的同构使得对任意 $a \in A_\theta$ 有 $h \circ a = f_\varphi(a) \circ h$; 所以 $\Delta_\varphi(K)$ 稳定 $h \circ z^{-1}$ (见 8.9.1) 从而 φ 是从 θ 到 ω 的 A-融合.

Proof: We already know that this condition is necessary. Conversely, if there exists f_φ as above and we have $\mathfrak{h}_\varphi = 0$, then there exists $z \in Z(A_\theta)^*$ such that, for any $y \in K$, we have $h^{-1} \circ h^{(y,\varphi(y))} = z^{-1} \circ z^y$, where $h: M_\theta \cong M_\omega$ is an A-module isomorphism such that $h \circ a = f_\varphi(a) \circ h$ for any $a \in A_\theta$; consequently, $\Delta_\varphi(K)$ stabilizes $h \circ z^{-1}$ (see 8.9.1), and therefore φ is a fusion from θ to ω in A.

推论 8.19. 设 H, K 是 G 的子群并且设 ω, θ 分别是 A 上的 H-除子与 K-除子. 如果 ξ 是一个 A 上的 G-除子使得 $\omega \subset \mathrm{res}_H^G(\xi)$, $\theta \subset \mathrm{res}_K^G(\xi)$ 那么

Corollary 8.19. *Let H and K be subgroups of G and consider respective divisors ω and θ of H and K on A. If ξ is a divisor of G on A such that $\omega \subset \mathrm{res}_H^G(\xi)$ and $\theta \subset \mathrm{res}_K^G(\xi)$ then we have*

8.19.1 $$融_{A_\xi}(K_\theta, H_\omega) = 融_A(K_\theta, H_\omega)\,.$$

证明: 首先, 假定 H, K 是相互同构的, 并且 $\mathrm{res}_1^H(\omega) = \mathrm{res}_1^K(\theta)$; 那么, 因为 (见同构 5.3.2)

Proof: First of all, assume that H and K are isomorphic, and that $\mathrm{res}_1^H(\omega) = \mathrm{res}_1^K(\theta)$; then, since we have (see isomorphism 5.3.2)

8.19.2 $$(A_\xi)_\omega \cong A_\omega\,, \quad (A_\xi)_\theta \cong A_\theta\,,$$

所以只要使用定理 8.18, 就得到等式 8.19.1. 一般来说, 由命题 8.13, 如果 $\varphi \in 融_A(K_\theta, H_\omega)$ 那么存在一个 A 上的 L-除子 ζ 使得

it suffices to apply Theorem 8.18, to get equality 8.19.1. In general, according to Proposition 8.13, if $\varphi \in 融_A(K_\theta, H_\omega)$ then there exists a divisor ζ of $L = \varphi(K)$ on A fulfilling

8.19.3 $$\mathrm{res}_1^K(\theta) = \mathrm{res}_1^L(\zeta)\,, \quad \zeta \subset \mathrm{res}_L^H(\omega)\,, \quad \psi \in 融_A(K_\theta, L_\zeta)\,,$$

其中 $L = \varphi(K)$; 特别是, 得到 $\zeta \subset \mathrm{res}_L^G(\xi)$ 并且有

in particular, we obtain $\zeta \subset \mathrm{res}_L^G(\xi)$ and moreover we get

8.19.4 $$融_{A_\xi}(K_\theta, L_\zeta) = 融_A(K_\theta, L_\zeta)\,,$$

从而也得到$\varphi \in$ 融$_{A_\xi}(K_\theta, H_\omega)$ (见 8.9 和 8.11).

and therefore we still get $\varphi \in$ 融$_{A_\xi}(K_\theta, H_\omega)$ (see 8.9 and 8.11).

类似地, 如果 φ 属于 融$_{A_\xi}(K_\theta, H_\omega)$ 那么存在 A_ξ 上的 L-除子 ζ 使得

Similarly, whenever $\varphi \in$ 融$_{A_\xi}(K_\theta, H_\omega)$, there exists a divisor ζ of $L = \varphi(K)$ on A_ξ fulfilling

8.19.5 $$\mathrm{res}_1^K(\theta) = \mathrm{res}_1^L(\zeta)\,, \quad \zeta \subset \mathrm{res}_L^H(\omega)\,, \quad \psi \in \text{融}_{A_\xi}(K_\theta, L_\zeta)\,,$$

其中 $L = \varphi(K)$; 再次, ζ 是一个 A 上的 L-除子并且有

and, once again, ζ is a divisor of L on A, and moreover we have

8.19.6 $$\text{融}_{A_\xi}(K_\theta, L_\zeta) = \text{融}_A(K_\theta, L_\zeta)\,;$$

所以得到 $\varphi \in$ 融$_A(K_\theta, H_\omega)$ (见 8.9 和 8.11).

consequently, we get $\varphi \in$ 融$_A(K_\theta, H_\omega)$ (see 8.9 and 8.11).

定理 8.20. 设 H, K 是 G 的相互同构的子群并设 i, j 分别是 A^H 与 A^K 的幂等元使得 $\mu_A^i = \mu_A^j$. 那么一个对任意 $y \in K$ 有 $\varphi(y) \in yN$ 的群同构 $\varphi\colon K \cong H$ 是一个从 $\mu^j_{A^K}$ 到 $\mu^i_{A^H}$ 的 A-融合当且仅当对 A^* 的适当的元素 a 并对任意 $y \in K$ 有

Theorem 8.20. *Let H and K be mutually isomorphic subgroups of G and consider respective idempotents i and j of A^H and A^K fulfilling $\mu_A^i = \mu_A^j$. Then, a group isomorphism $\varphi\colon K \cong H$ which fulfills $\varphi(y) \in yN$ for any $y \in K$ is a fusion from $\mu^j_{A^K}$ to $\mu^i_{A^H}$ in A if and only if, for a suitable element a of A^* and for any $y \in K$. we have*

8.20.1 $$j^a = i\,, \quad (ja)^y \cdot y^{-1}\varphi(y) = ja\,.$$

证明: 令 $\omega = \mu^i_{A^H}$, $\theta = \mu^j_{A^K}$; 这一次我们能选择 $M_\omega = Ai$ 与 $M_\theta = Aj$ (见 3.15). 首先假定存在 $a \in A^*$ 如上; 那么 Aj 的右边乘以 $ja = ai$ 显然决定一个 A-模同构 $h\colon Aj \cong Ai$; 确实 Ai 的右边乘以 $ia^{-1} = a^{-1}j$ 是其逆映射; 而且对任意 $y \in K$ 与 $b \in A$ 得到

Proof: Set $\omega = \mu^i_{A^H}$ and $\theta = \mu^j_{A^K}$; this time, we may choose $M_\omega = Ai$ and $M_\theta = Aj$ (see 3.15). First of all, assume that there is $a \in A^*$ as above; then the multiplication on the right by $ja = ai$ in Aj clearly determines an A-module isomorphism $h\colon Aj \cong Ai$; indeed, its inverse is the multiplication on the right by $ia^{-1} = a^{-1}j$ in Ai; moreover, for any $y \in K$ and any $b \in A$, we get

8.20.2 $$\begin{aligned} h^{(y,\varphi(y))}(bj) &= h\big(\varphi(y)y^{-1}\cdot(bj)^{y^{-1}}\big)^{\varphi(y)} \\ &= \big(\varphi(y)y^{-1}\cdot(bj)^{y^{-1}}ja\big)^{\varphi(y)} \\ &= y^{-1}\varphi(y)\cdot(bj)^{y^{-1}\varphi(y)}(ja)^{\varphi(y)} \\ &= bj\cdot\big(y^{-1}\varphi(y)\cdot(ja)^{\varphi(y)}\big) = bja = h(bj)\,, \end{aligned}$$

从而$\Delta_\varphi(K)$稳定h; 也就是说, φ是一个从θ到ω的A-融合.

反过来, 如果φ是一个从θ到ω的A-融合, 那么$\Delta_\varphi(K)$稳定一个A-模同构$h: Aj \cong Ai$; 而且, 因为$\mu_A^{1_A} - \mu_A^i = \mu_A^{1_A} - \mu_A^j$ 所以也存在一个A-模同构$g: A(1-j) \cong A(1-i)$. 现在, 考虑$h \oplus g: A \cong A$而且令$a = (h \oplus g)(1_A)$; 这样, 对任意$b \in A$有$(h \oplus g)(b) = ba$; 从而a是可逆的, 并且得到

and therefore $\Delta_\varphi(K)$ stabilizes h; in other words, φ is a fusion from θ to ω in A.

Conversely, if φ is a fusion from θ to ω in A then $\Delta_\varphi(K)$ stabilizes an A-module isomorphism $h: Aj \cong Ai$; moreover, since we have $\mu_A^{1_A} - \mu_A^i = \mu_A^{1_A} - \mu_A^j$, also exists an A-module isomorphism $g: A(1-j) \cong A(1-i)$. Now, consider $h \oplus g: A \cong A$ and moreover set $a = (h \oplus g)(1_A)$; consequently, we have $(h \oplus g)(b) = ba$ for any $b \in A$; hence, a is an inversible element, we get

8.20.3 $$Aja = Ai\,, \quad A(1-j)a = A(1-i)$$

还得到$(1-j^a)(1-i) = 1-j^a$与$j^a i = j^a$; 最后有$j^a = i$. 另一方面, 因为$\Delta_\varphi(K)$稳定h, 所以对任意$y \in K$有 (见 8.20.2)

and we still get $(1-j^a)(1-i) = 1-j^a$ and $j^a i = j^a$; finally, we have $j^a = i$. On the other hand, since $\Delta_\varphi(K)$ stabilizes h, for any $y \in K$ we also have (see 8.20.2)

8.20.4 $$\begin{aligned} ja = h(j) &= h^{(y,\varphi(y))}(j) = j\cdot\big(y^{-1}\varphi(y)\cdot(ja)^{\varphi(y)}\big) \\ &= (ja)^y\cdot y^{-1}\varphi(y)\,. \end{aligned}$$

命题 8.21. 设H与K是G的子群.

8.21.1. 设ω, ω'与θ, θ'分别是A上的H-除子与K-除子使得$\omega \cap \omega' = 0$与$\theta \cap \theta' = 0$. 那么有

Proposition 8.21. *Let H and K be subgroups of G.*

8.21.1. *Let ω, ω' and θ, θ' respectively be divisors of H and K on A fulfilling $\omega \cap \omega' = 0$ and $\theta \cap \theta' = 0$. Then, we have*

$$\text{融}_A(K_\theta, H_\omega) \cap \text{融}_A(K_{\theta'}, H_{\omega'}) \subset \text{融}_A(K_{\theta\cup\theta'}, H_{\omega\cup\omega'})\,.$$

8.21.2. 设β与γ分别是A上的H-除子与K-点. 对任意$0<n<m$有

8.21.2. *Let β and γ respectively be divisors of H and K on A. For any $0 < n < m$ we have*

$$\text{融}_A(K_{n\cdot\gamma}, H_{m\cdot\beta}) = \text{融}_A(K_\gamma, H_\beta)\,.$$

证明: 只要选择

Proof: It suffices to choose

8.21.3 $$M_{\omega\cup\omega'} = M_\omega \oplus M_{\omega'}\,, \quad M_{\theta\cup\theta'} = M_\theta \oplus M_{\theta'}$$

断言 8.21.1 就是显然的. 类似地, 可证明

and statement 8.21.1 becomes evident. Similarly, we prove that

8.21.4 $$\text{融}_A(K_\gamma, H_\beta) \subset \text{融}_A(K_{n\cdot\gamma}, H_{m\cdot\beta})\,.$$

最后, 我们假定 φ 是一个从 $n{\cdot}\gamma$ 到 $m{\cdot}\beta$ 的A-融合并且假定 $h\colon (Aj)^n \to (Ai)^m$ 是 $\Delta_\varphi(K)$-稳定的分离的单A-模同态使得$\mathrm{Im}(h)$ 是 AL-直和项, 其中 $i \in \beta$ 与 $j \in \gamma$ 并 $L = \varphi(K)$; 令 $f\colon Aj \to (Aj)^n$ 记下第一个自然分离的单同态. 此时, 因为 Aj 是不可分解 AK-模的, 所以存在 $g \in \mathrm{End}_{AL}((Ai)^m)^*$ 使得 $g(h(f(Aj)))$ 包含在第一个 $(Ai)^m$ 的自然直和项中 (见命题 3.19); 那么 $g \circ h \circ f$ 决定一个 $\Delta_\varphi(K)$-稳定的分离的单 A-模同态 $Aj \to Ai$ 使得其像是 AL-直和项, 从而 φ 也是一个从 γ 到 β 的 A-融合.

Finally, we assume that φ is a fusion from $n{\cdot}\gamma$ to $m{\cdot}\beta$ in A and that $\Delta_\varphi(K)$ stabilizes a direct injection $h\colon (Aj)^n \to (Ai)^m$ of A-modules such that $\mathrm{Im}(h)$ is a direct summand of $(Ai)^m$ as AL-modules, where $i \in \beta$, $j \in \gamma$ and we set $L = \varphi(K)$; denote by $f\colon Aj \to (Aj)^n$ the first canonical direct injection. In this situation, since Aj is an indecomposable AK-module, there exists g in $\mathrm{End}_{AL}((Ai)^m)^*$ such that $g(h(f(Aj)))$ is contained in the first canonical direct summand of $(Ai)^m$ (see Proposition 3.19); then, $g \circ h \circ f$ determines a $\Delta_\varphi(K)$-stable direct injection $Aj \to Ai$ of A-modules such that its image is a direct summand of Ai as AL-modules, and therefore φ is also a fusion from γ to β in A.

9. N-内 G-代数与 G-内代数相关

9. N-interior G-algebras through G-interior Algebras

9.1. 设G是一个有限群并设N是G的正规子群. 任意N-内 G-代数 A 决定一个G-内代数如下

9.1. Let G be a finite group and N a normal subgroup of G. Any N-interior G-algebra A determines a G-interior algebra as follows

9.1.1 $$A \otimes_N G = A \otimes_{\mathcal{O}N} \mathcal{O}G = \bigoplus_x A \otimes x$$

其中x跑遍一个$\bar{G} = G/N$的代表集合, 并且对任意 $a, b \in A$ 与$x, y \in G$ 我们令

where x runs over a set of representatives for $\bar{G} = G/N$ and, for any $a, b \in A$ and $x, y \in G$, we set

9.1.2 $$(a \otimes x)(b \otimes y) = ab^{x^{-1}} \otimes xy\,;$$

特别是, 有 $(1_A \otimes x)(1_A \otimes y) = 1_A \otimes xy$, 此时令 $A \otimes_N G$ 的 G-内结构的定义如下

in particular, we have $(1_A \otimes x)(1_A \otimes y) = 1_A \otimes xy$, so that we define the G-interior structure of $A \otimes_N G$ as follows

9.1.3 $$x\cdot(b \otimes y) = b^{x^{-1}} \otimes xy\,, \quad (a \otimes x)\cdot y = a \otimes xy\,.$$

而且, 还有 $(a \otimes 1)(b \otimes 1) = ab \otimes 1$, 从而$A \cong A \otimes 1$; 这样, 任意 A 上的点群 H_β 决定一个 $\hat{A} = A \otimes_N G$ 上的 H-除子 $\hat{\beta} = \mu_{\hat{A}^H}^{i\otimes 1}$, 其中$i \in \beta$ (见 3.21).

Moreover, we still have $(a\otimes 1)(b\otimes 1) = ab\otimes 1$, and therefore we have $A \cong A \otimes 1$; thus, any pointed group H_β on A determines a divisor $\hat{\beta} = \mu_{\hat{A}^H}^{i\otimes 1}$ of H on $\hat{A} = A \otimes_N G$, where $i \in \beta$ (see 3.21).

定理 9.2. **设P_γ与Q_δ是A上的局部点群. 那么有**

Theorem 9.2. *Let P_γ and Q_δ be local pointed groups on A. Then we have*

9.2.1 $$融_{\hat{A}}(Q_{\hat{\delta}}, P_{\hat{\gamma}}) = \bigcup_x 融_A\big((Q_\delta)^x, P_\gamma\big) \circ \kappa_x^Q$$

其中$x \in G$跑遍一个$\bar{G}$的代表集合而$\kappa_x^Q\colon Q \cong Q^x$是$x$-共轭同构.

where $x \in G$ runs over a set of representatives for $\bar{G}$, and $\kappa_x^Q\colon Q \cong Q^x$ is the isomorphism determined by the conjugation by x.

证明: 首先, 证明 $融_{\hat{A}}(Q_{\hat{\delta}}, P_{\hat{\gamma}})$ 包含 $融_A(Q_\delta, P_\gamma)$. 假定 φ 是 $融_A(Q_\delta, P_\gamma)$ 的元素; 由命题 8.13, 存在一个A上的点群R_ε与一个从δ到ε的双射A-融合ψ使得$R_\varepsilon \subset P_\gamma$与$\varphi = \psi \circ \iota$, 其中 $\iota\colon R \to P$ 是包含映射;

Proof: We firstly prove that $融_{\hat{A}}(Q_{\hat{\delta}}, P_{\hat{\gamma}})$ contains $融_A(Q_\delta, P_\gamma)$. Assume that φ is an element of $融_A(Q_\delta, P_\gamma)$; it follows from Proposition 8.13 that there exist a pointed group R_ε and a bijective fusion ψ from δ to ε in A such that $R_\varepsilon \subset P_\gamma$ and $\varphi = \psi \circ \iota$, where $\iota\colon R \to P$ is the inclusion map; then,

那么从 $\varepsilon \subset \mathrm{res}^P_R(\gamma)$ 显然可推出 $\hat{\varepsilon} \subset \mathrm{res}^P_R(\hat{\gamma})$，从而 ι 属于 融$_{\hat{A}}(R_{\hat{\varepsilon}}, P_{\hat{\gamma}})$；进一步，只要应用定理 8.20，就不难得到 $\psi \in$ 融$_{\hat{A}}(Q_{\hat{\delta}}, R_{\hat{\varepsilon}})$．所以，对任意 $x \in G$ 也得到 (见 8.10.1)

from $\varepsilon \subset \mathrm{res}^P_R(\gamma)$ we clearly can deduce that $\hat{\varepsilon} \subset \mathrm{res}^P_R(\hat{\gamma})$, and therefore ι belongs to 融$_{\hat{A}}(R_{\hat{\varepsilon}}, P_{\hat{\gamma}})$; moreover, by applying Theorem 8.20 to both A and $\hat{A}$, we easily get $\psi \in$ 融$_{\hat{A}}(Q_{\hat{\delta}}, R_{\hat{\varepsilon}})$. Consequently, for any $x \in G$, we still get

9.2.2 $$\text{融}_A\big((Q_\delta)^x, P_\gamma\big) \circ \kappa^Q_x \subset \text{融}_{\hat{A}}(Q_{\hat{\delta}}, P_{\hat{\gamma}})\,.$$

反过来,设 φ 是一个从 $\hat{\delta}$ 到 $\hat{\gamma}$ 的 $\hat{A}$-融合并且设

Conversely, let φ be a fusion from $\hat{\delta}$ to $\hat{\gamma}$ in $\hat{A}$ and let

9.2.3 $$\hat{h}\colon \hat{A}(j \otimes 1) \longrightarrow \hat{A}(i \otimes 1)$$

是一个 $\Delta_\varphi(Q)$-稳定的单 $\hat{A}$-模同态使得 $\mathrm{Im}(\hat{h})$ 是 $\hat{A}R$-模直和项，其中 $i \in \gamma$, $j \in \delta$, $R = \varphi(Q)$ (见 8.8). 也就是说，$\hat{a} = \hat{h}(j\otimes 1)$ 属于 $(j \otimes 1)\hat{A}(i \otimes 1)$ 并且有 $\hat{h}(\hat{a}') = \hat{a}'\hat{a}$，其中 $\hat{a}' \in \hat{A}(j\otimes 1)$．

be an injective $\Delta_\varphi(Q)$-stable A-module homomorphism such that $\mathrm{Im}(\hat{h})$ is a direct summand of $\hat{A}(i \otimes 1)$ as $\hat{A}R$-modules, where $i \in \gamma$, $j \in \delta$ and $R = \varphi(Q)$ (see 8.8). That is to say, $\hat{a} = \hat{h}(j\otimes 1)$ belongs to $(j\otimes 1)\hat{A}(i\otimes 1)$ and we have $\hat{h}(\hat{a}') = \hat{a}'\hat{a}$ for any $\hat{a}' \in \hat{A}(j\otimes 1)$.

进一步，$\hat{A}$ 的 G-内结构决定一个 $(j \otimes 1)\hat{A}(i \otimes 1)$ 上的 $\Delta_\varphi(Q)$-作用，并且对任意 $u \in Q$ 有 (见等式 8.6.1)

Moreover, the G-interior structure of $\hat{A}$ induces an action of $\Delta_\varphi(Q)$ on $(j \otimes 1)\hat{A}(i \otimes 1)$ and, for any $u \in Q$, we have (see equality 8.6.1)

9.2.4 $$\begin{aligned}\hat{a} &= \varphi(u)^{-1}(1_A \otimes \varphi(u)u^{-1})\cdot\hat{h}(j \otimes 1)\\ &= (1_A \otimes u^{-1})\hat{a}(1_A \otimes \varphi(u))\,,\end{aligned}$$

从而 $\hat{a} \in \big((j\otimes 1)\hat{A}(i\otimes 1)\big)^{\Delta_\varphi(Q)}$；进一步，如果 M 在 $\hat{A}(i\otimes 1)$ 里是 $\mathrm{Im}(\hat{h})$ 的补 $\hat{A}R$-模，那么不太难验证

so that $\hat{a} \in \big((j \otimes 1)\hat{A}(i \otimes 1)\big)^{\Delta_\varphi(Q)}$; furthermore, if M is a complement of $\mathrm{Im}(\hat{h})$ in $\hat{A}(i\otimes 1)$ as $\hat{A}R$-modules, it is not difficult to prove that

9.2.5 $$X = \Big(\sum_{x\in G-N} Aj^x \otimes x^{-1}\Big)\hat{a} \oplus M$$

在 $\hat{A}(i \otimes 1)$ 里是 $\hat{h}(Aj \otimes 1)$ 的补 $A\Delta_\varphi(Q)$-模. 令

is a complement of $\hat{h}(Aj \otimes 1)$ in $\hat{A}(i \otimes 1)$ as $A\Delta_\varphi(Q)$-modules. Set

9.2.6 $$E = \mathrm{End}_{A\Delta_\varphi(Q)}\big(\hat{A}(i \otimes 1)\big)\,.$$

这样，存在 E 的幂等元 e 使得 $e\big(\hat{A}(i\otimes 1)\big) = \hat{h}(Aj\otimes 1)$；请注意，因为 j 在 jA^Qj 中是

Thus, there exists an idempotent e in E such that $e\big(\hat{A}(i\otimes 1)\big) = \hat{h}(Aj\otimes 1)$; note that, since j is primitive in jA^Qj, the AQ-module

本原的, 所以 AQ-模 Aj 是不可分解的 (见例子 4.15), 从而 $A\Delta_\varphi(Q)$-模 $\hat{h}(Aj\otimes 1)$ 也是不可分解的; 也就是说, e 是本原的 (见例子 4.15), 从而只要考虑 $\Delta_\varphi(Q)$-稳定分解

Aj is indecomposable (see example 4.15), and therefore the $A\Delta_\varphi(Q)$-module $\hat{h}(Aj\otimes 1)$ is indecomposable too; in other words, e is primitive (see example 4.15), and therefore, by considering the following $\Delta_\varphi(Q)$-stable decomposition

9.2.7 $$\hat{A}(i\otimes 1)=\bigoplus_y Ai^y\otimes y^{-1}$$

其中$y\in G$跑遍$\bar{G}$的一个代表集合, 以及使用命题 3.19, 就可证明存在 $x\in G$ 并且 $g\in E^*$ 满足对任意 $y\in G-\Delta_\varphi(Q){\cdot}x$ 有 $e^{g^{-1}}(Ai^{y^{-1}}\otimes y)=\{0\}$, 还有

where $y\in G$ runs over a set of representatives for $\bar{G}$, and by applying Proposition 3.19, we can prove the existence of $x\in G$ and $g\in E^*$ such that, for any $y\in G-\Delta_\varphi(Q){\cdot}x$, we have $e^{g^{-1}}(Ai^{y^{-1}}\otimes y)=\{0\}$ and

9.2.8 $$g\big(\hat{h}(Aj\otimes 1)\big)\subset\bigoplus_{u\in U} Ai^{(ux)^{-1}}\otimes ux\varphi(u)^{-1}$$

其中G上的$\Delta_\varphi(Q)$-作用的定义是$(u,\varphi(u)){\cdot}y=ux\varphi(u)^{-1}$, 这里$u\in Q$, $y\in G$, 与$\Delta_\varphi(Q)_x$是xN的稳定化子并$U\subset Q$满足$\Delta_\varphi(U)$是$\Delta_\varphi(Q)/\Delta_\varphi(Q)_x$的代表集合. 请注意, 有

where the action of $\Delta_\varphi(Q)$ on G is defined by $(u,\varphi(u)){\cdot}y=ux\varphi(u)^{-1}$ for any $u\in Q$ and any $y\in G$, $\Delta_\varphi(Q)_x$ denotes the stabilizer of xN, and $U\subset Q$ is such that $\Delta_\varphi(U)$ is a set of representatives for $\Delta_\varphi(Q)/\Delta_\varphi(Q)_x$. Note that we have

9.2.9 $$\bigoplus_{u\in U} Ai^{(ux)^{-1}}\otimes ux\varphi(u)^{-1}\cong \mathrm{Ind}_{\Delta_\varphi(Q)_x}^{\Delta_\varphi(Q)}(Ai^{x^{-1}}\otimes x)=I_x\,.$$

这样, 更精解得到

Thus, more precisely we get

9.2.10 $$\mathrm{End}_\mathcal{O}\Big(g\big(\hat{h}(Aj\otimes 1)\big)\Big)\cong e^{g^{-1}}\mathrm{End}_\mathcal{O}(I_x)e^{g^{-1}}\,.$$

事实上, 从 Q_δ 是局部的条件可推出 $\Delta_\varphi(Q)_x=\Delta_\varphi(Q)$; 确实, 令

Actually, we claim that the hypothesis that Q_δ is local forces $\Delta_\varphi(Q)_x=\Delta_\varphi(Q)$; indeed, denote by

9.2.11 $$f: I_x\longrightarrow Ai^{x^{-1}}\otimes x\longrightarrow I_x$$

记自然满射和自然单射的合成; 那么, f 显然是 $A\Delta_\varphi(Q)_x$-模自同态满足

the composition of the canonical surjection and injection; then, f is an $A\Delta_\varphi(Q)_x$-module endomorphism which clearly fulfills

9.2.12 $$\mathrm{Tr}_{\Delta_\varphi(Q)_x}^{\Delta_\varphi(Q)}(f)=\mathrm{id}_{I_x}\,;$$

所以也得到

consequently, we still get

9.2.13 $$e^{g^{-1}} = \mathrm{Tr}_{\Delta_\varphi(Q)_x}^{\Delta_\varphi(Q)}(e^{g^{-1}} f e^{g^{-1}})$$

而且, 由同构式 9.2.10, 有

and moreover, by isomorphism 9.2.10, we have

9.2.14 $$e^{g^{-1}} f e^{g^{-1}} \in \mathrm{End}_{A\Delta_\varphi(Q)_x}\Big(g\big(\hat{h}(\hat{A}j \otimes 1)\big)\Big)\,;$$

特别是, 既然 g 是 $A\Delta_\varphi(Q)$-模同构, 对适当的 $\hat{h}(Aj \otimes 1)$ 的 $A\Delta_\varphi(Q)_x$-模自同态 d 也得到

in particular, since g is an $A\Delta_\varphi(Q)$-module isomorphism, for a suitable $A\Delta_\varphi(Q)_x$-module endomorphism d of $\hat{h}(Aj \otimes 1)$ we also have

9.2.15 $$e = \mathrm{Tr}_{\Delta_\varphi(Q)_x}^{\Delta_\varphi(Q)}(d)\,.$$

另一方面, 不难验证存在 jAj 的元素 c 使得对任意 $a \in A$ 有 $d\big(\hat{h}(aj \otimes 1)\big) = \hat{h}(ac \otimes 1)$; 这样, 对任意 $u \in Q$ 得到

On the other hand, it is quite easy to check that there is an element c of jAj fulfilling $d\big(\hat{h}(aj \otimes 1)\big) = \hat{h}(ac \otimes 1)$ for any $a \in A$; thus, for any $u \in Q$, we get

9.2.16 $$\begin{aligned} d\big(u^{-1}\cdot\hat{h}(aj \otimes 1)\cdot\varphi(u)\big) &= d\big(u^{-1}\cdot(aj \otimes 1)\hat{a}\cdot\varphi(u)\big) \\ &= d\big(\hat{h}(a^u j \otimes 1)\big) \\ &= \hat{h}(a^u c \otimes 1)\big) \\ &= u^{-1}\cdot\hat{h}(ac^{u^{-1}} \otimes 1)\cdot\varphi(u)\,. \end{aligned}$$

最后, 因为 $\hat{h}$ 是单射所以从等式 9.2.16 可推出 $c \in (jAj)^{Q_x}$ 并, 由等式 9.2.15, 有 $j = \mathrm{Tr}_{Q_x}^{Q}(c)$, 其中 $Q_x = \varphi^{-1}\big(\Delta_\varphi(Q)_x\big)$; 由于 Q_δ 是局部的, 得到 $Q_x = Q$ 与 $\Delta_\varphi(Q)_x = \Delta_\varphi(Q)$ (见命题 6.3)

Finally, since $\hat{h}$ is injective, from equality 9.2.16 we can deduce that $c \in (jAj)^{Q_x}$ and, by equality 9.2.15, we have $j = \mathrm{Tr}_{Q_x}^{Q}(c)$, where $Q_x = \varphi^{-1}\big(\Delta_\varphi(Q)_x\big)$; it follows that $Q_x = Q$ and $\Delta_\varphi(Q)_x = \Delta_\varphi(Q)$ since Q_δ is local (see Proposition 6.3).

特别是, 对任意 $u \in Q$ 得到 $\varphi(u)N = u^x N$, 并且 $\hat{h}(Aj \otimes 1)$ 与 $Ai^{x^{-1}} \otimes x$ 的某个 $A\Delta_\varphi(Q)$-模直和项是 $A\Delta_\varphi(Q)$-模同构的. 现在, 只要考虑 $h\colon Aj \to Ai^{x^{-1}}$ 使得

In particular, we get $\varphi(u)N = u^x N$ for any $u \in Q$, and the $A\Delta_\varphi(Q)$-module $\hat{h}(Aj \otimes 1)$ is isomorphic to some direct summand of the $A\Delta_\varphi(Q)$-module $Ai^{x^{-1}} \otimes x$. Now, considering $h\colon Aj \to Ai^{x^{-1}}$ such that, for any $a \in A$, we have

9.2.17 $$h(aj) \otimes x = g\big(\hat{h}(aj \otimes 1)\big)$$

其中 $a \in A$, 就可证明 $\kappa_{x^{-1}}^P \circ \varphi$ 是一个从 δ 到 $\gamma^{x^{-1}}$ 的 A-融合;

we can deduce that $\kappa_{x^{-1}}^P \circ \varphi$ is a fusion from δ to $\gamma^{x^{-1}}$ in A; indeed, h clearly is

确实, h 显然是 A-模同态而对任意 $u\in Q$ 只要记 $v=\varphi(u)^{x^{-1}}$ 以及使用等式 8.6.1 并且定义 9.1.3, 就我们得到

an A-module homomorphism and, for any $u \in Q$, setting $v = \varphi(u)^{x^{-1}}$ and applying equality 8.6.1 and definition 9.1.3, we get the following equalities

9.2.18
$$\begin{aligned} h^{(u,v)}(aj)\otimes x &= h(vu^{-1}\cdot a^{u^{-1}}j)^v\otimes x \\ &= \big(h(vu^{-1}\cdot a^{u^{-1}}j)\otimes x[\varphi(u^{-1}),x^{-1}]\big)^v \\ &= \Big(g\big(\hat h(vu^{-1}\cdot a^{u^{-1}}j\otimes 1)\big)\cdot[\varphi(u^{-1}),x^{-1}]\Big)^v \\ &= v^{-1}\cdot g\big(\hat h(vu^{-1}\cdot a^{u^{-1}}j\otimes 1)\big)\cdot\varphi(u) \\ &= v^{-1}u\cdot g\big(u^{-1}\cdot\hat h(vu^{-1}\cdot a^{u^{-1}}j\otimes 1)\cdot\varphi(u)\big) \\ &= v^{-1}u\cdot g\big(\hat h(u^{-1}v\cdot a^j\otimes 1)\big)\,, \end{aligned}$$

从而, 因为 $v^{-1}u$ 属于 N, 所以也得到

and therefore, since $v^{-1}u$ belongs to N, we still get

9.2.19
$$h^{(u,\varphi(u)^{x^{-1}})}(aj)\otimes x = g\big(\hat h(aj\otimes 1)\big) = h(aj)\otimes x\,;$$

也就是说, $\Delta_{\kappa^P_{x^{-1}}\circ\varphi}(Q)$ 稳定 h.

that is to say, $\Delta_{\kappa^P_{x^{-1}}\circ\varphi}(Q)$ stabilizes h.

类似地, 幂等元 $e^{g^{-1}}$ 决定 $\mathrm{End}_{\mathcal O}(Ai^x)$ 的幂等元 ℓ 使得

Similarly, the idempotent $e^{g^{-1}}$ determines an idempotent of $\mathrm{End}_{\mathcal O}(Ai^x)$ fulfilling

9.2.20
$$\ell(ai^x)\otimes x^{-1} = e^{g^{-1}}(ai^x\otimes x^{-1})$$

其中 $a\in A$; 特别是, 由于 e 的定义, 有

for any $a\in A$; in particular, according to the definition of e, we have

9.2.21
$$\begin{aligned} h(Aj)\otimes x = g\big(\hat h(Aj\otimes 1)\big) &= e^{g^{-1}}\big(\hat A(i\otimes 1)\big) \\ &= \ell(Ai^{x^{-1}})\otimes x \end{aligned}$$

并且 ℓ 属于 $\mathrm{End}_{AR^x}(Ai^x)$; 确实不难验证 ℓ 是 A-模同态; 而且对任意 $u\in Q$ 再次只要记 $v=\varphi(u)^{x^{-1}}$ 以及使用等式 8.6.1 并且定义 9.1.3, 就得到

and we claim that ℓ belongs to $\mathrm{End}_{AR^x}(Ai^x)$; indeed, it is clear that ℓ is an A-module homomorphism; moreover, for any $u\in Q$, once again setting $v=\varphi(u)^{x^{-1}}$ and applying equality 8.6.1 and definition 9.1.3, we get

9.2.22
$$\begin{aligned} \ell\big((ai^{x^{-1}})^{v^{-1}}\big)^v\otimes x &= \big(\ell(a^{v^{-1}}i^{x^{-1}})\otimes x[\varphi(u^{-1}),x^{-1}]\big)^v \\ &= \big(e^{g^{-1}}(a^{v^{-1}}i^{x^{-1}}\otimes x)[\varphi(u^{-1}),x^{-1}]\big)^v \\ &= v^{-1}\cdot e^{g^{-1}}(a^{v^{-1}}i^{x^{-1}}\otimes x)\cdot\varphi(u) \\ &= v^{-1}u\cdot e^{g^{-1}}\big(a^{v^{-1}u}i^{x^{-1}u}\otimes u^{-1}x\varphi(u)\big) \\ &= v^{-1}u\cdot e^{g^{-1}}\big((v^{-1}u)^{-1}\cdot ai^{x^{-1}}\cdot(v^{-1}u)\otimes(u^{-1}v)x\big) \end{aligned}$$

从而, 因为$v^{-1}u$属于N,所以也得到

and therefore, since $v^{-1}u$ belongs to N, we still get

9.2.23 $$\ell\big((ai^{x^{-1}})^{v^{-1}}\big)^v \otimes x = e^{g^{-1}}(ai^{x^{-1}} \otimes x) = \ell(ai^{x^{-1}}) \otimes x\,.$$

最后, 由等式 8.8.1, 有

Finally, it follows from equality 8.8.1 that

9.2.24 $$\varphi \circ \kappa^{Q^x}_{x^{-1}} = (\kappa^P_{x^{-1}} \circ \varphi)^x \in 融_A\big((Q_\delta)^x, P_\gamma\big)\,.$$

推论 9.3. **设 P_γ 是 A 上的局部点群. 那么 $融_A(P_\gamma)$ 是 $融_{\hat{A}}(P_\gamma)$ 的正规子群, 并商群 $融_{\hat{A}}(P_\gamma)/融_A(P_\gamma)$ 的次幂整除 $\bar{G}$ 的次幂.**

Corollary 9.3. *Let P_γ be a local pointed group on A. Then, $融_A(P_\gamma)$ is a normal subgroup of $融_{\hat{A}}(P_\gamma)$, and the exponent of the quotient $融_{\hat{A}}(P_\gamma)/融_A(P_\gamma)$ divises the exponent of $\bar{G}$.*

证明: 由定理 9.2,任意 $融_{\hat{A}}(P_\gamma)$ 的元素φ可分解为$\varphi = \psi \circ \kappa^P_x$ 其中$x \in G$, $\psi \in 融_A\big((P_\gamma)^x, P_\gamma\big)$; 这样, 对任意 $\eta \in 融_A(P_\gamma)$ 有

Proof: It follows from Theorem 9.2 that any element φ of $融_{\hat{A}}(P_\gamma)$ has a decomposition $\varphi = \psi \circ \kappa^P_x$, where $x \in G$, $\psi \in 融_A\big((P_\gamma)^x, P_\gamma\big)$; thus, for any $\eta \in 融_A(P_\gamma)$, we have

9.3.1 $$\begin{aligned}\varphi^{-1} \circ \eta \circ \varphi &= \kappa^{P^x}_{x^{-1}} \circ \psi^{-1} \circ \eta \circ \psi \circ \kappa^P_x \\ &= (\psi^{-1} \circ \eta \circ \psi)^{x^{-1}}\,,\end{aligned}$$

并且, 由等式 8.8.1, $\varphi^{-1}\circ\eta\circ\varphi$ 就属于 $融_A(P_\gamma)$. 而且, 对任意 $n \in \mathbb{N} - \{0\}$ 有 $融_A\big((P_\gamma)^{x^n}, P_\gamma\big)$ 的元素 ψ_n 使得 $\varphi^n = \psi_n \circ \kappa^P_{x^n}$; 确实, 只要使用归纳法, 就得到

and, by equality 8.8.1, $\varphi^{-1} \circ \eta \circ \varphi$ belongs to $融_A(P_\gamma)$. Moreover, for any $n \in \mathbb{N} - \{0\}$, there is an element ψ_n of $融_A\big((P_\gamma)^{x^n}, P_\gamma\big)$ fulfilling $\varphi^n = \psi_n \circ \kappa^P_{x^n}$; indeed, arguing by induction, we get

9.3.2 $$\varphi^n = (\psi_{n-1} \circ \kappa^P_{x^{n-1}}) \circ (\psi \circ \kappa^P_x) = (\psi_{n-1} \circ \psi^{x^{n-1}}) \circ \kappa^P_{x^n}\,;$$

所以, 如果 $x^n \in N$ 那么 $\kappa^P_{x^n}$ 属于 $融_A\big(P_\gamma, (P_\gamma)^{x^n}\big)$ (见 8.11), 而 φ^n 属于 $融_A(P_\gamma)$.

consequently, if $x^n \in N$ then $\kappa^P_{x^n}$ belongs to $融_A\big(P_\gamma, (P_\gamma)^{x^n}\big)$ (see 8.11), so that φ^n belongs to $融_A(P_\gamma)$.

9.4. 如果 k 是代数闭的, 而 $\bar{G}$ 是一个 p-群, 那么上面的断言可以更精确化; 确实在 $\hat{A}$ 上的局部点群之间的 $\hat{A}$-融合都已经明确了; 精确化的证明要应用引理 7.3. 请注意, 该引理的结论可以更清楚地描述如下: 设 B 与 B' 是 $\mathcal{O}$-代数并设 $g: B \to B'$ 是一个 $\mathcal{O}$-代数幺

9.4. If k is algebraically closed and $\bar{G}$ is a p-group, then we can be more precise on the statements above, namely by expliciting all the fusions between the local pointed groups on $\hat{A}$; the proof of these precisions will employ Lemma 7.3. Note that the conclusion of this lemma can be better understood as follows: let B and B' be $\mathcal{O}$-algebras and $g: B \to B'$ a unitary $\mathcal{O}$-algebra homomor-

同态; 令 $\mathfrak{C}_B$ 表示这样的范畴, 其元素是 B 的幂等元, 而两个元素 i,j 之间的同态集合就是 jBi; 类似地, $\mathfrak{C}_{B'}$ 记对应 B' 的范畴. 那么 g 显然决定一个函子 $\mathfrak{f}_g\colon \mathfrak{C}_B \to \mathfrak{C}_{B'}$ 而不难证明下面的条件是彼此等价的

phism; denote by $\mathfrak{C}_B$ the following category; its objects are the idempotents of B, whereas the set of morphisms between two objects j, i is just jBi; similarly, $\mathfrak{C}_{B'}$ denotes the corresponding category for B'. Then g induces a functor $\mathfrak{f}_g\colon \mathfrak{C}_B \to \mathfrak{C}_{B'}$ and it is easy to prove the equivalence of the following conditions:

9.4.1. **对任意 B 的本原幂等元 i, $g(i)$ 也是本原的并且有**

9.4.1. *For any primitive idempotent i of B, $g(i)$ still is primitive, and we have*

$$g\big(J(B)\big) \subset J(B')$$

9.4.2. **对任意 $\mathfrak{C}_B$ 的不可分解元素 i, $\mathfrak{f}_g(i)$ 也是不可分解的, 并且只要 $\mathfrak{f}_g(b)$ 在 $\mathfrak{C}_{B'}$ 中是同构, $\mathfrak{C}_B$ 的同态 b 就是同构.**

9.4.2. *For any indecomposable object i of $\mathfrak{C}_B$, $\mathfrak{f}_g(i)$ still is indecomposable, and a morphism b in $\mathfrak{C}_B$ is an isomorphism whenever $\mathfrak{f}_g(b)$ is an isomorphism in $\mathfrak{C}_{B'}$.*

定理 9.5. 假定 k 是代数闭的并假定 $\bar{G}$ 是 p-群. 那么对任意 G 的 p-子群 P 有

Theorem 9.5. *Assume that k is algebraically closed and that $\bar{G}$ is a p-group. Then, for any P-subgroup P of G, we have*

9.5.1 $$J(A^P) = A^P \cap J(\hat{A}^P)$$

并且只要 i 是 A^P 的本原幂等元 $i \otimes 1$ 就是 $\hat{A}^P$ 的本原幂等元. 特别是, 任意 A 上的 P-点 γ 决定唯一一个 $\hat{A}$ 上的 P-点 $\hat{\gamma}$ 使得 $\gamma \otimes 1 \subset \hat{\gamma}$; 那么 γ 是局部的当且仅当 $\hat{\gamma}$ 是局部的.

and if i is a primitive idempotent of A^P then $i \otimes 1$ is a primitive idempotent of $\hat{A}^P$. In particular, any point γ of P on A determines a unique point $\hat{\gamma}$ of P on $\hat{A}$ fulfilling $\gamma \otimes 1 \subset \hat{\gamma}$; then, γ is local if and only if $\hat{\gamma}$ is local.

证明: 我们对阶 $|P|$ 使用归纳法. 设 γ 是一个 A 上的 P-点; 如果 γ 不是局部的, 那么设 Q_δ 是一个 P_γ 的亏点群 (见 6.10); 由定理 7.2, 有 $\gamma = \mathrm{ind}_Q^P(\delta)$, 并且我们已经知道 $\hat{\delta}$ 是一个 $\hat{A}$ 上的 Q-点; 所以, 仍由定理 7.2, $\hat{\gamma} = \mathrm{ind}_Q^P(\hat{\delta})$ 是 $\hat{A}$ 上的非局部 P-点, 以及显然有 $\gamma \otimes 1 \subset \hat{\gamma}$.

Proof: We argue by induction on $|P|$. Let γ be a point of P on A; if γ is not local then let Q_δ be a defect pointed group of P_γ (see 6.10); by Theorem 7.2, we have $\gamma = \mathrm{ind}_Q^P(\delta)$, and we already know that $\hat{\delta}$ is a point of Q on $\hat{A}$; consequently, once again by Theorem 7.2, $\hat{\gamma} = \mathrm{ind}_Q^P(\hat{\delta})$ is a point of P on $\hat{A}$ which is not local, and clearly we have $\gamma \otimes 1 \subset \hat{\gamma}$.

假定 γ 是局部的. 因为 P 稳定直和分解 9.1.1, 所以有

Assume that γ is local. Since P stabilizes the direct sum decompostion 9.1.1, then

9.5.2 $$\hat{A}^P = \bigoplus_{\bar{X}} \Big(\bigoplus_{x} A \otimes x \Big)^P$$

其中 $\bar{X}$ 跑遍 $\bar{G}$ 上的 P-轨道的集合, 而x跑遍$\bar{X}$在G中的原像的一个代表集合; 如果 $|\bar{X}| \neq 1$ 那么任意 $(\oplus_x A\otimes x)^P$ 的元素显然是一个从P的某个真子群到P的相对迹, 特别是有 (见 6.4.1)

where $\bar{X}$ runs over the set of orbits of P on $\bar{G}$, and x on a set of representatives for $\bar{X}$ in the converse image of $\bar{X}$ in G; if $|\bar{X}| \neq 1$ then any element of $(\oplus_x A\otimes x)^P$ clearly is a trace from some proper subgroup of P to P, and in particular we have (see 6.4.1)

9.5.3 $$\mathrm{Br}_P\big((\bigoplus_x A\otimes x)^P\big) = \{0\}\,.$$

也就是说, 只要选择 $C_{\bar{G}}(\bar{P})$ 的原像的一个代表集合 $U\subset G$, 就不难证明

That is to say, choosing a set of representatives $U\subset G$ in the converse image of $C_{\bar{G}}(\bar{P})$, it is easy to check that we just get

9.5.4 $$\hat{A}(P) = \bigoplus_{u\in U}(A\otimes u)(P)\,;$$

而且, 只要使用等式 9.1.2, 在 $\hat{A}(P)$ 里就可推出

moreover, by applying equality 9.1.2, in $\hat{A}(P)$ we can deduce that

9.5.5 $$(A\otimes u)(P)\cdot(A\otimes v)(P) = (A\otimes w)(P)\,,$$

其中 $u,v,w\in U$ 满足 $\bar{w}=\bar{u}\bar{v}$. 最后, 对任意 $u\in U$ 令

for any $u,v,w\in U$ such that $\bar{w}=\bar{u}\bar{v}$. Finally, for any $u\in U$, set

9.5.6 $$\hat{A}(P)_{\bar{u}} = (A\otimes u)(P)\,;$$

那么 $\{\hat{A}(P)_{\bar{u}}\}_{\bar{u}\in C_{\bar{G}}(\bar{P})}$ 显然满足引理 7.3 的条件; 所以得到

then $\{\hat{A}(P)_{\bar{u}}\}_{\bar{u}\in C_{\bar{G}}(\bar{P})}$ clearly fulfills the hypothesis of Lemma 7.3; hence, we get

9.5.7 $$J\big((A)(P)\big) \cong J\big((A\otimes 1)(P)\big) = J\big(\hat{A}(P)\big)\cap(A\otimes 1)(P)$$

并且 $\mathrm{Br}_P(\hat{\gamma})$ 是一个 $\hat{A}(P)$ 的点.

and $\mathrm{Br}_P(\hat{\gamma})$ is a point of $\hat{A}(P)$.

我们使用反证法, 即假定 $\hat{\gamma}$ 不是 $\hat{A}$ 上的 P-点. 因为 $\mathrm{Br}_P(\hat{\gamma})$ 是 $\hat{A}(P)$ 的点, 所以存在 $\hat{A}$ 上的非局部 P-点 $\hat{\beta}$ 使得 $\hat{\beta}\subset\hat{\gamma}$; 设 $R_{\hat{\varepsilon}}$ 是一个 $P_{\hat{\beta}}$ 的亏点群; 仍由定理 7.2, 有 $\hat{\beta}=\mathrm{ind}_R^P(\hat{\varepsilon})$; 而且, 因为 $R\neq P$ 所以只要在等式 9.5.7 里使用 R 代替 P 就得到下面的 $\bar{N}_G(R_{\hat{\varepsilon}})$-代数幺同态

We argue by contradiction and assume that $\hat{\gamma}$ is not a point of P on $\hat{A}$. Since $\mathrm{Br}_P(\hat{\gamma})$ is a point of $\hat{A}(P)$, there exists a point $\hat{\beta}$ of P on $\hat{A}$ which is not local and fulfills $\hat{\beta}\subset\hat{\gamma}$; let $R_{\hat{\varepsilon}}$ be a defect pointed group of $P_{\hat{\beta}}$; by Theorem 7.2 again, we have $\hat{\beta}=\mathrm{ind}_R^P(\hat{\varepsilon})$; moreover, since $R\neq P$, it follows from equality 9.5.7 applied to R that we get the following unitary $\bar{N}_G(R_{\hat{\varepsilon}})$-algebra homomorphism

9.5.8 $$\prod_{\varepsilon\in\mathcal{P}_{\hat{\varepsilon}}} A(R_\varepsilon) \longrightarrow \hat{A}(R_{\hat{\varepsilon}})\,,$$

其中 $\mathcal{P}_{\hat{\varepsilon}}$ 是 A 上的满足 $\varepsilon \subset \hat{\varepsilon}$ 的 (局部) P-点的集合.

where $\mathcal{P}_{\hat{\varepsilon}}$ denotes the set of (local) points of P on A fulfilling $\varepsilon \subset \hat{\varepsilon}$.

因为k是代数闭的所以有 $\hat{A}(R_{\hat{\varepsilon}}) \cong \mathrm{End}_k(V_{\hat{\varepsilon}})$, 其中$V_{\hat{\varepsilon}}$是一个$k$-向量空间; 类似地, 对任意$\varepsilon \in \mathcal{P}_{\hat{\varepsilon}}$有$A(R_\varepsilon) \cong \mathrm{End}_k(V_\varepsilon)$ 其中 V_ε 是一个 k-向量空间, 并因为任意 $A(R_\varepsilon)$ 的本原幂等元在$\hat{A}(R_{\hat{\varepsilon}})$中也是本原的, 所以还有

Since k is algebraically closed, we have $\hat{A}(R_{\hat{\varepsilon}}) \cong \mathrm{End}_k(V_{\hat{\varepsilon}})$ for a suitable k-linear vector space $V_{\hat{\varepsilon}}$; similarly, for any $\varepsilon \in \mathcal{P}_{\hat{\varepsilon}}$, we have $A(R_\varepsilon) \cong \mathrm{End}_k(V_\varepsilon)$ for a suitable k-linear vector space V_ε, and moreover, since any primitive idempotent in $A(R_\varepsilon)$ remains primitive in $\hat{A}(R_{\hat{\varepsilon}})$, we still have

9.5.9 $$V_{\hat{\varepsilon}} \cong \bigoplus_{\varepsilon \in \mathcal{P}_{\hat{\varepsilon}}} V_\varepsilon .$$

已经知道任意$\mathrm{End}_k(V_{\hat{\varepsilon}})$ 的自同构可以被某个$f \in \mathrm{End}_k(V_{\hat{\varepsilon}})^*$的共轭决定; 这样, 对任意 $\bar{u} \in \bar{N}_P(R_{\hat{\varepsilon}})$ 我们可选择一个满足$f_{\bar{u}}$-共轭作用就是 $\bar{u}$-作用的$\mathrm{End}_k(V_{\hat{\varepsilon}})^*$的元素$f_{\bar{u}}$; 特别是, 对某个 $m \in \mathbb{N}$ 有

It is well-known that any automorphism of $\mathrm{End}_k(V_{\hat{\varepsilon}})$ is induced by conjugation by some $f \in \mathrm{End}_k(V_{\hat{\varepsilon}})^*$; thus, for any $\bar{u} \in \bar{N}_P(R_{\hat{\varepsilon}})$, we can choose an element $f_{\bar{u}}$ of $\mathrm{End}_k(V_{\hat{\varepsilon}})^*$ such that the $f_{\bar{u}}$-conjugation coincides with the action of $\bar{u}$; in particular, for some $m \in \mathbb{N}$, we have

9.5.10 $$(f_{\bar{u}})^{p^m} = \lambda \cdot \mathrm{id}_{V_{\hat{\varepsilon}}} ;$$

而且因为k是代数闭的, 所以存在唯一 $\mu \in k^*$ 使得 $\mu^{p^m} = \lambda$, 从而有

moreover, since k is algebraically closed, there is a unique $\mu \in k^*$ fulfilling $\mu^{p^m} = \lambda$ and therefore we get

9.5.11 $$(\mu^{-1} \cdot f_{\bar{u}})^{p^m} = \mathrm{id}_{V_{\hat{\varepsilon}}} .$$

这样, 我们能把 $A(R_{\hat{\varepsilon}})$ 上的 $\bar{N}_P(R_{\hat{\varepsilon}})$-作用唯一地提升到一个映射

Thus, we can lift the action of $\bar{N}_P(R_{\hat{\varepsilon}})$ on $A(R_{\hat{\varepsilon}})$ to a unique map

9.5.12 $$\rho : \bar{N}_P(R_{\hat{\varepsilon}}) \longrightarrow \mathrm{End}_k(V_{\hat{\varepsilon}})^*$$

使得$\rho(\bar{u})$是一个p-元素, 其中$\bar{u} \in \bar{N}_P(R_{\hat{\varepsilon}})$; 事实上, ρ 是群同态; 确实, 显然有$\rho(\bar{u}^n) = \rho(\bar{u})^n$ 其中$n \in \mathbb{N}$; 而且, 只要证明$\rho(\bar{u})\rho(\bar{v})$也是$p$-元素, 就我们有 $\rho(\bar{u})\rho(\bar{v}) = \rho(\overline{uv})$, 其中 $\bar{u}, \bar{v} \in \bar{N}_P(R_{\hat{\varepsilon}})$. 这样, 对任意$\bar{N}_P(R_{\hat{\varepsilon}})$ 的子群$\bar{U}$, 只要证明$\rho(\bar{U})$也是p-群; 对阶$|\bar{U}|$使用

such that $\rho(\bar{u})$ is a p-element for any element $\bar{u} \in \bar{N}_P(R_{\hat{\varepsilon}})$; actually, ρ is a group homomorphism; indeed, clearly $\rho(\bar{u}^n) = \rho(\bar{u})^n$ for any $n \in \mathbb{N}$; moreover, it suffices that we prove that $\rho(\bar{u})\rho(\bar{v})$ is also a p-element for any $\bar{u}, \bar{v} \in \bar{N}_P(R_{\hat{\varepsilon}})$, to get $\rho(\bar{u})\rho(\bar{v}) = \rho(\overline{uv})$. Thus, it suffices that we prove that $\rho(\bar{U})$ is a p-group for any subgroup $\bar{U}$ of $N_P(R_{\hat{\varepsilon}})$; we argue by induction on $|\bar{U}|$ and may

归纳法, 而可假定$\bar{U}\neq 1$; 此时, 设$\bar{V}$是$\bar{U}$的极大子群并选择$\bar{u}\in\bar{U}-\bar{V}$; 因为$\bar{V}\lhd\bar{U}$所以$\bar{u}$-作用稳定$\rho(\bar{V})$, 从而$\rho(\bar{V})\cdot\langle\rho(\bar{u})\rangle$是$p$-群; 特别是, $\rho(\bar{v})\rho(\bar{u})^n$是$p$-元素, 而$\rho(\bar{v})\rho(\bar{u})^n=\rho(\bar{v}\bar{u}^n)$其中$\bar{v}\in\bar{V}$与$n\in\mathbb{N}$; 最后有$\rho(\bar{U})=\rho(\bar{V})\cdot\langle\rho(\bar{u})\rangle$.

assume that $\bar{U}\neq 1$; then, let $\bar{V}$ be a maximal subgroup of $\bar{U}$ and choose $\bar{u}\in\bar{U}-\bar{V}$; since we have $\bar{V}\lhd\bar{U}$, the action of $\bar{u}$ stabilizes $\rho(\bar{V})$ and therefore $\rho(\bar{V})\cdot\langle\rho(\bar{u})\rangle$ is a p-group; in particular, $\rho(\bar{v})\rho(\bar{u})^n$ is a p-element for any $\bar{v}\in\bar{V}$ and any $n\in\mathbb{N}$, so that we get $\rho(\bar{v})\rho(\bar{u})^n=\rho(\bar{v}\bar{u}^n)$; finally, we obtain $\rho(\bar{U})=\rho(\bar{V})\cdot\langle\rho(\bar{u})\rangle$.

那么, $V_{\hat{\varepsilon}}$成为$k\bar{N}_P(R_{\hat{\varepsilon}})$-模; 而且, 因为同态9.5.8是$\bar{N}_G(R_{\hat{\varepsilon}})$-代数幺同态, 所以分解式9.5.9是$\bar{N}_P(R_{\hat{\varepsilon}})$-稳定的. 这样, 我们能应用下面的引理9.6. 更精确地, 令$\mathcal{Q}\subset\mathcal{P}_{\hat{\varepsilon}}$记$\mathcal{P}_{\hat{\varepsilon}}$上的$\bar{N}_P(R_{\hat{\varepsilon}})$-轨道的一个代表集合, 再令 (见定理 6.14)

Then, $V_{\hat{\varepsilon}}$ becomes a $k\bar{N}_P(R_{\hat{\varepsilon}})$-module; moreover, since homomorphism 9.5.8 is a unitary $k\bar{N}_G(R_{\hat{\varepsilon}})$-algebra homomorphism, $\bar{N}_P(R_{\hat{\varepsilon}})$ stabilizes decomposition 9.5.9. Thus, we are able to apply Lemma 9.6 below. More precisely, denote by $\mathcal{Q}\subset\mathcal{P}_{\hat{\varepsilon}}$ a set of representatives for the orbits of $\bar{N}_P(R_{\hat{\varepsilon}})$ in $\mathcal{P}_{\hat{\varepsilon}}$, and set (see Theorem 6.14)

9.5.13 $$\mathcal{P}_{\hat{\beta}}=\bigcup_{\varepsilon\in\mathcal{P}_{\hat{\varepsilon}}}\mathcal{P}_{A}(P)_{\varepsilon}\,;$$

同态9.5.8显然决定下面的k-代数同态

homomorphism 9.5.8 clearly determines the following k-algebra homomorphism

9.5.14 $$\prod_{\varepsilon\in\mathcal{Q}}A(R_\varepsilon)^{\bar{N}_P(R_\varepsilon)}\cong\Big(\prod_{\varepsilon\in\mathcal{P}_{\hat{\varepsilon}}}A(R_\varepsilon)\Big)^{\bar{N}_P(R_{\hat{\varepsilon}})}\longrightarrow\hat{A}(R_{\hat{\varepsilon}})^{\bar{N}_P(R_{\hat{\varepsilon}})},$$

从而, 由定理6.14与引理9.6, 我们得到下面的交换图

and so, by Theorem 6.14 and Lemma 9.6, we get the following commutative diagram

9.5.15 $$\begin{array}{ccc}\prod_{\varepsilon\in\mathcal{Q}}A(R_\varepsilon)^{\bar{N}_P(R_\varepsilon)} & \longrightarrow & \hat{A}(R_{\hat{\varepsilon}})^{\bar{N}_P(R_{\hat{\varepsilon}})}\\ \downarrow & & \downarrow\\ \prod_{\beta\in\mathcal{P}_{\hat{\beta}}}A(P_\beta) & \longrightarrow & \hat{A}(P_{\hat{\beta}})\end{array}$$

最后, 如果$i\in\gamma$那么在$\hat{A}(P_{\hat{\beta}})$里$i\otimes 1$的像不是零, 这是因为$\hat{\beta}\subset\hat{\gamma}$; 所以, 在$\prod_{\beta\in\mathcal{P}_{\hat{\beta}}}A(P_\beta)$中$i$的像也不是零; 也就是说, 存在$\beta\in\mathcal{P}_{\hat{\beta}}$使得$s_\beta(i)\neq 0$, 从而得到$\beta=\gamma$不是局部; 矛盾.

Finally, if $i\in\gamma$ then the image of $i\otimes 1$ in $\hat{A}(P_{\hat{\beta}})$ is not zero since we have $\hat{\beta}\subset\hat{\gamma}$; consequently, the image of i in $\prod_{\beta\in\mathcal{P}_{\hat{\beta}}}A(P_\beta)$ also is not zero; that is to say, there is $\beta\in\mathcal{P}_{\hat{\beta}}$ such that $s_\beta(i)\neq 0$ and therefore we get that $\beta=\gamma$ is local, a contradiction.

引理 9.6. 设 M 是一个 kG-模. 任意 $f \in \mathrm{End}_k(M)^G$ 都满足 $f(M_1^G) \subset M_1^G$; 特别是, 从 M 到 M_1^G 的限制给出下面的同态, 并它在 $\mathrm{End}_k(M)_1^G$ 上是满射

Lemma 9.6. *Let M be a kG-module. Any $f \in \mathrm{End}_k(M)^G$ fulfills $f(M_1^G) \subset M_1^G$; in particular, the restriction from M to M_1^G induces the following homomorphism which is already surjective over $\mathrm{End}_k(M)_1^G$.*

9.6.1 $$\mathrm{End}_k(M)_1^G \subset \mathrm{End}_k(M)^G \longrightarrow \mathrm{End}_k(M_1^G)\,.$$

进一步, 如果 $M = \oplus_{i\in I} M_i$ 是一个 G-稳定直和分解, 那么对任意 $i \in I$ 有

Moreover, if $M = \oplus_{i\in I} M_i$ is a G-stable direct sum decomposition then, for any $i \in I$, we have

9.6.2 $$\mathrm{End}_k(M_i)_1^{G_i} \cong \mathrm{Tr}_{G_i}^G\big(\mathrm{End}_k(M_i)_1^{G_i}\big) \subset \mathrm{End}_k(M)_1^G\,,$$

其中 G_i 是 i 的稳定化子, 而且 $\mathrm{Tr}_{G_i}^G\big(\mathrm{End}_k(M_i)_1^{G_i}\big)$ 的像是

where G_i denotes the stabilizer of i, and the image of $\mathrm{Tr}_{G_i}^G\big(\mathrm{End}_k(M_i)_1^{G_i}\big)$ is equal to

9.6.3 $$\mathrm{End}_k\Big(\mathrm{Tr}_{G_i}^G\big((M_i)_1^{G_i}\big)\Big) \cong \mathrm{End}_k\big((M_i)_1^{G_i}\big)\,.$$

特别是, 有下面的交换图

In particular, we have the following commutative diagram

9.6.4 $$\begin{array}{ccc} \prod_{j\in J} \mathrm{End}_k(M_j)^{G_j} & \longrightarrow & \mathrm{End}_k(M)^G \\ \downarrow & & \downarrow \\ \prod_{j\in J} \mathrm{End}_k\big((M_j)_1^{G_j}\big) & \longrightarrow & \mathrm{End}_k(M_1^G) \end{array}$$

其中 $J \subset I$ 是 I 上的 G-轨道的一个代表集合.

where $J \subset I$ is a set of representatives for the orbits of G on I.

证明: 对任意 $m \in M$, 显然有 $f\big(\mathrm{Tr}_1^G(m)\big) = \mathrm{Tr}_1^G\big(f(m)\big)$. 反过来, 设 g 是一个 M_1^G 的 k-向量空间自同态并设 T 是一个 M_1^G 的基; 这样, 对任意 $t \in T$ 有 $m_t \in M$ 使得 $g(t) = \mathrm{Tr}_1^G(m_t)$; 所以, 只要考虑对任意 $t \in T$ 满足 $h(t) = m_t$ 的唯一 k-向量空间同态 $h\colon M \to M$, 就得到 $\mathrm{Tr}_1^G(h)$ 在 M_1^G 上的限制等于 g. 进一步, 令 $\mathrm{id}_M = \sum_{i\in I} e_i$ 记被 $M = \oplus_{i\in I} M_i$ 决定的单位

Proof: For any $m \in M$, we clearly have $f\big(\mathrm{Tr}_1^G(m)\big) = \mathrm{Tr}_1^G\big(f(m)\big)$. Conversely, let g be a k-linear endomorphism of M_1^G and T a basis of M_1^G; thus, for any $t \subset T$, there exists $M_t \in M$ such that $g(t) = \mathrm{Tr}_1^G(m_t)$; consequently, considering the k-linear endomorphism $h\colon M \to M$ fulfilling $h(t) = m_t$ for any $t \in T$, we have that $\mathrm{Tr}_1^G(h)$ coincides with the restriction of g to M_1^G. Moreover, denote by $\mathrm{id}_M = \sum_{i\in I} e_i$ the decomposition of the identity map determined by the direct sum decomposition $M = \oplus_{i\in I} M_i$;

的分解; 我们已经知道有下面的 G_i-代数同构 (见例子 4.15)

we already know that we have the following G_i-algebra isomorphisms (see example 4.15)

9.6.5 $$\mathrm{End}_k(M_i) \cong e_i \mathrm{End}_k(M) e_i\,;$$

那么, 就不难完成证明了.

then, it is not difficult to finish the proof.

9.7. 当然, 如果 M 是 G 的正规子群使得 $M \subset N$, 那么有自然同态如下

9.7. Obviously, if M is a normal subgroup of G such that $M \subset N$, we have a canonical homomorphism as follows

9.7.1 $$\tilde{A} = \mathrm{Res}_M^N(A) \otimes_M G \longrightarrow A \otimes_N G = \hat{A}\,,$$

其中 $\mathrm{Res}_M^N(A)$ 记对应的 M-内 G-代数. 请注意, 如果 H 与 K 是 G 的子群, 并且 ω 与 θ 分别是 A 上的 H-除子与 K-除子, 显然 融$_{\mathrm{Res}_M^N(A)}(K_\theta, H_\omega)$ 是对任意 $y \in K$ 满足 $\varphi(y) \in yM$ 的 $\varphi \in$ 融$_A(K_\theta, H_\omega)$ 的集合. 假定 k 是代数闭的并假定 G/M 是 p-群;由定理 9.5, 对 G 的任意 p-子群 P, 任意 A^P 的本原幂等元 i 在 $\hat{A}^P$ 与 $\tilde{A}^P$ 中的像都是本原的, 从而在 $\hat{A}^P$ 中任意 $\tilde{A}^P$ 的本原幂等元的像是本原的. 事实上, 将要使用的这个结果就是下面的更精确的情况.

where $\mathrm{Res}_M^N(A)$ denotes the corresponding M-interior G-algebra. Note that, if H and K are subgroups of G and we consider respective divisors ω and θ of H and K on A, clearly 融$_{\mathrm{Res}_M^N(A)}(K_\theta, H_\omega)$ is the set of $\varphi \in$ 融$_A(K_\theta, H_\omega)$ such that $\varphi(y) \in yM$ for any $y \in K$. Assume that k is algebraically closed and that G/M is a P-group; it follows from Theorem 9.5 that, for any p-subgroup P of G, any primitive idempotent i of A^P is also primitive in both $\hat{A}^P$ and $\tilde{A}^P$, and therefore the image of any primitive idempotent of $\tilde{A}^P$ in $\hat{A}^P$ is primitive too. Actually, we will emply this result in the following more precise situation.

推论 9.8. 设 M 是 G 的正规子群使得 $M \subset N$. 假定 k 是代数闭的, 还假定 G/M 是 p-群. 令

Corollary 9.8. *Let M be a normal subgroup of G such that $M \subset N$. Assume that k is algebraically closed and that G/M is a p-group. Denote by*

9.8.1 $$f : \mathrm{Res}_M^N(A) \otimes_M G \longrightarrow A \otimes_N G$$

记自然的同态. 如果对任意 $\mathrm{Res}_M^N(A) \otimes_M G$ 上的局部点群 $P_{\tilde{\gamma}}$ 与 $Q_{\tilde{\delta}}$ 有

the canonical homomorphism. Whenever, for any local pointed groups $P_{\tilde{\gamma}}$ and $Q_{\tilde{\delta}}$ on $\mathrm{Res}_M^N(A) \otimes_M G$, we have

9.8.2 $$融_{\mathrm{Res}_M^N(A)\otimes_M G}(Q_{\tilde{\delta}}, P_{\tilde{\gamma}}) = 融_{A\otimes_N G}(Q_{\hat{\delta}}, P_{\hat{\gamma}})\,,$$

其中 $\hat{\gamma}$ 与 $\hat{\delta}$ 分别是 $A \otimes_N G$ 上的 P-点与 Q-点使得 $f(\tilde{\gamma}) \subset \hat{\gamma}$ 与 $f(\tilde{\delta}) \subset \hat{\delta}$, 那么 f 是严格覆盖的.

where $\hat{\gamma}$ and $\hat{\delta}$ respectively are the points of P and Q on $A \otimes_N G$ fulfilling $f(\tilde{\gamma}) \subset \hat{\gamma}$ and $f(\tilde{\delta}) \subset \hat{\delta}$, then f is a strict covering.

证明: 记 $\tilde{A} = \mathrm{Res}_M^N(A) \otimes_M G$ 与 $\hat{A} = A \otimes_N G$; 设 $P_{\hat{\gamma}}$ 是 $\hat{A}$ 上的局部点群, 而令 $\mathcal{P}_{\hat{\gamma}}$ 记满足 $\gamma \otimes_N 1 \subset \hat{\gamma}$ 的 $\gamma \in \mathcal{LP}_A(P)$ 的集合; 由定理 9.5, 包含单射 $A \otimes_N 1 \subset \hat{A}$ 决定一个 k-代数的单射的幺同态如下

Proof: Set $\tilde{A} = \mathrm{Res}_M^N(A) \otimes_M G$ and $\hat{A} = A \otimes_N G$; let $P_{\hat{\gamma}}$ be a local pointed group on $\hat{A}$, and denote by $\mathcal{P}_{\hat{\gamma}}$ the set of $\gamma \in \mathcal{LP}_A(P)$ fulfilling $\gamma \otimes_N 1 \subset \hat{\gamma}$; it follows from Theorem 9.5 that the inclusion map $A \otimes_N 1 \subset \hat{A}$ determines an injective unitary k-algebra homomorphism as follows

9.8.3 $$\prod_{\gamma \in \mathcal{P}_{\hat{\gamma}}} A(P_\gamma) \longrightarrow \hat{A}(P_{\hat{\gamma}}),$$

从而我们有

and therefore we have

9.8.4 $$\mu_{\hat{A}}^1(\hat{\gamma}) = \sum_{\gamma \in \mathcal{P}_{\hat{\gamma}}} \mu_A^1(\gamma).$$

类似地, 设 $\tilde{\gamma}$ 是 $\tilde{A}$ 的局部 P-点而令 $\mathcal{P}_{\tilde{\gamma}}$ 记满足 $\gamma \otimes_M 1 \subset \tilde{\gamma}$ 的 $\gamma \in \mathcal{LP}_A(P)$ 的集合; 那么还有

Similarly, let $\tilde{\gamma}$ be a local point of P on $\tilde{A}$, and denote by $\mathcal{P}_{\tilde{\gamma}}$ the set of $\gamma \in \mathcal{LP}_A(P)$ fulfilling $\gamma \otimes_M 1 \subset \tilde{\gamma}$; then, we still have

9.8.5 $$\mu_{\tilde{A}}^1(\tilde{\gamma}) = \sum_{\gamma \in \mathcal{P}_{\tilde{\gamma}}} \mu_A^1(\gamma).$$

最后, 令 $\tilde{\mathcal{P}}$ 记满足 $f(\tilde{\gamma}) \subset \hat{\gamma}$ 的 $\tilde{\gamma} \in \mathcal{LP}_{\tilde{A}}(P)$ 的集合; 显然有

Finally, denote by $\tilde{\mathcal{P}}$ the set of $\tilde{\gamma} \in \mathcal{LP}_{\tilde{A}}(P)$ fulfilling $f(\tilde{\gamma}) \subset \hat{\gamma}$; obviously, we have

9.8.6 $$\mathcal{P}_{\hat{\gamma}} = \bigsqcup_{\tilde{\gamma} \in \tilde{\mathcal{P}}} \mathcal{P}_{\tilde{\gamma}}.$$

而且, 如果 $\tilde{\gamma}, \tilde{\varepsilon} \in \mathcal{LP}_{\tilde{A}}(P)$ 是满足 $\hat{\gamma}$ 包含 $f(\tilde{\gamma})$ 与 $f(\tilde{\varepsilon})$ 的两个元素, 那么从等式 9.8.2 可推出 $\mathrm{id}_P \in$ 融$_{\tilde{A}}(P_{\tilde{\gamma}}, P_{\tilde{\varepsilon}})$, 从而 $\tilde{\gamma} = \tilde{\varepsilon}$ (见命题 8.10); 特别是, 有 $\mathcal{P}_{\hat{\gamma}} = \mathcal{P}_{\tilde{\gamma}}$ 与 $\mu_{\hat{A}}^1(\hat{\gamma}) = \mu_{\tilde{A}}^1(\tilde{\gamma})$. 也就是说, f 决定一个双射

Moreover, if $\tilde{\gamma}, \tilde{\varepsilon} \in \mathcal{LP}_{\tilde{A}}(P)$ are two elements such that $\hat{\gamma}$ contains both $f(\tilde{\gamma})$ and $f(\tilde{\varepsilon})$, then equality 9.8.2 implies that $\mathrm{id}_P \in$ 融$_{\tilde{A}}(P_{\tilde{\gamma}}, P_{\tilde{\varepsilon}})$, and therefore we have $\tilde{\gamma} = \tilde{\varepsilon}$ (see Proposition 8.10); in particular, we get $\mathcal{P}_{\hat{\gamma}} = \mathcal{P}_{\tilde{\gamma}}$ and $\mu_{\hat{A}}^1(\hat{\gamma}) = \mu_{\tilde{A}}^1(\tilde{\gamma})$. That is to say, f determines a bijective map

9.8.7 $$\mathrm{res}_{f^P} : \mathcal{LP}_{\tilde{A}}(P) \to \mathcal{LP}_{\hat{A}}(P)$$

使得对任意 $\tilde{\gamma} \in \mathcal{LP}_{\tilde{A}}(P)$ 有

such that, for any $\tilde{\gamma} \in \mathcal{LP}_{\tilde{A}}(P)$, we have

9.8.8 $$\tilde{A}(P_{\tilde{\gamma}}) \cong \hat{A}(P_{\hat{\gamma}})$$

其中 $\hat{\gamma} = \mathrm{res}_{f^P}(\tilde{\gamma})$; 此时, 只要应用推论 6.19 就证明完了.

where $\hat{\gamma} = \mathrm{res}_{f^P}(\tilde{\gamma})$; now, it suffices to apply Corollary 6.19 and we are done.

10. 群代数上的点群

10. Pointed Groups on the Group Algebra

10.1. 设 G 是有限群; 我们的主要目的是研究群代数 $\mathcal{O}G$; 事实上, 我们引进的工具都是为了分析这个 G-内代数. 请注意, $\mathcal{O}G$ 是对称的; 更精确地, 令 $\nu_G\colon \mathcal{O}G \to \mathcal{O}$ 记对任意 $x \in G$ 满足 $\nu_G(x) = \delta_{1,x}$ 的 $\mathcal{O}$-模同态 (如常 $\delta_{1,x} = 1$ 或 0 当 $x = 1$ 或 $x \neq 1$); 对任意 $\mathcal{O}G$ 的幂等元 i, i' 有一个 $\mathcal{O}$-模同构

10.1. Let G be a finite group; our main purpose is to study the group algebra $\mathcal{O}G$; actually, all the tools we are introducing are devoted to the analysis of this G-interior algebra. Note that $\mathcal{O}G$ is a symmetric $\mathcal{O}$-algebra; more precisely, denote by $\nu_G\colon \mathcal{O}G \to \mathcal{O}$ the $\mathcal{O}$-module homomorphism fulfilling $\nu_G(x) = \delta_{1,x}$ for any $x \in G$; for any idempotents i, i' of $\mathcal{O}G$, we have an $\mathcal{O}$-module homomorphism

10.1.1 $$\nu_G^{i,i'} : i(\mathcal{O}G)i' \cong \mathrm{Hom}_{\mathcal{O}}\big(i'(\mathcal{O}G)i, \mathcal{O}\big)$$

它对任意 $a, a' \in (\mathcal{O}G)$ 满足

which, for any $a, a' \in (\mathcal{O}G)$, fulfill

10.1.2 $$\big(\nu_G^{i,i'}(iai')\big)(i'a'i) = \nu_G(iai'a'i)\,;$$

确实, 不难证明我们能假定 $i = i' = 1$; 那么对任意 $x, y \in G$ 有

indeed, we easily reduce it to the case where $i = i' = 1$; then, for any $x, y \in G$, we have

10.1.3 $$\big(\nu_G^{1,1}(x)\big)(y) = \nu_G(xy) = \delta_{x^{-1},y}\,.$$

10.2. 首先, 因为 $(\mathcal{O}G)^G = Z(\mathcal{O}G)$ 所以 $\mathcal{O}G$ 上的 G-点 α 只包含唯一一个元素 b, 称为 G-块 (或者 $\mathcal{O}G$-块), 并且显然得到下面的 G-内代数

10.2. First of all, since $(\mathcal{O}G)^G = Z(\mathcal{O}G)$, a point α of G on $\mathcal{O}G$ contains a unique element b, called *block of* G (or *block of* $\mathcal{O}G$), and we clearly have the following G-interior algebra homomorphism

10.2.1 $$(\mathcal{O}G)_\alpha \cong \mathcal{O}Gb\,;$$

令 $\varepsilon_G\colon \mathcal{O}G \to \mathcal{O}$ 记增广映射; 显然存在唯一 G-块 $b_\circ$ 使得 $\varepsilon_G(b_\circ) = 1$; 它称为主块. 对任意 G 的子群 H, 不难验证

denote by $\varepsilon_G\colon \mathcal{O}G \to \mathcal{O}$ the augmentation map; clearly, there is a unique block $b_\circ$ of G fulfilling $\varepsilon_G(b_\circ) = 1$, called *principal block.* For any subgroup H of G, it is easy to check

10.2.2 $$(\mathcal{O}G)^H = \mathcal{O}C_G(H) \oplus \sum_{x \in G-H} \mathcal{O}\cdot\mathrm{Tr}_{C_H(x)}^H(x)\,;$$

特别是, 自然的同态从 $(\mathcal{O}G)^H$ 到 $(kG)^H$ 是满射.

in particular, the canonical homomorphism from $(\mathcal{O}G)^H$ to $(kG)^H$ is surjective.

命题 10.3. 对任意$\mathcal{O}G$上的点群H_β存在k的Galois扩张$Z^\beta=k(\theta)$,其中θ是p'-单位根,与一个Z^β-向量空间V_β使得

Proposition 10.3. *For any pointed group H_β on $\mathcal{O}G$, there exist a Galois extension $Z^\beta = k(\theta)$ of k by a p'-th root of unity θ, and a Z^β-linear vector space V_β fulfilling*

10.3.1 $$(\mathcal{O}G)(H_\beta) \cong \mathrm{End}_{Z^\beta}(V_\beta).$$

证明: 既然自然的同态从$(\mathcal{O}G)^H$到$(kG)^H$是满射,能假定$\mathcal{O}=k$. 令$k_\circ$记k的阶为p的子域, 再令$\bar{k}_\circ$记$k_\circ$的某个代数闭包; 特别是, 对$(\bar{k}_\circ G)^H$的任意本原幂等元i有

Proof: As far as the canonical homomorphism from $(\mathcal{O}G)^H$ to $(kG)^H$ is surjective, we may assume that $\mathcal{O} = k$. Let $k_\circ$ be the prime subfield of k and $\bar{k}_\circ$ an algebraic closure of $k_\circ$; in particular, for any primitive idempotent i of $(\bar{k}_\circ G)^H$, we have

10.3.2 $$i(\bar{k}_\circ G)^H i/J\big(i(\bar{k}_\circ G)^H i\big) \cong \bar{k}_\circ.$$

然后, 选择$(\bar{k}_\circ G)^H$的相互正交本原幂等元的集合I使得$\sum_{i\in I} i=1$,并令$i=\sum_{x\in G}\bar{\lambda}_{i,x}x$其中$i\in I$, $\bar{\lambda}_{i,x}\in\bar{k}_\circ$,再令$k_1$记由于$\{\bar{\lambda}_{i,x}\}_{i\in I,x\in G}$生成的子域.

then, choose a set I of pairwise orthogonal primitive idempotents of $(\bar{k}_\circ G)^H$ fulfilling $\sum_{i\in I} i = 1$, set $i = \sum_{x\in G} \bar{\lambda}_{i,x} x$, where $i \in I$ and $\bar{\lambda}_{i,x} \in \bar{k}_\circ$, and denote by k_1 the subfield generated by $\{\bar{\lambda}_{i,x}\}_{i\in I,\, x\in G}$.

也就是说, k_1是有限的, 并且I包含在$(k_1G)^H$中; 特别是商$i(k_1G)^H i/J(i(k_1G)^H i)$是一个有限的可除代数, 从而它是$k_1$的Galois扩张$k_1(\omega)$, 其中$\omega$是$p'$-单位根; 进一步, 因为

That is to say, k_1 is a finite field and I is contained in $(k_1G)^H$; in particular, the quotient $i(k_1G)^H i/J(i(k_1G)^H i)$ is a finite division algebra and therefore it is the Galois extension $k_1(\omega)$ of k_1 by a p'-th root of unity ω; moreover, since

10.3.3 $$(\bar{k}_\circ G)^H \cong \bar{k}_\circ \otimes_{k_1} (k_1G)^H,$$

所以$\bar{k}_\circ \otimes_{k_1} J\big((k_1G)^H\big)$包含在$J\big((\bar{k}_\circ G)^H\big)$里, 从而有

$\bar{k}_\circ \otimes_{k_1} J\big((k_1G)^H\big)$ is contained in $J\big((\bar{k}_\circ G)^H\big)$ and therefore we have

10.3.4 $$\bar{k}_\circ \otimes_{k_1} k_1(\omega)/J\big(\bar{k}_\circ \otimes_{k_1} k_1(\omega)\big) \cong \bar{k}_\circ.$$

这样, 对适当的$n\in\mathbb{N}$得到

Thus, for a suitable $n \in \mathbb{N}$, we get

10.3.5 $$0 = (1\otimes 1 - \omega\otimes\omega^{-1})^{p^n} = 1\otimes 1 - \omega^{p^n}\otimes\omega^{-p^n};$$

所以ω^{p^n}属于k_1,从而ω也属于k_1; 也就是说, 对任意$i\in I$还有

hence, ω^{p^n} belongs to k_1 and therefore ω belongs to k_1 too; that is to say, for any $i \in I$, we still have

10.3.6 $$i(k_1G)^H i/J\big(i(k_1G)^H i\big) \cong k_1;$$

更精确, 对任意$\beta_1 \in \mathcal{P}_{k_1G}(H)$, $(k_1G)(H_{\beta_1})$ 是一个 k_1 上的矩阵代数.

more precisely, for any point $\beta_1 \in \mathcal{P}_{k_1G}(H)$, $(k_1G)(H_{\beta_1})$ is a matrix algebra over k_1.

令k'记一个由 k, k_1生成的域; 请注意, 既然 k_1 是有限的, k' 是 k 的 Galois 扩张; 进一步, 显然有

Denote by k' a field generated by k and k_1; note that, as far as k_1 is finite, k' is a Galois extension of k; furthermore, we clearly have

10.3.7 $$(k'G)^H \cong k' \otimes_{k_1} (k_1G)^H\,, \quad (k'G)^H \cong k' \otimes_k (kG)^H\,;$$

所以一方面, 任意 $i \in I$ 在 $(k'G)^H$ 中也是本原的, 并得到

hence, on the one hand, any $i \in I$ is also primitive in $(k'G)^H$ and we get

10.3.8 $$k' \otimes_{k_1} S_1 \cong S'$$

其中我们令

where we are setting

10.3.9 $$S_1 = (k_1G)^H/J\big((k_1G)^H\big)\,, \quad S' = (k'G)^H/J\big((k'G)^H\big)\,.$$

另一方面, 我们有

On the other hand, we have

10.3.10 $$k' \otimes J\big((kG)^H\big) \subset J\big((k'G)^H\big)\,,$$

从而包含映射$(kG)^H \subset (k'G)^H$决定一个 k-代数同态

and therefore the inclusion $(kG)^H \subset (k'G)^H$ determines a k-algebra homomorphism

10.3.11 $$f : S = (kG)^H/J\big((kG)^H\big) \longrightarrow S'\,,$$

并且显然有

so that we clearly get

10.3.12 $$k' \otimes_k S/J(k' \otimes_k S) \cong S'\,;$$

特别是,还有$f(Z(S)) \subset Z(S')$.

in particular, we have $f(Z(S)) \subset Z(S')$.

令E与E'分别记$Z(S)$与$Z(S')$的本原幂等元的集合; 那么,得到$f(E) \subset \sum_{e' \in E'} k_\circ e'$. 而且, 令 I' 记 I在 S' 中的像; 因为任意 $i' \in I'$ 在S'中也是本原的 (见推论 2.14), 所以也得到 $E' \subset \sum_{i' \in I'} k_\circ i'$ (见命题 3.4); 也就是说, 显然得到 (见同构式 10.3.8)

Respectively denote by E and E' the set of primitive idempotents of $Z(S)$ and $Z(S')$; then, we get $f(E) \subset \sum_{e' \in E'} k_\circ e'$. Moreover, let I' be the image of I in S'; since any $i' \in I'$ is also primitive in S' (see Corollary 2.14), we still get $E' \subset \sum_{i' \in I'} k_\circ i'$ (see Proposition 3.4); that is to say, we clearly obtain (see isomorphism 10.3.8)

10.3.13 $$f(E) \subset \sum_{e' \in E'} k_\circ e' \subset Z(S_1)\,.$$

令 $k_2 = k \cap k_1$; 既然 k_1是有限的, k_1是k_2的 Galois 扩张,

Set $k_2 = k \cap k_1$; as far as k_1 is finite, k_1 is a Galois extension of k_2 and the Galois

以及 k 上的扩张 k' 的 Galois 群 Σ 与 k_2 上的扩张 k_1 的 Galois 群是相互同构的. 特别是, 有 $k_2=(k_1)_1^\Sigma$; 这样因为自然的同态

group Σ of k' over k is naturally isomorphic to the Galois group of k_1 over k_2. In particular, we have $k_2 = (k_1)_1^\Sigma$; thus, since the canonical homomorphism

10.3.14 $$\big((k_1G)^H\big)_1^\Sigma \longrightarrow (S_1)_1^\Sigma$$

是满射, 所以不难证明还有

is surjective, it is not difficult to check that

10.3.15 $$(S_1)^\Sigma \cong (k_2G)^H/J\big((k_2G)^H\big) = S_2\,,$$

从而得到 $f(E) \subset Z(S_2)$; 令一方面, 类似的上述, 包含映射 $(k_2G)^H \subset (kG)^H$ 决定一个 k_2-代数同态 $S_2 \to S$; 所以 $Z(S_2)$ 的任意本原幂等元在 $Z(S)$ 中也是本原的.

and therefore we get $f(E) \subset Z(S_2)$; on the other hand, arguing as above, the inclusion $(k_2G)^H \subset (kG)^H$ induces a k_2-algebra homomorphism $S_2 \longrightarrow S$; consequently, any primitive idempotent in $Z(S_2)$ remains primitive in $Z(S)$.

最后, 任意 $\beta \in \mathcal{P}_{kG}(H)$ 由适当的 $\beta_2 \in \mathcal{P}_{k_2G}(H)$ 一一决定, 并且有

Finally, any $\beta \in \mathcal{P}_{kG}(H)$ determines and is determined by a suitable $\beta_2 \in \mathcal{P}_{k_2G}(H)$, and then we have

10.3.16 $$k \otimes_{k_2} (k_2G)(H_{\beta_2})/J\big(k \otimes_{k_2} (k_2G)(H_{\beta_2})\big) \cong (kG)(H_\beta)\,;$$

由于 Wedderburn 定理, 对任意 $j_2 \in \beta_2$, 下面的 k_2-代数

it follows from Wedderburn's Theorem that the following k_2-algebra

10.3.17 $$Z^{\beta_2} = s_{\beta_2}(j_2)\big((k_2G)(H_{\beta_2})\big)s_{\beta_2}(j_2)$$

是一个有限的可除代数, 从而, 因为 k_2 是有限的, 所以 Z^{β_2} 是 k_2 的 Galois 扩张 $k_2(\theta)$ 其中 θ 是 p'-单位根, 并且对适当的 Z^{β_2}-向量空间 V_{β_2} 得到

is a division algebra for any $j_2 \in \beta_2$, and therefore, since k_2 is finite, Z^{β_2} is a Galois extension $k_2(\theta)$ of k_2 by a p'-th root of unity θ, and moreover, for a suitable Z^{β_2}-linear vector space V_{β_2}, we have

10.3.18 $$(k_2G)(H_{\beta_2}) \cong \mathrm{End}_{Z^{\beta_2}}(V_{\beta_2})\,.$$

这样, 由同构式 10.3.16 与同构式 10.3.18, 还得到

Thus, by isomorphisms 10.3.16 and 10.3.18, we still have

10.3.19 $$k \otimes_{k_2} Z^{\beta_2}/J(k \otimes_{k_2} Z^{\beta_2}) \cong s_\beta(j_2)\big((kG)(H_\beta)\big)s_\beta(j_2)\,;$$

因为这个同构式的左边是交换的, 那么只要仍应用 Wedderburn 定理, 就得到 $s_\beta(j_2)$ 也是本原的, 并且下面的 k-代数

since the left member of this isomorphism is commutative, applying Wedderburn's Theorem again, we get that $s_\beta(j_2)$ is primitive too, and moreover that the following k-algebra

10.3.20 $$Z^\beta = s_\beta(j_2)\big((kG)(H_\beta)\big)s_\beta(j_2)$$

是 k 的 Galois 扩张 $k(\theta)$，这是因为 Z^{β_2} 与 k 生成它；也得到

is the Galois extension $k(\theta)$ of k since Z^{β_2} and k generate it; we also get

10.3.21 $$(kG)(H_\beta) \cong \mathrm{End}_{Z^\beta}(V_\beta)\,.$$

10.4. 设 P 是 G 的 p-子群；别忘了，我们已经考虑 Brauer 商 $(\mathcal{O}G)(P)$ (见 6.4)；从等式 10.2.2 不难证明，同态 Br_P 决定下面的 $N_G(P)$-内 $C_G(P)$-代数同构

10.4. Let P be a p-subgroup of G; recall that we already have considered the Brauer quotient $(\mathcal{O}G)(P)$ (see 6.4); from equality 10.2.2, it is easily checked that the homomorphism Br_P induces the following $N_G(P)$-interior $C_G(P)$-algebra isomorphism

10.4.1 $$kC_G(P) \cong (\mathcal{O}G)(P)\,.$$

因为 $\varepsilon_G\big(\mathrm{Ker}(\mathrm{Br}_P)\big) \subset p\mathcal{O}$，所以有 $\mathrm{Br}_P(b_\circ) \neq 0$，从而存在 $\gamma \in \mathcal{LP}_{\mathcal{O}G}(P)$ 使得 $P_\gamma \subset G_{\{b_\circ\}}$ (见推论 6.6)；这样，任意 G 的 Sylow p-子群也是 $G_{\{b_\circ\}}$ 的亏点. 常常把 $kC_G(P)$ 与 $(\mathcal{O}G)(P)$ 等同一致.

Since $\varepsilon_G\big(\mathrm{Ker}(\mathrm{Br}_P)\big) \subset p\mathcal{O}$, we clearly get $\mathrm{Br}_P(b_\circ) \neq 0$ and therefore there exists $\gamma \in \mathcal{LP}_{\mathcal{O}G}(P)$ fulfilling $P_\gamma \subset G_{\{b_\circ\}}$ (see Corollary 6.6); thus, any Sylow p-subgroup of G is also a defect of $G_{\{b_\circ\}}$. Usually, we identify $kC_G(P)$ to $(\mathcal{O}G)(P)$ throughout isomorphism 10.4.1.

10.5. 由推论 6.6，这个同构决定一个从不可分解投射 $kC_G(P)$-模的同构类的集合到 $\mathcal{LP}_{\mathcal{O}G}(P)$ 的双射. 对任意 $\gamma \in \mathcal{LP}_{\mathcal{O}G}(P)$，令 W_γ 记一个对应的不可分解投射 $kC_G(P)$-模. 进一步，如果 $\mathrm{Br}_P(a) \neq 0$ 其中 $a \in (\mathcal{O}G)^P$，那么对任意 P 的子群 Q 显然有 $\mathrm{Br}_Q(a) \neq 0$，从而得到

10.5. By Corollary 6.6, this isomorphism induces a bijective map between the set of isomorphism classes of indecomposable projective $kC_G(P)$-modules and $\mathcal{LP}_{\mathcal{O}G}(P)$. For any $\gamma \in \mathcal{LP}_{\mathcal{O}G}(P)$, we denote by W_γ a corresponding indecomposable projective $kC_G(P)$-module. Moreover, if $\mathrm{Br}_P(a) \neq 0$ for some $a \in (\mathcal{O}G)^P$ then we clearly have $\mathrm{Br}_Q(a) \neq 0$ for any subgroup Q of P, and therefore we get

10.5.1. 对任意 $\gamma \in \mathcal{LP}_{\mathcal{O}G}(P)$ 存在 $\delta \in \mathcal{LP}_{\mathcal{O}G}(Q)$ 使得 $Q_\delta \subset P_\gamma$.

10.5.1. *For any $\gamma \in \mathcal{LP}_{\mathcal{O}G}(P)$ there exists $\delta \in \mathcal{LP}_{\mathcal{O}G}(Q)$ fulfilling $Q_\delta \subset P_\gamma$.*

确实，因为 $\mathrm{Br}_Q(\gamma) \neq \{0\}$，所以有 $\mathrm{Br}_Q(\gamma) \not\subset J\big((\mathcal{O}G)(Q)\big)$，从而存在 $\delta \in \mathcal{LP}_{\mathcal{O}G}(Q)$ 使得 $s_\delta(\gamma) \neq \{0\}$ (见推论 6.6).

Indeed, since we have $\mathrm{Br}_Q(\gamma) \neq \{0\}$, we still have $\mathrm{Br}_Q(\gamma) \not\subset J\big((\mathcal{O}G)(Q)\big)$ and therefore there is $\delta \in \mathcal{LP}_{\mathcal{O}G}(Q)$ such that $s_\delta(\gamma) \neq \{0\}$ (see Corollary 6.6).

10.6. 一般来说，如果 γ 是 $\mathcal{O}G$ 上的局部 P-点群而 Q 是 P 的子群，那么 $\mathcal{O}G$ 上的满足 $Q_\delta \subset P_\gamma$ 的局部 Q-点 δ 不是唯一的；可是，下面的结果说明被 $\mathrm{Br}_Q(\delta)$ 决定的 $C_G(Q)$-块是

10.6. Generally speaking, if γ is a point of P on $\mathcal{O}G$ and Q is a subgroup of P, there is more than one local point δ of Q on $\mathcal{O}G$ fulfilling $Q_\delta \subset P_\gamma$; however, the next result tells us that the block of $C_G(Q)$ determined by $\mathrm{Br}_Q(\delta)$ is unique. Note that,

唯一的. 请注意, 由命题 3.23 与同构式 10.4.1, Br_P 决定一个从 $\mathcal{P}\big(kC_G(P)\big)$ 到 $\mathcal{P}\big((\mathcal{O}G)^P\big)$ 的单射; 令

according to Proposition 3.23 and to isomorphism 10.4.1, Br_P induces an injective map from $\mathcal{P}\big(kC_G(P)\big)$ to $\mathcal{P}\big((\mathcal{O}G)^P\big)$; denote by

10.6.1 $$\mathrm{Br}_P^{\mathcal{D}}: \mathcal{D}\big(kC_G(P)\big) \longrightarrow \mathcal{D}_{\mathcal{O}G}(P)$$

记对应的线性单射, 再令 $b(\gamma)$ 记由 $\mathrm{Br}_P(\gamma)$ 决定的 $C_G(Q)$-块.

the corresponding additive map, and by $b(\gamma)$ the block of $C_G(Q)$ determined by $\mathrm{Br}_P(\gamma)$.

定理 10.7. 设 P 与 Q 是 G 的 p-子群使得 $Q \subset P$. 对任意 $Z\big(kC_G(P)\big)$ 的本原幂等元 e 存在唯一一个 $Z\big(kC_G(Q)\big)$ 的本原幂等元 f 使得

Theorem 10.7. *Let P and Q be p-subgroups of G such that $Q \subset P$. For any primitive idempotent e of $Z\big(kC_G(P)\big)$ there is a unique primitive idempotent f of $Z\big(kC_G(Q)\big)$ such that*

10.7.1 $$\mathrm{res}_{\mathrm{Br}_Q}\Big(\mathrm{res}_Q^P\big(\mathrm{Br}_P^{\mathcal{D}}(\mu^e_{kC_G(P)})\big)\Big) \subset \mu^f_{kC_G(Q)}.$$

注意 10.8. 对任意 $(\mathcal{O}G)^P$ 的本原幂等元 i 使得 $\mathrm{Br}_P(i)e = \mathrm{Br}_P(i) \neq 0$ 显然有

Remark 10.8. For any primitive idempotent i of $(\mathcal{O}G)^P$ such that $\mathrm{Br}_P(i)e = \mathrm{Br}_P(i) \neq 0$, we clearly have

10.8.1 $$\mu^i_{(\mathcal{O}G)^P} \subset \mathrm{Br}_P^{\mathcal{D}}(\mu^e_{kC_G(P)})$$

(见命题 3.19); 此时, 从 10.7.1 可推出 $\mathrm{Br}_Q(i)f = \mathrm{Br}_Q(i)$; 特别是, 假定 γ 与 δ 分别是 $\mathcal{O}G$ 上的局部 P-点与局部 Q-点使得 $Q_\delta \subset P_\gamma$; 那么从 $e = b(\gamma)$ 可推出 $f = b(\delta)$.

(see Proposition 3.19); then, from 10.7.1 we can deduce that $\mathrm{Br}_Q(i)f = \mathrm{Br}_Q(i)$; in particular, assume that γ and δ respectively are local points of P and Q on $\mathcal{O}G$ such that $Q_\delta \subset P_\gamma$; in this situation, the equality $e = b(\gamma)$ forces the equality $f = b(\delta)$.

反过来, 设 I 是 $(\mathcal{O}G)^P$ 的一个相互正交本原幂等元的集合满足 $\mathrm{Br}_P^{\mathcal{D}}(\mu^e_{kC_G(P)}) = \sum_{i\in I} i$; 如果对任意 $i \in I$ 有 $\mathrm{Br}_Q(i)f = \mathrm{Br}_Q(i)$ 那么当然有

Conversely, let I be a set of pairwise orthogonal primitive idempotents of $(\mathcal{O}G)^P$ fulfilling $\mathrm{Br}_P^{\mathcal{D}}(\mu^e_{kC_G(P)}) = \sum_{i\in I} i$; if we have $\mathrm{Br}_Q(i)f = \mathrm{Br}_Q(i)$ for any $i \in I$, obviously we still have

10.8.2 $$\mathrm{Br}_Q\Big(\sum_{i\in I} i\Big)f = \mathrm{Br}_Q\Big(\sum_{i\in I} i\Big)$$

从而得到10.7.1 (见命题 3.16).

and thus we get 10.7.1 (see Proposition 3.19).

证明: 对 $|P/Q|$ 使用归纳法, 当然可以假定 $Q \neq P$, 从而 $R = N_P(Q) \neq Q$; 这样, 存在唯一 $Z\big(kC_G(R)\big)$ 的本原幂等元

Proof: We argue by induction on $|P/Q|$ and we naturally may assume that $Q \neq P$ so that we have $R = N_P(Q) \neq Q$; thus, there exists a unique primitive idempotent g

g 使得如果 i 是满足 $\mathrm{Br}_P(i)e = \mathrm{Br}_P(i) \neq 0$ 的 $(\mathcal{O}G)^P$ 的本原幂等元那么 $\mathrm{Br}_R(i)g = \mathrm{Br}_R(i)$. 另一方面, 因为对任意 R 的子群 T 使得 $Q \subset T$ 有

of $Z\big(kC_G(R)\big)$ such that if i is a primitive idempotent of $(\mathcal{O}G)^P$ fulfilling $\mathrm{Br}_P(i)e = \mathrm{Br}_P(i) \neq 0$ then we have $\mathrm{Br}_R(i)g = \mathrm{Br}_R(i)$. On the other hand, since, for any subgroup T of R such that $Q \subset T$, we have

10.7.2 $$\mathrm{Br}_T^{kC_G(Q)}\big(kC_G(Q)^T\big) = kC_G(T),$$

所以还有

we still have

10.7.3 $$\mathrm{Br}_T^{kC_G(Q)}\Big(Z\big(kC_G(Q)\big)^T\Big) \subset Z\big(kC_G(T)\big),$$

从而存在唯一 $Z\big(kC_G(Q)\big)^R$ 的本原幂等元 f' 使得

and therefore there exists a unique primitive idempotent f' of $Z\big(kC_G(Q)\big)^R$ fulfilling

10.7.4 $$\mathrm{Br}_R^{kC_G(Q)}(f')g = g;$$

可是, 如果 f 是 $Z\big(kC_G(Q)\big)$ 的本原幂等元使得 $f'f = f$, 那么显然 $f' = \mathrm{Tr}_{R_f}^R(f)$, 其中 R_f 是 f 的稳定化子; 此时, 从 $\mathrm{Br}_R^{kC_G(Q)}(f') \neq 0$ 可推出 $R_f = R$ 与 $f' = f$.

however, if f is a primitive idempotent of $Z\big(kC_G(Q)\big)$ fulfilling $f'f = f$, then we clearly have $f' = \mathrm{Tr}_{R_f}^R(f)$, where R_f is the stabilizer of f; at that point, from the fact that $\mathrm{Br}_R^{kC_G(Q)}(f') \neq 0$, we can prove that $R_f = R$ and $f' = f$.

而且, 对 R 的任意真包含 Q 的子群 T 存在 $Z\big(kC_G(T)\big)$ 的本原幂等元 h 使得 $\mathrm{Br}_T(i) = \mathrm{Br}_T(i)h$, 其中 i 是如同上假设; 可是, 也存在 $Z\big(kC_G(T)\big)$ 的本原幂等元 h' 使得, 只要 j 是 $(\mathcal{O}G)^R$ 的本原幂等元满足 $\mathrm{Br}_R(j)g = \mathrm{Br}_R(j) \neq 0$, 就有 $\mathrm{Br}_T(j)h' = \mathrm{Br}_T(j)$; 因为对任意这些 i 能选择适当的上述的 j 使得 $ji = j = ij$ (见命题 3.19), 所以也有 $\mathrm{Br}_T(j)h = \mathrm{Br}_T(j)$, 从而得到 $h' = h$.

Moreover, for any subgroup T of R strictly containing Q there exists a primitive idempotent h of $Z\big(kC_G(T)\big)$ fulfilling $\mathrm{Br}_T(i) = \mathrm{Br}_T(i)h$ for any idempotent i as above; but, we also have the existence of a primitive idempotent h' of $Z\big(kC_G(T)\big)$ such that $\mathrm{Br}_T(j)h' = \mathrm{Br}_T(j)$ for any primitive idempotent j of $(\mathcal{O}G)^R$ fulfilling $\mathrm{Br}_R(j)g = \mathrm{Br}_R(j) \neq 0$; since, for any of those idempotents i, we may choose a suitable j as above fulfilling $ji = j = ij$ (see Proposition 3.19), we also have $\mathrm{Br}_T(j)h = \mathrm{Br}_T(j)$ and therefore we get $h' = h$.

进一步, 由于等式 10.7.4 与 $f' = f$, 如果 j 是 $(\mathcal{O}G)^R$ 的本原幂等元使得 $\mathrm{Br}_R(j)g = \mathrm{Br}_R(j) \neq 0$, 那么有

Furthermore, by equalities 10.7.4 and $f' = f$, if j is a prilmitive idempotent of $(\mathcal{O}G)^R$ fulfilling $\mathrm{Br}_R(j)g = \mathrm{Br}_R(j) \neq 0$, then we have

10.7.5 $$0 \neq \mathrm{Br}_R(j)\mathrm{Br}_R^{kC_G(Q)}(f) = \mathrm{Br}_R^{kC_G(Q)}\big(\mathrm{Br}_Q(j)f\big),$$

还有$\mathrm{Br}_Q(j)f=\mathrm{Br}_Q(j)$，这是因为$\mathrm{Br}_Q\big((\mathcal{O}G)^R\big)=kC_G(Q)^R$(见分解式10.2.2)，从而$\mathrm{Br}_Q(j)$在$kC_G(Q)^R$中是本原的；此时，得到

and thus still have $\mathrm{Br}_Q(j)f=\mathrm{Br}_Q(j)$ since $\mathrm{Br}_Q\big((\mathcal{O}G)^R\big)=kC_G(Q)^R$ (see the decomposition in 10.2.2) and therefore $\mathrm{Br}_Q(j)$ is primitive in $kC_G(Q)^R$; at that point, we get

10.7.6 $$\mathrm{Br}_T^{kC_G(Q)}\big(\mathrm{Br}_Q(j)f\big)h=\mathrm{Br}_T(j)h=\mathrm{Br}_T(j)\neq 0\,;$$

特别是，有$\mathrm{Br}_T^{kC_G(Q)}(f)h\neq 0$，从而$\mathrm{Br}_T^{kC_G(Q)}(f)h=h$ (见式子 10.7.3)．最后，因为对任意$(\mathcal{O}G)^P$的本原幂等元i使得$\mathrm{Br}_P(i)e=\mathrm{Br}_P(i)\neq 0$已经知道$\mathrm{Br}_T(i)h=\mathrm{Br}_T(i)$，所以得到

in particular, we have $\mathrm{Br}_T^{kC_G(Q)}(f)h\neq 0$ and therefore we also have $\mathrm{Br}_T^{kC_G(Q)}(f)h=h$ (see inclusion 10.7.3). Finally, since we already know that $\mathrm{Br}_T(i)h=\mathrm{Br}_T(i)$ for any primitive idempotent i of $(\mathcal{O}G)^P$ which fulfills $\mathrm{Br}_P(i)e=\mathrm{Br}_P(i)\neq 0$, we get

10.7.7 $$\mathrm{Br}_T^{kC_G(Q)}\big(\mathrm{Br}_Q(i)(f-1)\big)=\mathrm{Br}_T(i)h\mathrm{Br}_T^{kC_G(Q)}(f-1)=0\,.$$

也就是说，我们有

That is to say, we have

10.7.8 $$\mathrm{Br}_Q(i)(f-1)\in\bigcap_T\mathrm{Ker}(\mathrm{Br}_T^{kC_G(Q)})$$

其中T跑遍真包含Q的R的子群；可是，只要使用分解式10.2.2就可验证这个交是$kC_G(Q)_Q^R$；而且，只要把命题3.23应用到下面的满同态(见引理6.15)

where T runs over the set of subgroups of R strictly containing Q; but, by employing the decomposition 10.2.2, we can prove that this intersection is equal to $kC_G(Q)_Q^R$; moreover, by applying Proposition 3.23 to the following surjective homomorphim (see Lemma 6.15)

10.7.9 $$\mathrm{Br}_Q:\mathcal{O}\cdot i+i(\mathcal{O}G)_Q^P i\longrightarrow k\cdot\mathrm{Br}_Q(i)+\mathrm{Br}_Q(i)kC_G(Q)_Q^R\mathrm{Br}_Q(i)\,,$$

就得到存在一个$i(\mathcal{O}G)_Q^P i$的幂等元ℓ满足

we prove the existence of an idempotent ℓ of $i(\mathcal{O}G)_Q^P i$ fulfilling

10.7.10 $$\mathrm{Br}_Q(\ell)=\mathrm{Br}_Q(i)(f-1)\,.$$

最后，由推论 6.12，i不属于$(\mathcal{O}G)_Q^P$，并且因为它在$(\mathcal{O}G)^P$中是本原的，所以得到$\ell=0$；也就是说，$\mathrm{Br}_Q(i)f=\mathrm{Br}_Q(i)$．

Finally, according to Corollary 6.12, i does not belong to $(\mathcal{O}G)_Q^P$ and, since it is a primitive idempotent in $(\mathcal{O}G)^P$, we get $\ell=0$; that is to say, we have $\mathrm{Br}_Q(i)f=\mathrm{Br}_Q(i)$.

10.9. 设P_γ与Q_δ是$\mathcal{O}G$上的局部点群使得$Q\subset P$；请注意，为了解决P_γ包含Q_δ的有关问题，对应的不可分解投射$kC_G(P)$-模W_γ与

10.9. Let P_γ and Q_δ be local pointed groups on $\mathcal{O}G$ such that $Q\subset P$; as a matter of fact, the indecomposable projective $kC_G(P)$- and $kC_G(Q)$-modules W_γ and W_δ are not powerful enough to determine whe-

$kC_G(Q)$-模 W_δ 不够了. 可是, 由推论 6.17, 为了解决这个问题我们能假定 P 正规化 Q_δ; 此时, 下面的 $k(P{\cdot}C_G(Q))$-模

ther or not P_γ contains Q_δ. But, according to Corollary 6.17, in trying to answer this question, we may assume that P normalizes Q_δ; in this case, the $k(P{\cdot}C_G(Q))$-modules

10.9.1 $$\hat{W}_\gamma = \mathrm{Ind}_{C_G(P)}^{P\cdot C_G(Q)}(W_\gamma)\,, \quad \hat{W}_\delta = \mathrm{Ind}_{C_G(Q)}^{P\cdot C_G(Q)}(W_\delta)$$

分别允许 $k(P{\cdot}C_G(Q)\times P)$-模与 $k(P{\cdot}C_G(Q)\times Q)$-模的结构如下

respectively admit $k(P{\cdot}C_G(Q)\times P)$- and $k(P{\cdot}C_G(Q)\times Q)$-module structures as follows

10.9.2 $$u{\cdot}(z\otimes m) = zu^{-1}\otimes m\,, \quad v{\cdot}(z\otimes n) = zv^{-1}\otimes n\,,$$

其中 $z\in P{\cdot}C_G(Q)$ 与 $u\in P$ 与 $v\in Q$, 和 $m\in W_\gamma$ 与 $n\in W_\delta$; 事实上, 只要选择 $i\in\gamma$ 与 $j\in\delta$ 就得到

where $z\in P{\cdot}C_G(Q)$, $u\in P$, $v\in Q$, and $m\in W_\gamma$, $n\in W_\delta$; actually, choosing $i\in\gamma$ and $j\in\delta$, we have

10.9.3 $$W_\gamma \cong kC_G(P)\mathrm{Br}_P(i)\,, \quad W_\delta\cong kC_G(Q)\mathrm{Br}_Q(j)\,,$$

从而还得到

and therefore we get

10.9.4 $$\hat{W}_\gamma \cong k(P{\cdot}C_G(Q))\mathrm{Br}_P(i)\,, \quad \hat{W}_\delta \cong k(P{\cdot}C_G(Q))\mathrm{Br}_Q(j)\,.$$

10.10. 请注意,因为$\mathrm{Br}_Q(j)$是$kC_G(Q)$的本原幂等元, 所以只要假定k是代数闭的, 定理 9.5 就推出 $\mathrm{Br}_Q(j)$ 在 $k(P{\cdot}C_G(Q))$中也是本原的. 此时, 考虑下面的重数 (见 3.13)

10.10. Note that, since $\mathrm{Br}_Q(j)$ is a primitive idempotent of $kC_G(Q)$, assuming that k is algebraically closed, Theorem 9.5 implies that $\mathrm{Br}_Q(j)$ is also primitive in $k(P{\cdot}C_G(Q))$. At the same time, we consider the following multiplicity (see 3.13)

10.10.1 $$\mathrm{m}_\delta^\gamma = \big(\mathrm{res}_Q^P(\gamma)\big)(\delta)$$

(别忘了, δ 与 $\mathrm{res}_Q^P(\gamma)$ 分别是 A^Q 的点与除子). 而且, 令

(recall that δ and $\mathrm{res}_Q^P(\gamma)$ respectively are a point and a divisor of A^Q). Moreover, set

10.10.2 $$\hat{P} = P{\cdot}C_G(Q)\times P\,, \quad \hat{Q} = P{\cdot}C_G(Q)\times Q\,.$$

命题 10.11. 假定 k 是代数闭的. 设 P_γ 与 Q_δ 是 $\mathcal{O}G$ 上的局部点群使得 P 既包含 Q 也正规化 Q_δ. 那么存在唯一满足 $M(\Delta(P))\neq\{0\}$ 的 $k\hat{P}$-模 $\hat{W}_\gamma$ 的不可分解直和项 M 的同构类, 并且 m_δ^γ 是 $\hat{W}_\delta$ 在 m_δ^γ 里的重数.

Proposition 10.11. *Assume that k is algebraically closed. Let P_γ and Q_δ local pointed groups on $\mathcal{O}G$ such that P contains Q and normalizes Q_δ. Then, there exists a unique isomorphism class of indecomposable direct summands M of the $k\hat{P}$-module $\hat{W}_\gamma$ fulfilling $M(\Delta(P))\neq\{0\}$, and moreover m_δ^γ is the multiplicity of $\hat{W}_\delta$ in m_δ^γ.*

证明: 令 $\bar{\imath} \in kG$ 记 $i \in \gamma$ 的像; 请注意, 我们能选择 i 使得 $\bar{\imath}\mathrm{Br}_P(i) = \mathrm{Br}_P(i)\bar{\imath}$; 确实, 只要用满足 $\bar{\imath}' \in \mathrm{Br}_P(i)(kG)^P\mathrm{Br}_P(i)$ 的 $i' \in \gamma$ (见 3.21) 代替 i 就得到上述的等式; 此时, 特别是有

Proof: Denote by $\bar{\imath} \in kG$ the image of $i \in \gamma$; note that we may choose i in such a way that we have $\bar{\imath}\mathrm{Br}_P(i) = \mathrm{Br}_P(i)\bar{\imath}$; indeed, it suffices to replace i by $i' \in \gamma$ such that $\bar{\imath}'$ belongs to $\mathrm{Br}_P(i)(kG)^P\mathrm{Br}_P(i)$ (see 3.21), to get the equality above; in this case, we have

10.11.1 $$\mathrm{Br}_Q(i)\mathrm{Br}_P(i) = \mathrm{Br}_P(i)\mathrm{Br}_Q(i)\,.$$

而且, 因为 $\mathrm{Br}_Q\big((\mathcal{O}G)^P\big) = \big(kC_G(Q)\big)^P$ (见分解式 10.2.2) 所以 $\mathrm{Br}_Q(i)$ 在 $\big(kC_G(Q)\big)^P$ 中是本原的 (见命题 3.23); 这样, 由定理 9.5, 它在 $k\big(P{\cdot}C_G(Q)\big)^P$ 中也是本原的; 特别是, 因为

Moreover, since we have $\mathrm{Br}_Q\big((\mathcal{O}G)^P\big) = \big(kC_G(Q)\big)^P$ (see the decomposition 10.2.2), $\mathrm{Br}_Q(i)$ is primitive in $\big(kC_G(Q)\big)^P$ (see Proposition 3.23); thus, it follows from Theorem 9.5 that it is primitive in $k\big(P{\cdot}C_G(Q)\big)^P$ too; in particular, since we have

10.11.2 $$\mathrm{End}_{k\hat{P}}\Big(k\big(P{\cdot}C_G(Q)\big)\mathrm{Br}_Q(i)\Big) \cong \mathrm{Br}_Q(i)k\big(P{\cdot}C_G(Q)\big)^P\mathrm{Br}_Q(i)\,,$$

所以 $M = k\big(P{\cdot}C_G(Q)\big)\mathrm{Br}_Q(i)$ 是不可分解 $k\hat{P}$-模; 而且, 既然

$M = k\big(P{\cdot}C_G(Q)\big)\mathrm{Br}_Q(i)$ is an indecomposable $k\hat{P}$-module; moreover, as far as we have

10.11.3 $$\mathrm{Br}_P^{k(P\cdot C_G(Q))}\big(\mathrm{Br}_Q(i)\big) = \mathrm{Br}_P(i)\,,$$

不难验证 (见分解式 10.2.2)

it is easy to check (see decomposition 10.2.2)

10.11.4 $$\begin{aligned} M\big(\Delta(P)\big) &\cong \Big(k\big(P{\cdot}C_G(Q)\big)\Big)(P)\,\mathrm{Br}_P(i) \\ &\cong kC_G(P)\mathrm{Br}_P(i) \neq \{0\}\,. \end{aligned}$$

进一步, 设 J 是一个 $(\mathcal{O}G)^Q$ 的相互正交本原幂等元使得 $\sum_{j\in J} j = i$; 类似如上, 由定理 9.5, 对任意满足 $\mathrm{Br}_Q(j) \neq 0$ 的 $j \in J$, 幂等元 $\mathrm{Br}_Q(j)$ 在 $k\big(P{\cdot}C_G(Q)\big)^P$ 中是本原的, 并且得到

Furthermore, let J be a set of pairwise orthogonal primitive idempotents of $(\mathcal{O}G)^Q$ fulfilling $\sum_{j\in J} j = i$; as above, it follows from Theorem 9.5 that, for any $j \in J$ such that $\mathrm{Br}_Q(j) \neq 0$, the idempotent $\mathrm{Br}_Q(j)$ is primitive in $k\big(P{\cdot}C_G(Q)\big)^P$, and moreover we get

10.11.5 $$\mathrm{End}_{k\hat{Q}}(M) \cong \mathrm{Br}_Q(i)k\big(P{\cdot}C_G(Q)\big)^Q\mathrm{Br}_Q(i)\,;$$

所以下面的直和项分解

consequently, in the following decomposition

10.11.6 $$M = \bigoplus_{j\in J'} k\big(P{\cdot}C_G(Q)\big)\mathrm{Br}_Q(j)\,,$$

其中$J' \subset J$是满足$\mathrm{Br}_Q(j) \neq 0$的$j \in J$的集合，是一个不可分的$k\hat{Q}$-模分解，并且不难验证δ包含$j \in J$当且仅当存在一个$k\hat{Q}$-模同构如下

where $J' \subset J$ is the set of $j \in J$ fulfilling $\mathrm{Br}_Q(j) \neq 0$, all the direct summands are indecomposable $k\hat{Q}$-modules, and it is easy to prove that δ contains $j \in J$ if and only if there is a $k\hat{Q}$-module isomorphism as follows

10.11.7 $$k\big(P{\cdot}C_G(Q)\big)\mathrm{Br}_Q(j) \cong \hat{W}_\delta\,;$$

也就是说，$\mathrm{m}_\delta^\gamma = |\delta \cap J|$是$\hat{W}_\delta$在$\mathrm{Res}_{\hat{Q}}^{\hat{P}}(M)$里的重数.

that is to say, $\mathrm{m}_\delta^\gamma = |\delta \cap J|$ is the multiplicity of $\hat{W}_\delta$ in $\mathrm{Res}_{\hat{Q}}^{\hat{P}}(M)$.

最后,由等式10.11.1只要令

Finally, from equality 10.11.1, setting

10.11.8 $$M' = k\big(P{\cdot}C_G(Q)\big)\big(\mathrm{Br}_P(i) - \mathrm{Br}_Q(i)\big)\,,$$

就得到$\hat{W}_\gamma \cong M \oplus M'$；以及，由等式10.11.3，也得到

we get $\hat{W}_\gamma \cong M \oplus M'$; moreover, from equality 10.11.3, we still get

10.11.9 $$M'\big(\Delta(P)\big) = \{0\}\,.$$

10.12. 设P_γ是$\mathcal{O}G$上的局部点群并设H是G的子群使得$\tilde{C}_G(P) = P{\cdot}C_G(P) \subset H$. 令$Z^\gamma$记$(\mathcal{O}G)(P_\gamma)$的中心，它是$k$上的Galois扩张（见命题10.3）. 由同构式10.4.1有

10.12. Let P_γ be a local pointed group on $\mathcal{O}G$ and H a subgroup of G such that $\tilde{C}_G(P) = P{\cdot}C_G(P) \subset H$. Denote by Z^γ the center of $(\mathcal{O}G)(P_\gamma)$ which is a Galois extension of k (see Proposition 10.3). From isomorphism 10.4.1 we have

10.12.1 $$s_\gamma\big(\mathcal{O}C_G(P)\big) = (\mathcal{O}G)(P_\gamma)$$

从而，由推论6.6，还有

and then, from Corollary 6.6, we still have

10.12.2 $$\begin{aligned} s_\gamma\big((\mathcal{O}G)^H\big) &\subset s_\gamma\big((\mathcal{O}G)^{N_H(P_\gamma)}\big) \\ &\subset (\mathcal{O}G)(P_\gamma)^{C_G(P)} = Z^\gamma\,; \end{aligned}$$

所以存在唯一$\mathcal{O}G$上的H-点β使得$P_\gamma \subset H_\beta$（见5.2），并且$s_\gamma(\beta) = \{1\}$；更精确地，因为$\mathrm{Br}_P(\beta) \subset Z\big(kC_G(P)\big)$所以有$\mathrm{Br}_P(\beta) = \{e\}$，其中$e$是一个$kC_G(P)$-块的和. 特别是，对任意$\mathcal{O}G$上的点群$L_\lambda$使得$P_\gamma \subset L_\lambda$与$L \subset H$不难验证$L_\lambda \subset H_\beta$（见5.2）.

consequently, there exists a unique point β of H on $\mathcal{O}G$ fulfilling $P_\gamma \subset H_\beta$ (see 5.2), and we have $s_\gamma(\beta) = \{1\}$; more precisely, since $\mathrm{Br}_P(\beta) \subset Z\big(kC_G(P)\big)$, we have $\mathrm{Br}_P(\beta) = \{e\}$, where e is a sum of blocks of $kC_G(P)$. In particular, for any pointed group L_λ on $\mathcal{O}G$ such that $P_\gamma \subset L_\lambda$ and $L \subset H$, it is easily checked that $L_\lambda \subset H_\beta$ (see 5.2).

10.13. 进一步, 如果Q_δ与R_ε是$\mathcal{O}G$上的局部点群使得$P_\gamma \subset Q_\delta \supset R_\varepsilon$与$P \subset R \subset H$, 那么有 $R_\varepsilon \subset H_\beta$; 确实, 由定理10.7 和注意 10.8, $b(\delta)$ 唯一地决定 $b(\gamma)$ 与 $b(\varepsilon)$, 从而 $b(\varepsilon)$ 唯一地决定 $b(\gamma)$; 所以有 $\mathrm{Br}_P(\varepsilon)b(\gamma) = \mathrm{Br}_P(\varepsilon)$, 从而还有 $\mathrm{Br}_P(\varepsilon)e = \mathrm{Br}_P(\varepsilon)$, 这是因为显然 $b(\gamma)e = b(\gamma)$; 特别是, 得到 $s_\varepsilon(\beta) \neq \{0\}$. 别忘了(见 6.13.1 和 8.11)

10.13. Furthermore, if Q_δ and R_ε are local pointed groups on $\mathcal{O}G$ fulfilling $P_\gamma \subset Q_\delta \supset R_\varepsilon$ and $P \subset R \subset H$, then we have $R_\varepsilon \subset H_\beta$; indeed, it follows from Theorem 10.7 and Remark 10.8 that $b(\gamma)$ and $b(\varepsilon)$ are the blocks uniquely determined by $b(\delta)$ and therefore $b(\gamma)$ is the block uniquely determined by $b(\varepsilon)$; hence, we get $\mathrm{Br}_P(\varepsilon)b(\gamma) = \mathrm{Br}_P(\varepsilon)$, and thus still get $\mathrm{Br}_P(\varepsilon)e = \mathrm{Br}_P(\varepsilon)$ since we clearly have $b(\gamma)e = b(\gamma)$; in particular, $s_\varepsilon(\beta) \neq \{0\}$. Recall that (see 6.13.1 and 8.11)

10.13.1 $$\bar{C}_G(P) = C_G(P)/Z(P), \quad E_H(P_\gamma) \cong N_H(P_\gamma)/\tilde{C}_G(P);$$

这样, 有一个自然群同态

thus, there is a natural group homomorphism

10.13.2 $$E_H(P_\gamma) \longrightarrow \mathfrak{Gal}(Z^\gamma/k)$$

其中$\mathfrak{Gal}(Z^\gamma/k)$ 记 Z^γ 在 k 上的 Galois 群; 令 $K_H(P_\gamma)$ 记其核, 再令 $\bar{b}(\gamma)$ 记 $b(\gamma)$ 在 $k\bar{C}_G(P)$中的像.

where $\mathfrak{Gal}(Z^\gamma/k)$ denotes the Galois group of Z^γ over k; denote by $K_H(P_\gamma)$ the kernel of this homomorphism, and by $\bar{b}(\gamma)$ the image of $b(\gamma)$.

定理 10.14. 设P_γ是$\mathcal{O}G$上的局部点群并设 H 是 G 的子群使得 H 包含 $\tilde{C}_G(P)$. 令 β 记 $\mathcal{O}G$ 上的 H-点使得 $P_\gamma \subset H_\beta$. 那么 P_γ 是 H_β 的亏点群当且仅当

Theorem 10.14. *Let P_γ be a local pointed group on $\mathcal{O}G$ and H a subgroup of G which contains $\tilde{C}_G(P)$. Denote by β the point of H on $\mathcal{O}G$ such that $P_\gamma \subset H_\beta$. Then, P_γ is a defect pointed group of H_β if and only if*

10.14.1. 对任意 $\mathcal{O}G$ 上的局部点群 Q_δ 使得 $P_\gamma \subset Q_\delta$ 有 $C_Q(P) = Z(P)$.

10.14.1. *For any local pointed group Q_δ on $\mathcal{O}G$ such that $P_\gamma \subset Q_\delta$, we have* $C_Q(P) = Z(P)$.

10.14.2. p 不整除 $|K_H(P_\gamma)|$.

10.14.2. *p does not divide $|K_H(P_\gamma)|$.*

而且, 如果 P_γ 是 H_β 的亏点群, 那么Br_P 决定下面的同构

Moreover, if P_γ is a defect pointed group of H_β then Br_P *induces a isomorphism*

10.14.3 $$k\bar{C}_G(P)\bar{b}(\gamma) \cong (\mathcal{O}G)(P_\gamma).$$

证明: 设ζ是$\mathcal{O}G$上的$\tilde{C}_G(P)$-点使得 $P_\gamma \subset \tilde{C}_G(P)_\zeta$; 从分解式 10.2.2 不难验证

Proof: Let ζ be the point of $\tilde{C}_G(P)$ on $\mathcal{O}G$ such that $P_\gamma \subset \tilde{C}_G(P)_\zeta$; from the decomposition 10.2.2, it is easily checked that

10.14.4 $$\mathrm{Br}_P\big((\mathcal{O}G)^{\tilde{C}_G(P)}\big) = Z\big(kC_G(P)\big);$$

所以有 $\mathrm{Br}_P(\zeta) = \{b(\gamma)\}$，从而还有 $\tilde{C}_G(P)_\zeta \subset H_\beta$ (见10.12). 另一方面, 令 K 记自然群同态 10.13.2 的核; 请注意, 由命题 10.3 与等式 10.12.1, 不难验证

hence we have $\mathrm{Br}_P(\zeta) = \{b(\gamma)\}$ and therefore we still have $\tilde{C}_G(P)_\zeta \subset H_\beta$ (see 10.12). On the other hand, denote by K the kernel of homomorphism 10.13.2; from Proposition 10.3 and equality 10.12.1, it is easy to get

10.14.5 $$(Z^\gamma)_1^{E_H(P_\gamma)} = |K|\cdot(Z^\gamma)^{E_H(P_\gamma)} .$$

首先, 假定 P_γ 是 H_β 的亏点群; 所以 P_γ 也是 $\tilde{C}_G(P)_\zeta$ 的亏点群 (见10.12). 设 Q_δ 是 $\mathcal{O}G$ 上的局部点群使得 $P_\gamma \subset Q_\delta$，而令 $R = P\cdot C_Q(P)$；由陈述 10.5.1, 存在$\mathcal{O}G$上的一个局部 R-点 ε 使得 $R_\varepsilon \subset Q_\delta$；此时, 由 10.12, 得到 $R_\varepsilon \subset \tilde{C}_G(P)_\zeta$. 那么, 只要应用定理 6.8, 就得到 $R_\varepsilon \subset P_\gamma$，特别是 $C_Q(P) \subset P$. 另一方面, 由定理 6.14, 有

First of all, assume that P_γ is a defect pointed group of H_β; hence, P_γ still is a defect pointed group of $\tilde{C}_G(P)_\zeta$ (see 10.12). Let Q_δ be a local pointed group on $\mathcal{O}G$ such that $P_\gamma \subset Q_\delta$, and set $R = P\cdot C_Q(P)$; according to statement 10.5.1, there exists a local point ε of R on $\mathcal{O}G$ such that $R_\varepsilon \subset Q_\delta$; at that point, from 10.12, we get $R_\varepsilon \subset \tilde{C}_G(P)_\zeta$. Then, it follows from Theorem 6.8 that $R_\varepsilon \subset P_\gamma$, so that $C_Q(P) \subset P$. On the other hand, from Theorem 6.14, we get

10.14.6 $$(Z^\gamma)_1^{E_H(P_\gamma)} \neq \{0\}$$

(见10.12.1); 这样, p 不整除 $|K|$ (见等式10.14.5).

(see 10.12.1); thus, p does not divide $|K|$ (see equality 10.14.5).

反过来, 假定 P_γ 满足条件 10.14.1 和 10.14.2; 如果 Q_δ 既是 $\tilde{C}_G(P)_\zeta$ 的亏点群也满足 $P_\gamma \subset Q_\delta$，那么显然有 $Q = P\cdot C_Q(P) = P$；此时, 由定理 6.14, 得到

Conversely, assume that P_γ fulfills conditions 10.14.1 and 10.14.2; if Q_δ is a defect pointed group of $\tilde{C}_G(P)_\zeta$ which fulfills $P_\gamma \subset Q_\delta$, then we clearly have $Q = P\cdot C_Q(P) = P$; at that point, from Theorem 6.14, we get

10.14.7 $$(\mathcal{O}G)(P_\gamma)_1^{\bar{C}_G(P)} \neq \{0\} ;$$

那么, 因为 $(\mathcal{O}G)(P_\gamma)_1^{\bar{C}_G(P)}$ 是 Z^γ 的理想 (见 10.12.1), 所以

then, since $(\mathcal{O}G)(P_\gamma)_1^{\bar{C}_G(P)}$ is an ideal of Z^γ (see 10.12.1), we get

10.14.8 $$(\mathcal{O}G)(P_\gamma)_1^{\bar{C}_G(P)} = Z^\gamma ,$$

从而得到 (见等式 10.14.5)

and therefore we obtain (see equality 10.14.5)

10.14.9 $$(\mathcal{O}G)(P_\gamma)_1^{\bar{N}_H(P_\gamma)} = (Z^\gamma)_1^{E_H(P_\gamma)} = (Z^\gamma)^{E_H(P_\gamma)} ;$$

这样, 仍由定理 6.14, P_γ 是 H_β 的亏点群.

thus, once again by Theorem 6.14, P_γ is a defect pointed group of H_β.

而且, 由命题 10.3, 有

Moreover, by Proposition 10.3, we have

10.14.10
$$s_\gamma\big(kC_G(P)\,b(\gamma)\big) = (\mathcal{O}G)(P_\gamma) \cong \mathrm{End}_{Z^\gamma}(V_\gamma)\,,$$
$$s_\gamma\big(Z(P)\big) = \{\mathrm{id}_{V_\gamma}\}\,,$$

其中 V_γ 是 Z^γ-向量空间; 显然 V_γ 成为单 $k\bar{C}_G(P)$-模并且, 由等式 10.14.8, 对适当的 $f \in \mathrm{End}_{Z_\gamma}(V_\gamma)$ 有

where V_γ is a Z^γ-linear vector space; clearly, V_γ becomes a simple $k\bar{C}_G(P)$-module and, by equality 10.14.8, for some $f \in \mathrm{End}_{Z_\gamma}(V_\gamma)$ we have

10.14.11
$$\mathrm{id}_{V_\gamma} = \mathrm{Tr}_1^{\bar{C}_G(P)}(f)\,.$$

也就是说, 选择满足 $\mathrm{End}_k(M)$ 是完备诱导 $\bar{C}_G(P)$-内代数并 V_γ 是 M 的直和项的 $k\bar{C}_G(P)$-模 M (见定理 5.11); 特别是, 对 $\mathrm{End}_{k\bar{C}_G(P)}(M)$ 的某个幂等元 ℓ 有

In other terms, consider a $k\bar{C}_G(P)$-module M such that the $\bar{C}_G(P)$-interior algebra $\mathrm{End}_k(M)$ is inductively complete and that it admits V_γ as a direct summand (see Theorem 5.11); in particular, for some idempotent ℓ of $\mathrm{End}_{k\bar{C}_G(P)}(M)$ we have

10.14.12
$$V_\gamma \cong \ell(M)\,, \quad \ell = \mathrm{Tr}_1^{\bar{C}_G(P)}(g)\,,$$

其中 $g \in \mathrm{End}_k(M)$; 那么只要把 $\bar{C}_G(P)$-内代数 $\mathrm{End}_k(M)$ 应用定理 5.12 就得到单 $k\bar{C}_G(P)$-模 V_γ 也是投射的 (见同构式 5.8.6). 所以有

where $g \in \mathrm{End}_k(M)$; then, it suffices to apply Theorem 5.12 to the $\bar{C}_G(P)$-interior algebra $\mathrm{End}_k(M)$, to conclude that the simple $k\bar{C}_G(P)$-module V_γ is projective too (see isomorphism 5.8.6). Consequently, we have

10.14.13
$$k\bar{C}_G(P)\bar{\imath} \cong V_\gamma\,,$$

其中 $\bar{\imath}$ 是某个 $i \in \gamma$ 的像 (见同构式 10.4.1).

for the image $\bar{\imath}$ of some $i \in \gamma$ (see isomorphism 10.4.1).

特别是, 因为对任意与 i 不共轭的 $(\mathcal{O}G)^P$ 的本原幂等元 i' 有 $s_\gamma(i') = 0$, 所以还有 $\bar{\imath}'k\bar{C}_G(P)\bar{\imath} = \{0\}$, 其中 $\bar{\imath}'$ 是 i' 的像; 此时, 由同构式 10.1.1, 也得到 $\bar{\imath}\bar{C}_G(P)\bar{\imath}' = \{0\}$. 也就是说, 如果 $\bar{I}$ 是一个 $k\bar{C}_G(P)$ 的相互正交本原幂等元的集合使得 $\sum_{\bar{\imath}''\in\bar{I}} \bar{\imath}'' = 1$, 那么下面的和

In particular, since $s_\gamma(i') = 0$ for any primitive idempotent i' of $(\mathcal{O}G)^P$ which is not conjugate to i, we get $\bar{\imath}'k\bar{C}_G(P)\bar{\imath} = \{0\}$, where $\bar{\imath}'$ denotes the image of i'; at this point, by isomorphism 10.1.1, we also get $\bar{\imath}\bar{C}_G(P)\bar{\imath}' = \{0\}$. That is to say, if $\bar{I}$ is a set of pairwise orthogonal primitive idempotents of $k\bar{C}_G(P)$ fulfilling $\sum_{\bar{\imath}''\in\bar{I}} \bar{\imath}'' = 1$, then the following sum

10.14.14
$$\bar{e} = \sum_{\bar{\imath}''\in\bar{\gamma}\cap\bar{I}} \bar{\imath}''$$

是 $k\bar{C}_G(P)$ 的中心幂等元使得 $\bar{e}\bar{b}(\gamma) = \bar{e}$，其中 $\bar{\gamma}$ 是 γ 的像；所以得到

is a central idempotent of $k\bar{C}_G(P)$ which fulfills $\bar{e}\bar{b}(\gamma) = \bar{e}$, where $\bar{\gamma}$ is the image of γ; consequently, we get

10.14.15 $$k\bar{C}_G(P)\bar{e} \cong (\mathcal{O}G)(P_\gamma).$$

而且，因为 $\left(k\bar{C}_G(P)\right)_1^{\bar{C}_G(P)}$ 就是 $\left(kC_G(P)\right)_1^{\bar{C}_G(P)}$ 的像，而 $\bar{e}$ 属于 $\left(k\bar{C}_G(P)\right)_1^{\bar{C}_G(P)}$，所以也得到 $\bar{e} = \bar{b}(\gamma)$ (见命题 3.23).

Moreover, since $\left(k\bar{C}_G(P)\right)_1^{\bar{C}_G(P)}$ is the image of $\left(kC_G(P)\right)_1^{\bar{C}_G(P)}$ and $\bar{e}$ belongs to $\left(k\bar{C}_G(P)\right)_1^{\bar{C}_G(P)}$, we also get $\bar{e} = \bar{b}(\gamma)$ (see Proposition 3.23).

10.15. 我们把满足条件 10.14.1 的 $\mathcal{O}G$ 上的局部点群称为自中心化的点群；请注意，$\mathcal{O}G$ 上的局部点群只要包含一个 $\mathcal{O}G$ 上的自中心化的点群就是自中心化的.

10.15. We call *self-centralizing pointed groups* the local pointed groups on $\mathcal{O}G$ which fulfill condition 10.14.1; note that a local pointed group on $\mathcal{O}G$ which contains a self-centralizing pointed group on $\mathcal{O}G$ is self-centralizing too.

推论 10.16. 设 P_γ 是 $\mathcal{O}G$ 上的局部点群. 令 ζ 记 $\mathcal{O}G$ 上的 $\tilde{C}_G(P)$-点使得 $P_\gamma \subset \tilde{C}_G(P)_\zeta$. 下面的论断是相互等价.

10.16.1. P_γ 是自中心化的.

10.16.2. P_γ 是 $\tilde{C}_G(P)_\zeta$ 的亏点群.

10.16.3. Br_P 决定同构如下

Corollary 10.16. *Let P_γ be a local pointed group on $\mathcal{O}G$. Denote by ζ the point of $\tilde{C}_G(P)$ on $\mathcal{O}G$ fulfilling $P_\gamma \subset \tilde{C}_G(P)_\zeta$. The following statements are mutually equivalent*

10.16.1. *P_γ is self-centralizing.*

10.16.2. *P_γ is a defect pointed group of $\tilde{C}_G(P)_\zeta$.*

10.16.3. Br_P *induces an isomorphism*

$$k\bar{C}_G(P)\bar{b}(\gamma) \cong (\mathcal{O}G)(P_\gamma).$$

那么，对任意包含 P_γ 的 $\mathcal{O}G$ 上的局部点群 Q_δ，γ 是唯一一个满足 $P_\gamma \subset Q_\delta$ 的 $\mathcal{O}G$ 的局部 P-点.

Then, for any local pointed group Q_δ on $\mathcal{O}G$ which contains P_γ, γ is the unique local point of P on $\mathcal{O}G$ fulfilling $P_\gamma \subset Q_\delta$.

证明: 既然显然 $E_{\tilde{C}_G(P)}(P_\gamma) = 1$，从定理 10.14 可推出论断 10.16.1 与论断 10.16.2 是等价的. 类似地，如果 P_γ 是 $\tilde{C}_G(P)_\zeta$ 的亏点群，那么从定理 10.14 可推出论断 10.16.3. 假定 P_γ 满足论断 10.16.3; 如果 Q_δ 是包含 P_γ 的 $\mathcal{O}G$ 的局部点群

Proof: As far as clearly $E_{\tilde{C}_G(P)}(P_\gamma) = 1$, it follows from Theorem 10.14 that statements 10.16.1 and 10.16.2 are equivalent each other. Similarly, if P_γ is a defect pointed group of $\tilde{C}_G(P)_\zeta$, statement 10.16.3 follows from Theorem 10.14. Assume that P_γ fulfills statement 10.16.3; if Q_δ is a local pointed group on $\mathcal{O}G$ which contains P_γ and γ' is a lo-

并且 γ' 是满足 $P_{\gamma'} \subset Q_\delta$ 的 $\mathcal{O}G$ 上的局部 P-点, 那么从注意 10.8 可推出 $b(\gamma') = b(\gamma)$; 此时, 从上述的同构可推出 γ' 在 $(\mathcal{O}G)(P_\gamma)$ 是的像不是零; 所以得到 $\gamma' = \gamma$. (见推论 6.6).

进一步, 从上述的同构并且命题 10.3, 我们已经知道

cal point of P on $\mathcal{O}G$ fulfilling $P_{\gamma'} \subset O_\delta$, then we can deduce from Remark 10.8 that $b(\gamma') = b(\gamma)$; at this point, from the isomorphism above, we can deduce that the image of γ' in $(\mathcal{O}G)(P_\gamma)$ is not zero; consequently, we get $\gamma' = \gamma$ (see Corollary 6.6).

Further, from the isomorphism above and Proposition 10.3, we already know that

10.16.4 $$k\bar{C}_G(P)\bar{b}(\gamma) \cong (\mathcal{O}G)(P_\gamma) \cong \mathrm{End}_{Z^\gamma}(V_\gamma),$$

其中 V_γ 是一个单 $(\mathcal{O}G)(P_\gamma)$-模, 并 $Z^\gamma = k(\theta)$, 这里 θ 是一个 p'-单位根; 此时, 因为 V_γ 是正则 $(\mathcal{O}G)(P_\gamma)$-模 $\mathrm{End}_{Z^\gamma}(V_\gamma)$ 的直和项, 所以 V_γ 也是投射 $k\bar{C}_G(P)$-模, 从而已经知道它是 $\mathrm{Ind}_1^{\bar{C}_G(P)}(k)$ 的直和项.

这样, 由定理 5.12 和例子 4.15, 得到 $\mathrm{id}_{V_\gamma} = \mathrm{Tr}_1^{\bar{C}_G(P)}(f)$, 其中 $f \in \mathrm{End}_k(V_\gamma)$; 可是, 因为 p 不整除由 θ 在 $\mathrm{End}_k(V_\gamma)^*$ 中的像生成的子群 U 的阶, 所以

where V_γ is a simple $(\mathcal{O}G)(P_\gamma)$-module and $Z^\gamma = k(\theta)$, θ being a p'-th root of unity; moreover, since V_γ is a direct summand of $\mathrm{End}_{Z^\gamma}(V_\gamma)$ considered as an $(\mathcal{O}G)(P_\gamma)$-module, V_γ becomes a projective $k\bar{C}_G(P)$-module and therefore, as it is well-known, is a direct summand of $\mathrm{Ind}_1^{\bar{C}_G(P)}(k)$.

Thus, by Theorem 5.12 and example 4.15, we get $\mathrm{id}_{V_\gamma} = \mathrm{Tr}_1^{\bar{C}_G(P)}(f)$ for some $f \in \mathrm{End}_k(V_\gamma)$; but, since p does not divide the order of the subgroup U of $\mathrm{End}_k(V_\gamma)^*$ generated by the image of θ, the sum

10.16.5 $$g = \frac{1}{|U|} \sum_{u \in U} f^u$$

属于 $\mathrm{End}_{Z^\gamma}(V_\gamma)$; 此时, 既然 $U \subset \mathrm{End}_{k\bar{C}_G(P)}(V_\gamma)$, 还得到 $\mathrm{id}_{V_\gamma} = \mathrm{Tr}_1^{\bar{C}_G(P)}(g)$, 从而显然

belongs to $\mathrm{End}_{Z^\gamma}(V_\gamma)$; at this point, as far as we have $U \subset \mathrm{End}_{k\bar{C}_G(P)}(V_\gamma)$, we also get $\mathrm{id}_{V_\gamma} = \mathrm{Tr}_1^{\bar{C}_G(P)}(g)$ and therefore clearly

10.16.6 $$\mathrm{Br}_P(\zeta) = \{\bar{b}(\gamma)\} \subset (\mathcal{O}G)(P_\gamma)_1^{\bar{C}_G(P)} - \{0\};$$

最后, 由定理 6.14, P_γ 就是 $\tilde{C}_G(P)_\zeta$ 的亏点群.

finally, it follows from Theorem 6.14 that P_γ is a defect pointed group of $\tilde{C}_G(P)_\zeta$.

推论 10.17. **假定 k 是代数闭的. 设 P_γ 是 $\mathcal{O}G$ 上的局部点群, 并设 $\tilde{Q}$ 是 $E_G(P_\gamma)$ 的 p-子群. 令 $H \subset N_G(P_\gamma)$ 记 $\tilde{Q}$ 的原像, 并令 β 记 $\mathcal{O}G$ 上的 H-点使得 $P_\gamma \subset H_\beta$. 那么, 任意 H_β 的亏点群 Q_δ 满足 $Q{\cdot}C_G(P) = H$; 特别是, $\tilde{Q}$ 就是 Q 的像.**

Corollary 10.17. *Assume that k is algebraically closed. Let P_γ be a local pointed group on $\mathcal{O}G$ and $\tilde{Q}$ a p-subgroup of $E_G(P_\gamma)$. Denote by $H \subset N_G(P_\gamma)$ the converse image of $\tilde{Q}$ and by β the point of H on $\mathcal{O}G$ such that $P_\gamma \subset H_\beta$. Then, any defect pointed group Q_δ of H_β fulfills $Q{\cdot}C_G(P) = H$; in particular, the image of Q coincides with $\tilde{Q}$.*

证明: 请注意,有$P_\gamma \subset Q_\delta$;确实,由定理 6.8, 存在 $y \in H$ 使得 $(P_\gamma)^y \subset Q_\delta$, 可是 $(P_\gamma)^y = P_\gamma$. 记 $L = Q{\cdot}C_G(P)$ 与 $N = N_H(L)$ 和令 λ 与 ν 分别记 $\mathcal{O}G$ 上的 L-点与 N-点使得 (见 10.12)

Proof: Note that we have $P_\gamma \subset Q_\delta$; indeed, by Theorem 6.8, there is $y \in H$ such that $(P_\gamma)^y \subset Q_\delta$, but we have $(P_\gamma)^y = P_\gamma$. Set $L = Q{\cdot}C_G(P)$ and $N = N_H(L)$, and denote by λ and ν the respective points of L and of N on $\mathcal{O}G$ fulfilling (see 10.12)

10.17.1 $$P_\gamma \subset Q_\delta \subset L_\lambda \subset N_\nu \subset H_\beta\,;$$

事实上, λ 也是唯一 $\mathcal{O}G$ 上的 L-点使得 $P_\gamma \subset L_\lambda$ (见 10.12), 从而有 $L_\lambda \lhd N_\nu$, 这是因为 N 稳定 P_γ 与 L; 特别是, 对任意 $x \in N$ 有 $(Q_\delta)^x \subset L_\lambda$; 这样, Q_δ 与 $(Q_\delta)^x$ 是 L_λ 的亏点群, 从而有 $x = ny$, 其中 n 属于 $N_N(Q_\delta)$ 与 y 属于 L; 也就是说, 我们得到 $N = L{\cdot}N_N(Q_\delta)$, 从而 $N/L \cong N_N(Q_\delta)/N_L(Q_\delta)$.

actually, λ also is the unique point of L on $\mathcal{O}G$ fulfilling $P_\gamma \subset L_\lambda$ (see 10.12), and therefore we have $L_\lambda \lhd N_\nu$ since n stabilizes P_γ and L; in particular, for any $x \in N$ we have $(Q_\delta)^x \subset L_\lambda$; thus, Q_δ and $(Q_\delta)^x$ are defect pointed groups of L_λ and therefore we have $x = ny$ for suitable elements n of $N_N(Q_\delta)$ and y of L; in other terms, we get $N = L{\cdot}N_N(Q_\delta)$ and therefore we also get $N/L \cong N_N(Q_\delta)/N_L(Q_\delta)$.

而且, 由定理 10.14, p 不整除 $|E_H(Q_\delta)|$; 既然 $\tilde{C}_G(Q) \subset L$, p 也不整除 $|N/L|$. 最后, 令 $\tilde{L}$ 与 $\tilde{N}$ 分别记 L 与 N 在 $E_G(P_\gamma)$ 里的像; 显然 $\tilde{N} = N_{\tilde{Q}}(\tilde{L})$, 并且 $\tilde{N}/\tilde{L} \cong N/L$; 因为 $\tilde{Q}$ 是 p-子群, 所以得到 $L = N$ 与 $\tilde{L} = \tilde{Q}$.

Moreover, it follows from Theorem 10.14 that p does not divide $|E_H(Q_\delta)|$; since $\tilde{C}_G(Q) \subset L$, p does not divide $|N/L|$ too. Finally, respectively denote by $\tilde{L}$ and $\tilde{N}$ the images of L and N in $E_G(P_\gamma)$; clearly we have $\tilde{N} = N_{\tilde{Q}}(\tilde{L})$ and $\tilde{N}/\tilde{L} \cong N/L$; since $\tilde{Q}$ is a p-group, we get $L = N$ and $\tilde{L} = \tilde{Q}$.

10.18. 别忘了, 如果 H_β 与 K_ξ 是 $\mathcal{O}G$ 上的点群, 那么对任意从 K_ξ 到 H_β 的 $\mathcal{O}G$-融合 φ 存在 $\mathcal{O}G$ 上的点群 L_ε 使得 $L_\varepsilon \subset H_\beta$ 与 $\varphi(K) = L$, 并且 φ 的限制 $\psi: K \to L$ 是从 K_ξ 到 L_ε 的 $\mathcal{O}G$-融合 (见命题 8.13); 进一步, 如果 $H \cong K$ 那么, 由定理 8.20, $\varphi: K \cong H$ 是从 K_ξ 到 H_β 的 $\mathcal{O}G$-融合当且仅当, 对某个 $i \in \beta$ 与 $j \in \xi$, 存在 $a \in (\mathcal{O}G)^*$ 使得 $ja{\cdot}\varphi(y) = y{\cdot}ai$, 其中 $y \in K$ (这里, 只要假定 融$_{\mathcal{O}G}(K_\xi, H_\beta) \neq \emptyset$, 命题 8.13 就推出$\mathrm{res}_1^K(\xi) = \mathrm{res}_1^H(\beta)$).

10.18 Recall that, if H_β and K_ξ are pointed groups on $\mathcal{O}G$, then, for any fusion φ from K_ξ to H_β in $\mathcal{O}G$, there exists a pointed group L_ε on $\mathcal{O}G$ such that $L_\varepsilon \subset H_\beta$, that $\varphi(K) = L$ and that the restriction $\psi: K \to L$ of φ is a fusion from K_ξ to L_ε in $\mathcal{O}G$ (see Proposition 8.13); moreover, if $H \cong K$ then, according to Theorem 8.20, $\varphi: K \cong H$ is a fusion from K_ξ to H_β in $\mathcal{O}G$ if and only if, for some $i \in \beta$ and some $j \in \xi$, there exists $a \in (\mathcal{O}G)^*$ fulfilling $ja{\cdot}\varphi(y) = y{\cdot}ai$ for any $y \in K$ (here, it suffices to assume that 融$_{\mathcal{O}G}(K_\xi, H_\beta) \neq \emptyset$ and Proposition 8.13 implies that we have $\mathrm{res}_1^K(\xi) = \mathrm{res}_1^H(\beta)$).

定理 10.19. 设 P_γ, Q_δ 是 $\mathcal{O}G$ 上的局部点群. 那么有

Theorem 10.19. *Let P_γ and Q_δ be local pointed groups on $\mathcal{O}G$. Then we have*

10.19.1 $$E_G(Q_\delta, P_\gamma) = F_{\mathcal{O}G}(Q_\delta, P_\gamma)\,.$$

证明: 只要证明如果 φ 属于 融$_{\mathcal{O}G}(Q_\delta, P_\gamma)$, 那么 $\tilde{\varphi}$ 属于 $E_G(Q_\delta, P_\gamma)$, 这个定理的证明就完成了 (见包含式 8.11.1); 而且能假定 $\varphi(Q) = P$ (见 10.18). 此时, 对某个 $i \in \gamma$ 与 $j \in \delta$, 存在 $a \in (\mathcal{O}G)^*$ 使得 $u{\cdot}ai = ja{\cdot}\varphi(u)$, 其中 $u \in Q$. 考虑 $(\mathcal{O}G)_\delta$ 的自然 $\mathcal{O}(Q \times Q)$-模结构; 由下面的引理 10.24, $(\mathcal{O}Q)j = j(\mathcal{O}Q)$ 是 $j(\mathcal{O}G)j$ 的 $\mathcal{O}(Q \times Q)$-模直和项. 这样, $aj(\mathcal{O}Q)$ 显然是 $aj(\mathcal{O}G)j$ 的 $\mathcal{O}(P \times Q)$-模直和项, 以及有

Proof: In order to prove this equality it suffices to prove that if φ belongs to 融$_{\mathcal{O}G}(Q_\delta, P_\gamma)$ then $\tilde{\varphi}$ belongs to $E_G(Q_\delta, P_\gamma)$ (see inclusion 8.11.1); moreover, we may assume that $\varphi(Q) = P$ (see 10.18). In this case, for some $i \in \gamma$ and some $j \in \delta$ there is $a \in (\mathcal{O}G)^*$ fulfilling $u{\cdot}ai = ja{\cdot}\varphi(u)$ for any $u \in Q$. Consider the natural $\mathcal{O}(Q \times Q)$-module structure of $(\mathcal{O}G)_\delta$; according to Lemma 10.24 below, $(\mathcal{O}Q)j = j(\mathcal{O}Q)$ is a direct summand of $j(\mathcal{O}G)j$ as $\mathcal{O}(Q \times Q)$-modules. Thus, $aj(\mathcal{O}Q)$ clearly is a direct summand of $aj(\mathcal{O}G)j$ as $\mathcal{O}(P \times Q)$-modules, and moreover we have

10.19.2 $$aj(\mathcal{O}G)j = aja^{-1}(\mathcal{O}G)j = i(\mathcal{O}G)j\,.$$

另一方面, 我们显然有 $\mathcal{O}Q \cong \mathrm{Ind}_{\Delta(Q)}^{Q\times Q}(\mathcal{O})$ 从而, 由下面的引理 10.20, 这个 $\mathcal{O}(Q \times Q)$-模是不可分解. 令 (见 8.7.1)

On the other hand, it is clear that $\mathcal{O}Q \cong \mathrm{Ind}_{\Delta(Q)}^{Q\times Q}(\mathcal{O})$ and therefore, according to Lemma 10.20 below, this $\mathcal{O}(Q \times Q)$-module is indecomposable. Set (see 8.7.1)

10.19.3 $$\Delta_\varphi(Q) = (\varphi \times \mathrm{id}_Q)(\Delta(Q))\,;$$

类似地, $aj(\mathcal{O}Q) \cong \mathrm{Ind}_{\Delta_\varphi(Q)}^{P\times Q}(\mathcal{O})$ 并且这个不可分解 $\mathcal{O}(P \times Q)$-模是 $i(\mathcal{O}G)j$ 的直和项. 所以, 由下面的引理 10.22, 存在 $x \in G$ 使得有一个 $\mathcal{O}(P \times Q)$-模同构如下

similarly, we have $aj(\mathcal{O}Q) \cong \mathrm{Ind}_{\Delta_\varphi(Q)}^{P\times Q}(\mathcal{O})$ and this indecomposable $\mathcal{O}(P\times Q)$-module is a direct summand of $i(\mathcal{O}G)j$. Consequently, it follows from Lemma 10.22 below that there exists $x \in G$ such that we have an $\mathcal{O}(P\times Q)$-module isomorphism as follows

10.19.4 $$\mathcal{O}(PxQ) \cong \mathrm{Ind}_{\Delta_\varphi(Q)}^{P\times Q}(\mathcal{O})\,;$$

特别是, $\mathcal{O}(PxQ)$ 的 $\mathcal{O}$-秩是等于 $|P|$, 从而 $PxQ = Px = xQ$, 等价的 $P^x = Q$.

in particular, the $\mathcal{O}$-rank of $\mathcal{O}(PxQ)$ is equal to $|P|$ and thus we have $PxQ = Px = xQ$, so that $P^x = Q$.

而且, 不难验证 $P\times Q$ 的子群 $\Delta_x(Q) = \{(u, u^x)\}_{u\in P}$ 是 $\mathcal{O}(P\times Q)$-模 $\mathcal{O}(PxQ)$ 的顶点; 因为 $\Delta_\varphi(Q)$ 也是 $\mathcal{O}(PxQ)$ 的

Moreover, it is easy to check that the subgroup $\Delta_x(Q) = \{(u, u^x)\}_{u\in P}$ of $P \times Q$ is a vertex of the $\mathcal{O}(P \times Q)$-module $\mathcal{O}(PxQ)$; since $\Delta_\varphi(Q)$ also is a vertex of $\mathcal{O}(PxQ)$,

顶点，所以存在 $(u,v)\in P\times Q$ 使得

there exists $(u,v)\in P\times Q$ fulfilling

10.19.5 $$\Delta_x(Q)^{(u,v)}=\Delta_\varphi(Q)\,,$$

从而对任意 $w\in Q$ 得到 $\varphi(w)=w^y$，其中 $vy=x^{-1}u$；也就是说，被φ与x-共轭决定的群外同态是一样的；这样，$\tilde\varphi$ 属于 $E_G(Q_\delta,P_{\delta^{x^{-1}}})$ (见 8.11)。特别是，因为 φ^{-1} 是从 P_γ 到 Q_δ 的 $\mathcal{O}G$-融合 (见 8.12)，所以 $\text{融}_{\mathcal{O}G}(P_\gamma,P_{\delta^{x^{-1}}})$ 包含 id_P；此时，由命题 8.10，得到 $\gamma=\delta^{x^{-1}}$.

and therefore, for any $w\in Q$, we get $\varphi(w)=w^y$, where $vy=x^{-1}u$; that is to say, the group exomorphisms determined by φ and by the conjugation by x coincide; thus, $\tilde\varphi$ belongs to $E_G(Q_\delta,P_{\delta^{x^{-1}}})$ (see 8.11). In particular, since φ^{-1} is a fusion from P_γ to Q_δ in $\mathcal{O}G$ (see 8.12), $\text{融}_{\mathcal{O}G}(P_\gamma,P_{\delta^{x^{-1}}})$ contains id_P; at this point, it follows from Proposition 8.10 that $\gamma=\delta^{x^{-1}}$.

引理 10.20. **设 P 是一个有限 p-群并设 Q 是 P 的子群. 那么 $\mathrm{Ind}_Q^P(\mathcal{O})$ 是不可分解 $\mathcal{O}P$-模.**

Lemma 10.20. *Let P be a finite p-group and Q a subgroup of P. Then, $\mathrm{Ind}_Q^P(\mathcal{O})$ is an indecomposable $\mathcal{O}P$-module.*

注意 10.21. 因为不假定 k 是代数闭，所以不可以应用定理 7.2.

Remark 10.21. Since k need not be algebraically closed, we can not use Theorem 7.2.

证明: 因为有 $\mathrm{Ind}_1^Q(k)=kQ$ 与 $kQ/J(kQ)\cong k$，所以得到下面的满同态

Proof: Since we have $\mathrm{Ind}_1^Q(k)=kQ$ and $kQ/J(kQ)\cong k$, we get the following surjective homomorphism

10.20.1 $$kP=\mathrm{Ind}_1^P(k)\cong\mathrm{Ind}_Q^P\big(\mathrm{Ind}_1^Q(k)\big)\longrightarrow\mathrm{Ind}_Q^P(k)\,;$$

可是从 $kP/J(kP)\cong k$ 可推出 kP 有唯一单商 kP-模；这样，$\mathrm{Ind}_Q^P(k)$ 也有唯一单商 kP-模，从而它是不可分解；所以，$\mathrm{Ind}_Q^P(\mathcal{O})$ 也是不可分解.

but, from $kP/J(kP)\cong k$, it follows that the kP-module kP has a unique simple quotient; thus, $\mathrm{Ind}_Q^P(k)$ also has a unique simple quotient and therefore it is indecomposable; hence, $\mathrm{Ind}_Q^P(\mathcal{O})$ is indecomposable too.

引理 10.22. **设 P, Q 是 G 的 p-子群并设 i, j 分别是 $(\mathcal{O}G)^P$, $(\mathcal{O}G)^Q$ 的幂等元. 存在一个满足 $P{\cdot}T{\cdot}Q=T$ 并 $|P{\cdot}t|=|P|$ 与 $|t{\cdot}Q|=|Q|$，其中 $t\in T$，的 $i(\mathcal{O}G)j$ 的 $\mathcal{O}$-基 T. 特别是，对 $k(P\times Q)$-模 $i(\mathcal{O}G)j$ 的任意不可分解直和项 M 存在 $x\in G$ 使得**

Lemma 10.22. *Let P and Q be p-subgroups of G and consider respective idempotents i and j of $(\mathcal{O}G)^P$ and $(\mathcal{O}G)^Q$. There exists an $\mathcal{O}$-basis T of $i(\mathcal{O}G)j$ such that $P{\cdot}T{\cdot}Q=T$ and that we have $|P{\cdot}t|=|P|$ and $|t{\cdot}Q|=|Q|$ for any $t\in T$. In particular, for any indecomposable direct summand M of $i(\mathcal{O}G)j$ as $k(P\times Q)$-module, there exists $x\in G$ such that*

10.22.1 $$M\cong\mathrm{Ind}_{\Delta_x(P^x\cap Q)}^{P\times Q}(\mathcal{O})\,.$$

Remark 10.23. 上述的如同，对任意$x \in G$，令$\Delta_x(P^x \cap Q)$记x在$P \times Q$中的稳定化子，并且令$\mathcal{O}(PxQ)$记被x生成的$\mathcal{O}G$的子$\mathcal{O}(P \times Q)$-模.不难验证

Remark 10.23. As above, for any $x \in G$, we denote by $\Delta_x(P^x \cap Q)$ the stabilizer of x in $P \times Q$, and by $\mathcal{O}(PxQ)$ the $\mathcal{O}(P \times Q)$-submodule of $\mathcal{O}G$ generated by x. It is quite easy to check that

10.23.1 $$\mathcal{O}(PxQ) \cong \mathrm{Ind}_{\Delta_x(P^x \cap Q)}^{P \times Q}(\mathcal{O}).$$

证明: 如果$X \subset G$是$P\backslash G/Q$的代表集合，那么显然

Proof: If $X \subset G$ is a set of representatives for $P\backslash G/Q$, we clearly have

10.22.2 $$\mathcal{O}G = \bigoplus_{x \in X} \mathcal{O}(PxQ);$$

而且，对任意$x \in G$有

moreover, for any $x \in G$, we get

10.22.3 $$|Px| = |P|, \quad |xQ| = |Q|.$$

这样，由引理10.20，上述的$\mathcal{O}G$的$\mathcal{O}(P \times Q)$-模分解就是不可分的直和分解. 此时，因为$i(\mathcal{O}G)j$是$\mathcal{O}G$的$\mathcal{O}(P \times Q)$-模直和项，所以存在$Y \subset X$使得有一个$\mathcal{O}(P \times Q)$-模同构如下

Thus, by Lemma 10.20, in the direct sum decomposition of $\mathcal{O}G$ all the terms above are indecomposable $\mathcal{O}(P \times Q)$-modules. Then, since $i(\mathcal{O}G)j$ is a direct summand of $\mathcal{O}G$ as $\mathcal{O}(P \times Q)$-modules, there is $Y \subset X$ such that we have an $\mathcal{O}(P \times Q)$-module isomorphism

10.22.4 $$f : i(\mathcal{O}G)j \cong \bigoplus_{y \in Y} \mathcal{O}(PyQ);$$

那么，下面的集合满足上述条件

then, the following set fulfills the conditions

10.22.5 $$T = \bigcup_{y \in Y} f^{-1}(PyQ).$$

引理 10.24. 设P_γ是$\mathcal{O}G$上的局部点群. 对任意$i \in \gamma$，有$(\mathcal{O}P)i \cong \mathcal{O}P$并存在$(\mathcal{O}P)i$在$i(\mathcal{O}G)i$中的$\mathcal{O}(P \times P)$-模补.

Lemma 10.24. *Let P_γ be a local pointed group on $\mathcal{O}G$. For any $i \in \gamma$, we have $(\mathcal{O}P)i \cong \mathcal{O}P$ and there is a complement of $(\mathcal{O}P)i$ in $i(\mathcal{O}G)i$ as $\mathcal{O}(P \times P)$-modules.*

证明: 由引理 10.22，存在一个满足$P{\cdot}T{\cdot}P = T$，并且对任意$t \in T$有$|P{\cdot}t| = |P|$，的$i(\mathcal{O}G)i$的$\mathcal{O}$-基T；特别是，不难验证$\{\mathrm{Tr}_{P_t}^P(t)\}_{t \in S}$是$i(\mathcal{O}G)^P i$的$\mathcal{O}$-基，其中$S \subset T$是$T$的$P$-共轭作用的轨道的代表集合，而$P_t$是$t \in S$的稳定化子.

Proof: By Lemma 10.22, there exists an $\mathcal{O}$-basis T of $i(\mathcal{O}G)i$ such that $P{\cdot}T{\cdot}P = T$ and that $|P{\cdot}t| = |P|$ for any $t \in T$; in particular, it is easy to check that $\{\mathrm{Tr}_{P_t}^P(t)\}_{t \in S}$ is an $\mathcal{O}$-basis of $i(\mathcal{O}G)^P i$, where $S \subset T$ is a set of representatives for the orbits of the action by conjugation of P on T, and P_t is the stabilizer of $t \in S$. Consequently, since

这样，因为 $\mathrm{Br}_P(i) \neq 0$ (见推论 6.6) 所以存在 $t \in T \cap (\mathcal{O}G)^P$ 使得 $\mathrm{Br}_P(t) \neq 0$; 那么，因为 i 是 $(\mathcal{O}G)^P$ 的本原幂等元，所以 t 在 $i(\mathcal{O}G)i$ 中是可逆的; 因此 $t^{-1}T$ 也是 $i(\mathcal{O}G)i$ 的 $\mathcal{O}$-基，并且 i 属于 $t^{-1}T$. 最后，得到下面的 $\mathcal{O}(P \times P)$-模分解

$\mathrm{Br}_P(i) \neq 0$ (see Corollary 6.6), there exists $t \in T \cap (\mathcal{O}G)^P$ fulfilling $\mathrm{Br}_P(t) \neq 0$; then, since i is a primitive idempotent in $(\mathcal{O}G)^P$, t is inversible in $i(\mathcal{O}G)i$; it follows that $t^{-1}T$ is also an $\mathcal{O}$-basis of $i(\mathcal{O}G)i$, and this time i belongs to $t^{-1}T$. Finally, we obtain the following $\mathcal{O}(P \times P)$-module decomposition

10.24.1 $$i(\mathcal{O}G)i = (\mathcal{O}P)i \oplus \Big(\bigoplus_{t' \in T - Pt} \mathcal{O}t^{-1}t'\Big).$$

11. G-块的融合$\mathbb{Z}$-代数

11. Fusion $\mathbb{Z}$-algebra of a Block

11.1. 设 G 是有限群并设 b 是一个 G-块; 令 $\alpha = \{b\}$ 记对应的 $\mathcal{O}G$ 上的 G-点; 选定 G_α 的亏点群 P_γ . 由定理 6.8, 对 $\mathcal{O}G$ 上的任意含于 G_α 的局部点群 Q_δ (也就是说, 对任意$\mathcal{O}Gb$上的局部点群Q_δ) 存在 $x \in G$ 使得 $(Q_\delta)^x \subset P_\gamma$, 从而 $E_G(Q_\delta, P_\gamma) \neq \emptyset$; 一般来说, 虽然 $|E_G(Q_\delta, G_\alpha)| = 1$ 但 $E_G(Q_\delta, P_\gamma)$ 有数个元素. 这里, 我们要研究 $E_G(Q_\delta, P_\gamma)$ 的元素之间的"差".

11.1. Let G be a finite group and b a block of G; denote by $\alpha = \{b\}$ the corresponding point of G on $\mathcal{O}G$; choose a defect pointed group P_γ of G_α. By Theorem 6.8, for any local pointed group Q_δ on $\mathcal{O}G$ contained in G_α (or, equivalently, for any local pointed group Q_δ on $\mathcal{O}Gb$), there is $x \in G$ such that $(Q_\delta)^x \subset P_\gamma$ and therefore we have $E_G(Q_\delta, P_\gamma) \neq \emptyset$; generally speaking, although certainly $|E_G(Q_\delta, G_\alpha)| = 1$, $E_G(Q_\delta, P_\gamma)$ has more than one element. In this section, we study the "differences" between elements of $E_G(Q_\delta, P_\gamma)$.

11.2. 更精确地, 我们考虑下面的 $\mathbb{Z}$-代数

11.2. More precisely, we consider the following $\mathbb{Z}$-algebra

11.2.1 $$\mathfrak{F}_G(b) = \bigoplus_{Q_\delta, R_\varepsilon} \mathbb{Z}E_G(Q_\delta, R_\varepsilon),$$

其中 Q_δ 与 R_ε 跑遍 $\mathcal{O}Gb$ 上的局部点群, 而 $\mathbb{Z}E_G(Q_\delta, R_\varepsilon)$ 是集合$E_G(Q_\delta,R_\varepsilon)$上的自由$\mathbb{Z}$-模, 并且为了定义 $\mathfrak{F}_G(b)$ 的乘法, 利用融合的合成如下: 对$\mathcal{O}Gb$ 上的任意局部点群 $Q_\delta, Q'_{\delta'}$, R_ε $R'_{\varepsilon'}$, 以及对任意 $E_G(Q_\delta, R_\varepsilon)$与$E_G(Q'_{\delta'}, R'_{\varepsilon'})$ 的分别元素$\tilde{\varphi}$ 与 $\tilde{\varphi}'$我们规定 $\tilde{\varphi}'\,\tilde{\varphi} = \tilde{\varphi}' \circ \tilde{\varphi}$ 或0 当 $R_\varepsilon = Q'_{\delta'}$ 或 $R_\varepsilon \neq Q'_{\delta'}$.

where Q_δ and R_ε run over the set of local pointed groups on $\mathcal{O}G$, $\mathbb{Z}E_G(Q_\delta, R_\varepsilon)$ is the free $\mathbb{Z}$-module over the set $E_G(Q_\delta, R_\varepsilon)$ and, in order to define the multiplication in $\mathfrak{F}_G(b)$, we use the composition of fusions as follows: for any local pointed groups $Q_\delta, Q'_{\delta'}, R_\varepsilon, R'_{\varepsilon'}$ on $\mathcal{O}G$ and any respective elements $\tilde{\varphi}$ and $\tilde{\varphi}'$ of $E_G(Q_\delta, R_\varepsilon)$ and $E_G(Q'_{\delta'}, R'_{\varepsilon'})$, we set $\tilde{\varphi}'\,\tilde{\varphi} = \tilde{\varphi}' \circ \tilde{\varphi}$ or $= 0$ according to $R_\varepsilon = Q'_{\delta'}$ or $R_\varepsilon \neq Q'_{\delta'}$.

11.3. 在分解式 11.2.1 中, 直和项 $\mathbb{Z}E_G(Q_\delta, R_\varepsilon)$ 的元素$\tilde{\varphi}$称为 $\mathfrak{F}_G(b)$ 的齐性元素, 并记 $\ell(\tilde{\varphi}) = |R|/|Q|$, 称为 $\tilde{\varphi}$ 的长度. 这样, 在这个 $\mathbb{Z}$-代数中一定能考虑 $E_G(Q_\delta, P_\gamma)$ 的元素之间的差, 并记

11.3. In the decomposition 11.2.1, the elements $\tilde{\varphi}$ of all the direct summands $\mathbb{Z}E_G(Q_\delta, R_\varepsilon)$ are called *homogeneous elements* of $\mathfrak{F}_G(b)$ and we set $\ell(\tilde{\varphi}) = |R|/|Q|$, called the *length* of $\tilde{\varphi}$. Now, in this $\mathbb{Z}$-algebra, we certainly can consider the differences between elements of $E_G(Q_\delta, P_\gamma)$, and we set

11.3.1 $$\mathfrak{D}_G(Q_\delta, R_\varepsilon) = \sum_{\tilde{\varphi}, \tilde{\psi} \in E_G(Q_\delta, R_\varepsilon)} \mathbb{Z}(\tilde{\varphi} - \tilde{\psi}).$$

进一步，我们也令

Moreover, we still set

11.3.2
$$\mathfrak{D}_G(b) = \bigoplus_{Q_\delta\,,\,x} \mathfrak{D}_G\big(Q_\delta\,,(P_\gamma)^x\big)$$

其中 Q_δ 跑遍 $\mathcal{O}Gb$ 上的局部点群，而 x 跑遍 G；请注意，$\mathfrak{D}_G(b)$ 在 $\mathfrak{F}_G(b)$ 里显然是一个理想：我们要研究这个理想的生成集合 $\mathfrak{G}$. 当然，齐性元素在 $\mathfrak{G}$ 的元素的齐性分解中的集合也生成 $\mathfrak{D}_G(b)$. 这种生成集合称为 $\mathfrak{D}_G(b)$ 的齐性的生成集.

where Q_δ runs over the set of local pointed groups on $\mathcal{O}G$ and x runs over G; note that, in $\mathfrak{F}_G(b)$, $\mathfrak{D}_G(b)$ clearly is an ideal: we want to study the *generator* sets $\mathfrak{G}$ of this ideal. Obviously, the set of homogeneous elements in the homogeneous decomposition of all the elements of $\mathfrak{G}$ generates $\mathfrak{D}_G(b)$ too. We call it *homogeneous generator set.*

11.4. 只要引进所谓 $\mathfrak{D}_G(b)$ 的可约元素，就能解答上述的问题；对任意 $x,y\in G$, 令

11.4. In order to solve this question, it is useful to introduce the so-called *reducible* elements of $\mathfrak{D}_G(b)$; for any $x,y\in G$, set

11.4.1
$$\mathfrak{R}_G\big((P_\gamma)^y,(P_\gamma)^x\big) = \mathfrak{D}_G\big((P_\gamma)^y,(P_\gamma)^x\big)\,,$$

再对 $\mathcal{O}Gb$ 上的任意非极大的局部点群 Q_δ 令

and also, for any nonmaximal local pointed group Q_δ on $\mathcal{O}G$, set

11.4.2
$$\begin{aligned}\mathfrak{R}_G\big(Q_\delta\,,(P_\gamma)^x\big) &= \sum_{R_\varepsilon} \mathfrak{D}_G\big(R_\varepsilon\,,(P_\gamma)^x\big)\cdot\mathbb{Z}E_G(Q_\delta\,,R_\varepsilon)\\ &\subset \mathfrak{D}_G\big(Q_\delta\,,(P_\gamma)^x\big)\end{aligned}$$

其中 R_ε 跑遍满足 $|R|>|Q|$ 的 $\mathcal{O}Gb$ 上的局部点群的集合；那么下面的直和在 $\mathfrak{F}_G(b)$ 里就是 $\mathfrak{D}_G(b)$ 的可约元素的集合,

where R_ε runs over the set of local pointed groups on $\mathcal{O}G$ fulfilling $|R|>|Q|$; then, the set of *reducible elements* of $\mathfrak{F}_G(b)$ is the following direct sum

11.4.3
$$\mathfrak{R}_G(b) = \bigoplus_{Q_\delta\,,\,x} \mathfrak{R}_G\big(Q_\delta\,,(P_\gamma)^x\big)\,,$$

其中 Q_δ 跑遍 $\mathcal{O}Gb$ 上的局部点群，而 x 跑遍 G；请注意，它也是理想.

where Q_δ runs over the set of local pointed groups on $\mathcal{O}G$ and x runs over G; note that it is an ideal too.

11.5. 进一步，如果存在 $x\in G$ 使得

11.5. Moreover, whenever there exists $x\in G$ fulfilling

11.5.1
$$\mathfrak{R}_G\big(Q_\delta\,,(P_\gamma)^x\big) \neq \mathfrak{D}_G\big(Q_\delta\,,(P_\gamma)^x\big)\,,$$

那么这个 $\mathcal{O}Gb$ 上的局部点群称为关键点群；事实上，$\mathcal{O}Gb$上的

we call *essential pointed group* such a local pointed group on $\mathcal{O}G$; actually, a local

局部点群 Q_δ 是关键的当且仅当

pointed group Q_δ is essential if and only if

11.5.2 $$\mathfrak{R}_G(Q_\delta\,, P_\gamma) \neq \mathfrak{D}_G(Q_\delta\,, P_\gamma)\,.$$

请注意, 如果 Q_δ 是关键的, 那么从右边乘以 $E_G(Q_\delta)$ 的作用在

Note that, if Q_δ is essential then, the multiplication on the right by $E_G(Q_\delta)$ in the quotient

11.5.3 $$\mathfrak{W}_{\delta,\gamma} = \mathbb{Z}E_G(Q_\delta\,, P_\gamma)/\mathfrak{R}_G(Q_\delta\,, P_\gamma)$$

里显然稳定 $E_G(Q_\delta\,, P_\gamma)$ 的像. 对任意 $\tilde{\varphi} \in E_G(Q_\delta\,, P_\gamma)$ 令 $E_G(Q_\delta)_{\tilde{\varphi}}$ 记 $\tilde{\varphi}$ 的像的稳定化子, 再令 $\mathbb{Z}\big(E_G(Q_\delta)_{\tilde{\varphi}}\backslash E_G(Q_\delta)\big)$ 记 $E_G(Q_\delta)_{\tilde{\varphi}}\backslash E_G(Q_\delta)$ 上的自由 $\mathbb{Z}$-模. 事实上, 要证明 $E_G(Q_\delta)$ 在 $E_G(Q_\delta\,, P_\gamma)$ 的像上可迁作用.

stabilizes the image of $E_G(Q_\delta\,, P_\gamma)$. For any element $\tilde{\varphi} \in E_G(Q_\delta\,, P_\gamma)$, denote by $E_G(Q_\delta)_{\tilde{\varphi}}$ the stabilizer of $\tilde{\varphi}$, and by $\mathbb{Z}\big(E_G(Q_\delta)_{\tilde{\varphi}}\backslash E_G(Q_\delta)\big)$ the free $\mathbb{Z}$-module over $E_G(Q_\delta)_{\tilde{\varphi}}\backslash E_G(Q_\delta)$. Actually, we will prove that $E_G(Q_\delta)$ acts transitively on the image of $E_G(Q_\delta\,, P_\gamma)$.

命题 11.6. 设 $\mathfrak{S}$ 是一个 $\mathfrak{D}_G(b)$ 的齐性的生成集. 那么, $\mathfrak{S}$ 的不可约元素的集合也生成 $\mathfrak{D}_G(b)$. 而且, 对任意 $\mathcal{O}Gb$ 上的关键点群 Q_δ 存在 $x, y \in G$ 使得 $\mathfrak{S}$ 有某个 $\mathfrak{D}_G\big((Q_\delta)^y, (P_\gamma)^x\big)$ 的不可约元素.

Proposition 11.6. *Let $\mathfrak{S}$ be a homogeneous generator set of $\mathfrak{D}_G(b)$. Then, the set of nonreducible elements of $\mathfrak{S}$ still generates $\mathfrak{D}_G(b)$. Moreover, for any essential pointed group Q_δ on $\mathcal{O}G$ there exist $x, y \in G$ such that $\mathfrak{S}$ contains some nonreducible element of $\mathfrak{D}_G\big((Q_\delta)^y, (P_\gamma)^x\big)$.*

证明: 如果 $\tilde{\sigma} \in \mathfrak{S}$ 是可约, 那么有 $\tilde{\sigma} = \sum_{i\in I} \tilde{\eta}_i\,\tilde{\rho}_i\,\tilde{\nu}_i$, 其中 $\tilde{\eta}_i$ $\tilde{\rho}_i$, $\tilde{\nu}_i$ 是 $\mathfrak{F}_G(b)$ 的齐性元素使得 $\ell(\tilde{\eta}_i) = 1$ 与 $\ell(\tilde{\rho}_i) < \ell(\tilde{\sigma})$; 进一步, 对任意 $i \in I$ 有 $\tilde{\rho}_i = \sum_{\tilde{\tau}} \tilde{\theta}_{i,\tilde{\tau}}\,\tilde{\tau}\,\tilde{\mu}_{i,\tilde{\tau}}$, 其中 $\tilde{\tau}$ 跑遍满足 $\ell(\tilde{\tau}) < \ell(\tilde{\sigma})$ 的 $\mathfrak{S}$ 的元素, 并且 $\tilde{\theta}_{i,\tilde{\tau}}$ 与 $\tilde{\mu}_{i,\tilde{\tau}}$ 是 $\mathfrak{F}_G(b)$ 的齐性的元素; 特别是, 得到

Proof: If $\tilde{\sigma} \in \mathfrak{S}$ is reducible then we have $\tilde{\sigma} = \sum_{i\in I} \tilde{\eta}_i\,\tilde{\rho}_i\,\tilde{\nu}_i$, where $\tilde{\eta}_i$ $\tilde{\rho}_i$ and $\tilde{\nu}_i$ are homogeneous elements of $\mathfrak{F}_G(b)$ such that $\ell(\tilde{\eta}_i) = 1$ and $\ell(\tilde{\rho}_i) < \ell(\tilde{\sigma})$; furthermore, for any $i \in I$, we have $\tilde{\rho}_i = \sum_{\tilde{\tau}} \tilde{\theta}_{i,\tilde{\tau}}\,\tilde{\tau}\,\tilde{\mu}_{i,\tilde{\tau}}$, where $\tilde{\tau}$ runs over the set of elements of $\mathfrak{S}$ fulfilling $\ell(\tilde{\tau}) < \ell(\tilde{\sigma})$, and where $\tilde{\theta}_{i,\tilde{\tau}}$ and $\tilde{\mu}_{i,\tilde{\tau}}$ are homogeneous elements of $\mathfrak{F}_G(b)$; in particular, we get

11.6.1 $$\tilde{\sigma} = \sum_{\tilde{\tau}\in\mathfrak{S}-\{\tilde{\sigma}\}} \tilde{\eta}'_i\,\tilde{\tau}\,\tilde{\nu}'_i$$

其中 $\tilde{\eta}'_i$, $\tilde{\nu}'_i$ 是 $\mathfrak{F}_G(b)$ 的齐性的元素, 从而 $\mathfrak{S} - \{\tilde{\sigma}\}$ 也是一个 $\mathfrak{D}_G(b)$ 的齐性的生成集.

where $\tilde{\eta}'_i$ and $\tilde{\nu}'_i$ are homogeneous elements of $\mathfrak{F}_G(b)$ and therefore $\mathfrak{S} - \{\tilde{\sigma}\}$ is also a homogeneous generator set of $\mathfrak{D}_G(b)$.

如果 Q_δ 是 $\mathcal{O}Gb$ 上的关键点群, 那么存在某个不可约元素 $\tilde{\chi} \in \mathfrak{D}_G(Q_\delta, P_\gamma)$; 因为

If Q_δ is an essential pointed group on $\mathcal{O}G$ then there is some nonreducible element $\tilde{\chi} \in \mathfrak{D}_G(Q_\delta, P_\gamma)$; since we have

11.6.2
$$\tilde{\chi} = \sum_{\tilde{\sigma} \in \mathfrak{S}} \tilde{\xi}_{\tilde{\sigma}}\, \tilde{\sigma}\, \tilde{\lambda}_{\tilde{\sigma}}$$

其中 $\tilde{\xi}_{\tilde{\sigma}}$, $\tilde{\lambda}_{\tilde{\sigma}}$ 是 $\mathfrak{F}_G(b)$ 的齐性的元素使得 $\ell(\tilde{\xi}_{\tilde{\sigma}}) = 1$, 所以既然 $\tilde{\chi}$ 非可约, 存在某个不可约 $\tilde{\sigma} \in \mathfrak{S}$ 使得 $\ell(\tilde{\sigma}) = \ell(\tilde{\chi})$ 与 $\tilde{\xi}_{\tilde{\sigma}}\, \tilde{\sigma}\, \tilde{\lambda}_{\tilde{\sigma}} \neq 0$. 这样, 既然 $\tilde{\sigma}$ 是齐性的, 对适当的 $\mathcal{O}Gb$ 上的局部点群 R_ε 与 $x \in G$, 有 $\tilde{\sigma} \in \mathfrak{D}_G\big(R_\varepsilon, (P_\gamma)^x\big)$, 并且从 $\ell(\tilde{\sigma}) = \ell(\tilde{\chi})$ 可得到 $|Q| = |R|$; 进一步, 因为 $\tilde{\lambda}_{\tilde{\sigma}} \in \mathbb{Z}E_G(Q_\delta, R_\varepsilon)$ 与 $\tilde{\lambda}_{\tilde{\sigma}} \neq 0$, 所以我们得到 $E_G(Q_\delta, R_\varepsilon) \neq \emptyset$; 从而存在 $y \in G$ 使得 $(Q_\delta)^y = R_\varepsilon$.

where $\tilde{\xi}_{\tilde{\sigma}}$ and $\tilde{\lambda}_{\tilde{\sigma}}$ are homogeneous elements of $\mathfrak{F}_G(b)$ such that $\ell(\tilde{\xi}_{\tilde{\sigma}}) = 1$, and since $\tilde{\chi}$ is nonreductible, there exists some nonreducible $\tilde{\sigma} \in \mathfrak{S}$ fulfilling $\ell(\tilde{\sigma}) = \ell(\tilde{\chi})$ and $\tilde{\xi}_{\tilde{\sigma}}\, \tilde{\sigma}\, \tilde{\lambda}_{\tilde{\sigma}} \neq 0$. Thus, as far as $\tilde{\sigma}$ is a homogeneous element, for a suitable local pointed group R_ε on $\mathcal{O}G$ and some $x \in G$, we have $\tilde{\sigma} \in \mathfrak{D}_G\big(R_\varepsilon, (P_\gamma)^x\big)$, and then from $\ell(\tilde{\sigma}) = \ell(\tilde{\chi})$ we get $|Q| = |R|$; moreover, since $\tilde{\lambda}_{\tilde{\sigma}} \in \mathbb{Z}E_G(Q_\delta, R_\varepsilon)$ and $\tilde{\lambda}_{\tilde{\sigma}} \neq 0$, we still get $E_G(Q_\delta, R_\varepsilon) \neq \emptyset$; hence, there exists $y \in G$ such that $(Q_\delta)^y = R_\varepsilon$.

定理 11.7. **假定 k 是代数闭的. $\mathcal{O}Gb$ 上的一个局部点群 Q_δ 是关键的当且仅当它满足如下两个条件**

11.7.1. **Q_δ 是自中心化的.**

11.7.2. **存在一个 $E_G(Q_\delta)$ 的真子群 $\tilde{M}$ 使得 p 整除 $|\tilde{M}|$, 但是不整除任意 $|M \cap \tilde{M}^{\tilde{\sigma}}|$, 其中 $\tilde{\sigma} \in E_G(Q_\delta) - \tilde{M}$.**

此时, 对任意 $\tilde{\varphi} \in E_G(Q_\delta, P_\gamma)$, $E_G(Q_\delta)_{\tilde{\varphi}}$ 是极小的满足条件 11.7.2 的 $E_G(Q_\delta)$ 的子群; 特别是, 它包含 $E_G(Q_\delta)$ 的某个 Sylow p-子群. 而且, 右边合成以 $\tilde{\varphi}$ 决定下面的 $\mathbb{Z}$-模同构

Theorem 11.7. *Assume that k is algebraically closed. A local pointed group Q_δ on $\mathcal{O}G$ is essential if and only if it fullfills the following two conditions*

11.7.1. *Q_δ is self-centralizing.*

11.7.2. *There exists a proper subgroup $\tilde{M}$ of $E_G(Q_\delta)$ such that p divides $|\tilde{M}|$, whereas it does not divide $|M \cap \tilde{M}^{\tilde{\sigma}}|$ for any $\tilde{\sigma} \in E_G(Q_\delta) - \tilde{M}$.*

Then, for any $\tilde{\varphi} \in E_G(Q_\delta, P_\gamma)$, $E_G(Q_\delta)_{\tilde{\varphi}}$ is a minimal subgroup of $E_G(Q_\delta)$ fulfilling condition 11.7.2; in particular, it contains a Sylow p-subgroup of $E_G(Q_\delta)$. Moreover, the composition on the right with $\tilde{\varphi}$ induces the following $\mathbb{Z}$-module isomorphism

11.7.3
$$\mathbb{Z}\big(E_G(Q_\delta)_{\tilde{\varphi}} \backslash E_G(Q_\delta)\big) \cong \mathfrak{W}_{\delta,\gamma}\,.$$

特别是, $E_G(Q_\delta)$ 在 $E_G(Q_\delta, P_\gamma)$ 的像上可迁作用.

In particular, $E_G(Q_\delta)$ acts transitively on the image of $E_G(Q_\delta, P_\gamma)$.

注意 11.8. 事实上, 任意包含 $E_G(Q_\delta)_{\tilde\varphi}$ 的 $E_G(Q_\delta)$ 的真子群 $\tilde M$ 满足上述的条件; 确实, 考虑 $E_G(Q_\delta, P_\gamma)$ 在 $\mathfrak{W}_{\delta,\gamma}$ 中的像; 由条件 11.7.2, 任意 $E_G(Q_\delta)$ 的非平凡的 p-子群在这个像上稳定唯一一个元素; 可是, p 不整除 $|E_G(Q_\delta)_{\tilde\varphi}\backslash\tilde M|$; 所以对任意 $\tilde\sigma \in E_G(Q_\delta) - \tilde M$, 任意 $\tilde M$ 的非平凡的 p-子群不包含在 $\tilde M^{\tilde\sigma}$ 里.

Remark 11.8. Actually, any subgroup $\tilde M$ of $E_G(Q_\delta)$ containing $E_G(Q_\delta)_{\tilde\varphi}$ fulfills the condition above; indeed, consider the image of $E_G(Q_\delta, P_\gamma)$ in $\mathfrak{W}_{\delta,\gamma}$; according to condition 11.7.2, any nontrivial p-subgroup of $E_G(Q_\delta)$ stabilizes a unique element in this image; but, p does not divide $|E_G(Q_\delta)_{\tilde\varphi}\backslash\tilde M|$; consequently, for any $\tilde\sigma \in E_G(Q_\delta) - \tilde M$, all the nontrivial p-subgroups of $\tilde M$ are not contained in $\tilde M^{\tilde\sigma}$.

证明: 设 Q_δ 是 $\mathcal{O}Gb$ 上的局部点群, 并且设 ν 是 $\mathcal{O}Gb$ 上的 $N_G(Q_\delta)$-点使得 $Q_\delta \subset N_G(Q_\delta)_\nu$; 选择 $N_G(Q_\delta)_\nu$ 的亏点群 R_ε 并选择 $x \in G$ 使得 $R_\varepsilon \subset (P_\gamma)^x$. 请注意, R_ε 包含 Q_δ (见定理 6.8), 并且由推论 10.17, R 在 $E_G(Q_\delta)$ 里的像 $\tilde R$ 是 $E_G(Q_\delta)$ 的 Sylow p-子群. 令 $\varphi: Q \to P$ 记被 x-共轭决定的群同态, 这样 $\tilde\varphi \in E_G(Q_\delta, P_\gamma)$.

Proof: Let Q_δ be a local pointed group on $\mathcal{O}Gb$ and ν the point of $N_G(Q_\delta)$ on $\mathcal{O}Gb$ fulfilling $Q_\delta \in N_G(Q_\delta)_\nu$; choose a defect pointed group R_ε of $N_G(Q_\delta)_\nu$ and an element $x \in G$ such that $R_\varepsilon \subset (P_\gamma)^x$. Note that R_ε contains Q_δ (see Theorem 6.8) and that, by Corollary 10.17, the image of R in $E_G(Q_\delta)$ is a Sylow p-subgroup. Denote by $\varphi: Q \to P$ the group homomorphism determined by the conjugation by x, so that $\tilde\varphi \in E_G(Q_\delta, P_\gamma)$.

由推论 6.17, 如果 Q_δ 非极大的局部点群, 那么对任意 $x' \in G$ 使得 $Q_\delta \subset (P_\gamma)^{x'}$ 存在 $\mathcal{O}Gb$ 上的局部点群 $R'_{\varepsilon'}$ 使得

It follows from Corollary 6.17 that, if Q_δ is not a maximal local pointed group, for any $x' \in G$ fulfilling $Q_\delta \subset (P_\gamma)^{x'}$ there exists a local pointed group $R'_{\varepsilon'}$ on $\mathcal{O}G$ fulfilling

11.7.4 $$Q_\delta \lhd R'_{\varepsilon'} \subset (P_\gamma)^{x'}, \quad |R'| > |Q|;$$

此时, 既然 ν 是 $N_G(Q_\delta)_\nu$ 包含 Q_δ 的 $\mathcal{O}Gb$ 上的唯一 $N_G(Q_\delta)$-点 (见 10.12), $N_G(Q_\delta)_\nu$ 就包含 $R'_{\varepsilon'}$, 从而存在 $n \in N_G(Q_\delta)$ 使得 $R'_{\varepsilon'} \subset (R_\varepsilon)^n$. 那么, 令 $\iota': Q \to R'$ 记包含映射, 再令 $\varphi': Q \to P$, $\rho': R' \to P$, $\psi': R' \to P$, $\sigma: Q \to Q$ 分别记被 x'-共轭, xn-共轭, x'-共轭, n-共轭决定的群同态; 显然有

moreover, as far as ν is the unique point of $N_G(Q_\delta)_\nu$ on $\mathcal{O}Gb$ such that $N_G(Q_\delta)_\nu$ contains Q_δ (see 10.12), $N_G(Q_\delta)_\nu$ still contains $R'_{\varepsilon'}$ and therefore there is $n \in N_G(Q_\delta)$ fulfilling $R'_{\varepsilon'} \subset (R_\varepsilon)^n$. Then, denote by $\iota': Q \to R'$ the inclusion homomorphism, and by $\varphi': Q \to P$, $\rho': R' \to P$, $\psi': R' \to P$ and $\sigma: Q \to Q$ the group homomorphisms respectively determined by the conjugation by the elements x', xn, x', and n; we clearly have

11.7.5 $$\tilde\varphi' - \tilde\varphi \circ \tilde\sigma = (\tilde\psi' - \tilde\rho') \circ \tilde\iota'.$$

现在, 我们假定 Q_δ 是关键的, 从而它非极大的局部点群. 那么, 既然 $(\tilde{\psi}'-\tilde{\rho}')\circ\tilde{\iota}'$ 显然是可约的, $\tilde{\varphi}'$ 与 $\tilde{\varphi}\circ\tilde{\sigma}$ 在 $\mathfrak{W}_{\delta,\gamma}$ 中的像是一样的; 也就是说, $E_G(Q_\delta)$ 在 $E_G(Q_\delta, P_\gamma)$ 上的像可迁作用, 而右边合成以 $\tilde{\varphi}$ 决定满$\mathbb{Z}$-模同态如下

Now, we assume that Q_δ is essential and thus it is not a maximal local pointed group. Then, as far as $(\tilde{\psi}' - \tilde{\rho}')\circ\tilde{\iota}'$ is certainly reducible, $\tilde{\varphi}'$ and $\tilde{\varphi}\circ\tilde{\sigma}$ have the same image in $\mathfrak{W}_{\delta,\gamma}$; in other terms, $E_G(Q_\delta)$ acts transitively on the image of $E_G(Q_\delta, P_\gamma)$, and the composition by $\tilde{\varphi}$ on the right induces a surjective $\mathbb{Z}$-module homomorphism

11.7.6 $$\mathbb{Z}E_G(Q_\delta) \longrightarrow \mathfrak{W}_{\delta,\gamma}\,;$$

事实上, 由 $E_G(Q_\delta)_{\tilde{\varphi}}$ 的定义, 可以把这个同态提升到下面的满 $\mathbb{Z}$-模同态

actually, by the definition of $E_G(Q_\delta)_{\tilde{\varphi}}$, this homomorphism can be factorized *via* the following surjective $\mathbb{Z}$-module homomorphism

11.7.7 $$\mathbb{Z}\big(E_G(Q_\delta)_{\tilde{\varphi}}\backslash E_G(Q_\delta)\big) \longrightarrow \mathfrak{W}_{\delta,\gamma}\,,$$

这样, 得到$E_G(Q_\delta)_{\tilde{\varphi}} \neq E_G(Q_\delta)$.

in particular, we get $E_G(Q_\delta)_{\tilde{\varphi}} \neq E_G(Q_\delta)$.

令ζ记满足$Q_\delta \subset \tilde{C}_G(Q_\delta)_\zeta$ 的 $\mathcal{O}Gb$ 上的唯一 $\tilde{C}_G(Q)$-点 (见 10.12), 并且选择$\tilde{C}_G(Q)_\zeta$ 的亏点群 T_θ; 既然 $(T_\theta)^n$ 也是 $\tilde{C}_G(Q)_\zeta$ 的亏点群, 对适当的 $m \in N_G(T_\theta)$ 与 $z \in C_G(Q)$ 有 $n = mz$. 那么, 令 $\iota{:}\,Q \to T$ 记包含映射, 再令 $\psi{:}\,T \to P$ 与 $\rho{:}\,T \to T$ 分别记被 x-共轭与 m-共轭决定的群同态; 显然得到

Denote by ζ the unique point of $\tilde{C}_G(Q)$ on $\mathcal{O}G$ fulfilling $Q_\delta \subset \tilde{C}_G(Q_\delta)_\zeta$ (see 10.12), and choose a defect pointed group T_θ of $\tilde{C}_G(Q_\delta)_\zeta$; as far as $(T_\theta)^n$ also is a defect pointed group of $\tilde{C}_G(Q_\delta)_\zeta$, we have $n = mz$ for suitable $m \in N_G(T_\theta)$ and $z \in C_G(Q)$. Then, denote by $\iota{:}\,Q \to T$ the inclusion map and by $\psi{:}\,T \to P$ and $\rho{:}\,T \to T$ the group homomorphisms respectively induced by conjugation by the elements x and m; we clearly get

11.7.8 $$\tilde{\varphi} - \tilde{\varphi}\circ\tilde{\sigma} = \tilde{\psi}\circ(\widetilde{\mathrm{id}}_T - \tilde{\rho})\circ\tilde{\iota}\,.$$

既然 $E_G(Q_\delta)_{\tilde{\varphi}} \neq E_G(Q_\delta)$, 可取$\tilde{\sigma} \notin E_G(Q_\delta)_{\tilde{\varphi}}$;那么$\tilde{\varphi} - \tilde{\varphi}\circ\tilde{\sigma}$ 不是可约的, 所以从等式 11.7.8 可推出 $T = Q$; 此时, 由推论 10.16, Q_δ 是自中心化的.

since we have $E_G(Q_\delta)_{\tilde{\varphi}} \neq E_G(Q_\delta)$, we can choose $\tilde{\sigma} \notin E_G(Q_\delta)_{\tilde{\varphi}}$; then $\tilde{\varphi} - \tilde{\varphi}\circ\tilde{\sigma}$ is not reducible, and therefore from equality 11.7.8 we conclude that $T = Q$; at this point, by Corollary 10.16, Q_δ is self-centralizing.

进一步, 从 $Q_\delta \neq P_\gamma$ 可推出 $Q_\delta \neq R_\varepsilon$ (见推论 6.16); 此时, 只要选择$x' = x$, $R'_{\varepsilon'} = R_\varepsilon$ $n \in R$, 就从等式 11.7.5 可推出$\tilde{R}$稳定$\tilde{\varphi}$; 这样,$E_G(Q_\delta)_{\tilde{\varphi}}$包含 $E_G(Q_\delta)$ 的 Sylow p-子群 $\tilde{R}$.

Furthermore, $Q_\delta \neq P_\gamma$ implies that $Q_\delta \neq R_\varepsilon$ (see Corollary 6.16); then, choosing $x' = x$, $R'_{\varepsilon'} = R_\varepsilon$ and $n \in R$, the corresponding equalities 11.7.5 imply that $\tilde{R}$ stablizes $\tilde{\varphi}$; thus, $E_G(Q_\delta)_{\tilde{\varphi}}$ contains a Sylow p-subgroup of $E_G(Q_\delta)$.

另一方面, 假定 p 整除 $|E_G(Q_\delta)_{\tilde\varphi'} \cap E_G(Q_\delta)_{\tilde\varphi}|$; 由推论 10.17, 存在 $\mathcal{O}Gb$ 上的一个局部点群 T_θ 使得 $Q_\delta \triangleleft T_\theta$ 并 T 在 $E_G(Q_\delta)$ 中的像 $\tilde T$ 是 $E_G(Q_\delta)_{\tilde\varphi'} \cap E_G(Q_\delta)_{\tilde\varphi}$ 的非平凡子群; 令 M 与 M' 分别记 $E_G(Q_\delta)_{\tilde\varphi}$ 与 $E_G(Q_\delta)_{\tilde\varphi'}$ 在 $N_G(Q_\delta)$ 里的原像, 再令 μ 与 μ' 分别记 $\mathcal{O}Gb$ 上的 M-点与 M'-点使得 $M_\mu \supset Q_\delta \subset M'_{\mu'}$ (见 10.12); 只要仍应用 10.12, 并使用$E_G(Q_\delta)_{\tilde\varphi\circ\tilde\sigma} = E_G(Q_\delta)_{\tilde\varphi'}$ 就得到

On the other hand, assume that p divides $|E_G(Q_\delta)_{\tilde\varphi'} \cap E_G(Q_\delta)_{\tilde\varphi}|$; according to Corollary 10.17, there exists a local pointed group T_θ on $\mathcal{O}G$ such that we have $Q_\delta \triangleleft T_\theta$ and that the image $\tilde T$ of T in $E_G(Q_\delta)$ is a nontrivial subgroup of $E_G(Q_\delta)_{\tilde\varphi'} \cap E_G(Q_\delta)_{\tilde\varphi}$; denote by M and M' the respective converse images of $E_G(Q_\delta)_{\tilde\varphi}$ and $E_G(Q_\delta)_{\tilde\varphi'}$ in $N_G(Q_\delta)$, and by μ and μ' the respective points of M and M' on $\mathcal{O}Gb$ which fulfill $M_\mu \supset Q_\delta \subset M'_{\mu'}$ (see 10.12); it follows from the uniqueness in 10.12 and from the obvious equality $E_G(Q_\delta)_{\tilde\varphi\circ\tilde\sigma} = E_G(Q_\delta)_{\tilde\varphi'}$ that we get

11.7.9 $$R_\varepsilon \subset M_\mu \supset T_\theta \subset M'_{\mu'} \supset (R_\varepsilon)^n\,,$$

从而存在 $m \in M$ 与 $m' \in M'$ 使得

and therefore there are $m \in M$ and $m' \in M'$ fulfilling

11.7.10 $$(R_\varepsilon)^m \supset T_\theta \subset (R_\varepsilon)^{nm'}\,.$$

那么令 $\iota{:}\,Q \to T$ 记包含映射,再令 $\xi{:}\,T \to P$, $\xi'{:}\,T \to P$, $\tau{:}\,Q \to Q$, $\tau'{:}\,Q \to Q$ 分别记被 xm-共轭, xnm'-共轭, m^{-1}-共轭, m'^{-1}-共轭决定的群同态; 显然得到

Then, denote by $\iota{:}\,Q \to T$ the inclusion homomorphism, and by $\xi{:}\,T \to P$, $\xi'{:}\,T \to P$, $\tau{:}\,Q \to Q$ and $\tau'{:}\,Q \to Q$ the group homomorphisms respectively induced by the conjugation by the elements xm, xnm', m^{-1} and m'^{-1}; we clearly get

11.7.11 $$\tilde\xi \circ \tilde\iota = \tilde\varphi \circ \tilde\tau\,, \quad \tilde\xi' \circ \tilde\iota = \tilde\varphi \circ \tilde\sigma \circ \tilde\tau'\,;$$

所以 $\mathfrak{R}_G(Q_\delta, P_\gamma)$ 包含 $\tilde\varphi - \tilde\xi \circ \tilde\iota$, $(\tilde\xi - \tilde\xi') \circ \tilde\iota$, $\tilde\xi' \circ \tilde\iota - \tilde\varphi \circ \tilde\sigma$, 从而它还包含 $\tilde\varphi - \tilde\varphi \circ \tilde\sigma$; 最后, 得到

consequently, $\mathfrak{R}_G(Q_\delta, P_\gamma)$ contains $\tilde\varphi - \tilde\xi \circ \tilde\iota$, $(\tilde\xi - \tilde\xi') \circ \tilde\iota$ and $\tilde\xi' \circ \tilde\iota - \tilde\varphi \circ \tilde\sigma$, and therefore it also contains $\tilde\varphi - \tilde\varphi \circ \tilde\sigma$; finally, we get

11.7.12 $$E_G(Q_\delta)_{\tilde\varphi} = E_G(Q_\delta)_{\tilde\varphi\circ\tilde\sigma} = E_G(Q_\delta)_{\tilde\varphi'}\,.$$

反过来, 假定 Q_δ 是自中心化的, 再假定 $\tilde M$ 是 $E_G(Q_\delta)$ 的真子群使得 p 整除 $|\tilde M|$, 以及不整除 $|M \cap \tilde M^{\tilde\sigma}|$, 其中 $\tilde\sigma$ 属于 $E_G(Q_\delta) - \tilde M$. 由此可见, 对 $\tilde M$ 的任意非平凡 p-子群 $\tilde T$, $\tilde M$ 包含 $N_{E_G(Q_\delta)}(\tilde T)$; 这样, 如果 $\tilde T$ 是 $\tilde M$ 的 Sylow p-子群, 那么 $\tilde T$ 也

Conversely, assume that Q_δ is self-centralizing, and that $\tilde M$ is a proper subgroup of $E_G(Q_\delta)$ such that p divides $|\tilde M|$ and does not divide $|M \cap \tilde M^{\tilde\sigma}|$ for any $\tilde\sigma$ in $E_G(Q_\delta) - \tilde M$. Accordingly, for any nontrivial p-subgroup $\tilde T$ of $\tilde M$, $\tilde M$ contains $N_{E_G(Q_\delta)}(\tilde T)$; thus, if $\tilde T$ is a Sylow p-subgroup of $\tilde M$, then $\tilde T$ is also a Sylow p-subgroup

是$N_{E_G(Q_\delta)}(\tilde{T})$的Sylow p-子群，从而$\tilde{T}$就是$E_G(Q_\delta)$的Sylow p-子群. 由此能假定$\tilde{M}$包含R'在$E_G(Q_\delta)$中的像$\tilde{R}'$；进一步，因为$|R'|>|Q|$与$C_{R'}(Q)=Z(Q)$ (见 10.15)，所以$\tilde{R}'$非平凡；因此，对任意$\tilde{\sigma}'\in E_G(Q_\delta)$使得$\tilde{R}'\subset \tilde{M}^{\tilde{\sigma}'}$，有$\tilde{M}\tilde{\sigma}'=\tilde{M}\tilde{\sigma}$.

这样，任意$\tilde{\varphi}'\in E_G(Q_\delta,P_\gamma)$决定$\tilde{M}\backslash E_G(Q_\delta)$的元素$\tilde{M}\tilde{\sigma}$；令

of $N_{E_G(Q_\delta)}(\tilde{T})$ and therefore $\tilde{T}$ is a Sylow p-subgroup of $E_G(Q_\delta)$. Hence, we may assume that $\tilde{M}$ contains the image $\tilde{R}'$ of R' in $E_G(Q_\delta)$; moreover, since $|R'|>|Q|$ and $C_{R'}(Q)=Z(Q)$ (see 10.15), R' is not trivial; consequently, for any $\tilde{\sigma}'\in E_G(Q_\delta)$ such that $\tilde{R}'\subset \tilde{M}^{\tilde{\sigma}'}$, we have $\tilde{M}\tilde{\sigma}'=\tilde{M}\tilde{\sigma}$.

Thus, any $\tilde{\varphi}'\in E_G(Q_\delta,P_\gamma)$ determines an element $\tilde{M}\tilde{\sigma}$ of $\tilde{M}\backslash E_G(Q_\delta)$; denote by

11.7.13 $$\mathbb{Z}E_G(Q_\delta,P_\gamma)\longrightarrow \mathbb{Z}\big(\tilde{M}\backslash E_G(Q_\delta)\big)$$

记对应的$\mathbb{Z}$-模同态；显然，这个同态是满射. 而且$\mathfrak{R}_G(Q_\delta,P_\gamma)$的像是零；确实，由$\mathfrak{R}_G(Q_\delta,P_\gamma)$的定义，只要表示$(\tilde{\eta}'-\tilde{\eta})\circ\tilde{\omega}$在$\mathbb{Z}\big(\tilde{M}\backslash E_G(Q_\delta)\big)$中的像是零，其中$\tilde{\omega}\in E_G(Q_\delta,T_\theta)$与$\tilde{\eta}',\tilde{\eta}$属于$E_G(T_\theta,P_\gamma)$，这里$T_\theta$是满足$|T|>|Q|$的$\mathcal{O}Gb$上的局部点群，就证明上述的主张. 已经知道，如果$s\in G$决定$\tilde{\omega}$，那么有$Q_\delta\subset (T_\theta)^s$；只要仍应用推论 6.17，就得到

the corresponding $\mathbb{Z}$-module homomorphism; clearly, this homomorphism is surjective. Moreover, the image of $\mathfrak{R}_G(Q_\delta,P_\gamma)$ is zero; indeed, by the very definition of $\mathfrak{R}_G(Q_\delta,P_\gamma)$, in order to prove our claim, it suffices to show that, for any $\tilde{\omega}\in E_G(Q_\delta,T_\theta)$ and any $\tilde{\eta}'$, $\tilde{\eta}$ in $E_G(T_\theta,P_\gamma)$, where T_θ is a local pointed group on $\mathcal{O}G$ such that $|T|>|Q|$, the image of $(\tilde{\eta}'-\tilde{\eta})\circ\tilde{\omega}$ in $\mathbb{Z}\big(\tilde{M}\backslash E_G(Q_\delta)\big)$ is zero. As we already know, if $s\in G$ induces $\tilde{\omega}$ then we have $Q_\delta\subset (T_\theta)^s$; thus, applying again Corollary 6.17, we get

11.7.14 $$Q_\delta\triangleleft U_\upsilon\subset (T_\theta)^s\,,\quad |U|>|Q|$$

其中U_υ是$\mathcal{O}Gb$上的局部点群.

我们能假定$s=1$与$U_\upsilon=T_\theta$；此时有$T_\theta\subset N_G(Q_\delta)_\nu$，并且$T$在$E_G(Q_\delta)$里的像$\tilde{T}$非平凡，这是因为$C_T(Q)=Z(Q)$ (见 10.15)；而且，如果$y,y'\in G$分别决定$\tilde{\eta}$与$\tilde{\eta}'$，那么存在$\tilde{\tau}\in E_G(Q_\delta)$与$\tilde{\tau}'\in E_G(Q_\delta)$使得$N_{P^y}(Q_\delta)$与$N_{P^{y'}}(Q_\delta)$在$E_G(Q_\delta)$的像分别包含在$\tilde{M}^{\tilde{\tau}}$与$\tilde{M}^{\tilde{\tau}'}$中；此时，因为$(P_\gamma)^y$与$(P_\gamma)^{y'}$包含$T_\theta$，所以我们有$T_\theta\subset \tilde{M}^{\tilde{\tau}}\cap\tilde{M}^{\tilde{\tau}'}$，从而还有$\tilde{M}^{\tilde{\tau}}=\tilde{M}^{\tilde{\tau}'}$，以及$\tilde{M}\tilde{\tau}=\tilde{M}\tilde{\tau}'$.

where U_υ is a local pointed group on $\mathcal{O}Gb$.

We may assume that $s=1$ and $U_\upsilon=T_\theta$; then we have $T_\theta\subset N_G(Q_\delta)_\nu$ and the image $\tilde{T}$ of T in $E_G(Q_\delta)$ is not trivial since we have $C_T(Q)=Z(Q)$ (see 10.15); furthermore, if $y,y'\in G$ respectively determine $\tilde{\eta}$ and $\tilde{\eta}'$, then there are $\tilde{\tau}\in E_G(Q_\delta)$ and $\tilde{\tau}'\in E_G(Q_\delta)$ such that the images of $N_{P^y}(Q_\delta)$ and $N_{P^{y'}}(Q_\delta)$ in $E_G(Q_\delta)$ are respectively contained in $\tilde{M}^{\tilde{\tau}}$ and $\tilde{M}^{\tilde{\tau}'}$; at this point, since $(P_\gamma)^y$ and $(P_\gamma)^{y'}$ contain T_θ, we have $T_\theta\subset \tilde{M}^{\tilde{\tau}}\cap\tilde{M}^{\tilde{\tau}'}$ and therefore we still have $\tilde{M}^{\tilde{\tau}}=\tilde{M}^{\tilde{\tau}'}$, so that $\tilde{M}\tilde{\tau}=\tilde{M}\tilde{\tau}'$.

最后, 我们得到下面的满ℤ-模同态 (见 11.7.7 和 11.7.13)

Finally, we get the surjective ℤ-module homomorphism (see 11.7.7, 11.7.13)

11.7.15 $$\mathbb{Z}\big(E_G(Q_\delta)_{\tilde\varphi}\backslash E_G(Q_\delta)\big) \longrightarrow \mathfrak{W}_{\delta,\gamma} \longrightarrow \mathbb{Z}\big(\tilde M\backslash E_G(Q_\delta)\big)\,,$$

其中 $E_G(Q_\delta)_{\tilde\varphi}\tilde\sigma$ 的像就是 $\tilde M\tilde\sigma$; 特别是, Q_δ 是关键的, 并且有 $E_G(Q_\delta)_{\tilde\varphi}\subset\tilde M$. 此时, 我们能选择 $\tilde M = E_G(Q_\delta)_{\tilde\varphi}$; 从这个选择可证明 ℤ-模同态 11.7.7 也是单射, 从而它是双射.

where $\tilde M\tilde\sigma$ is the image of $E_G(Q_\delta)_{\tilde\varphi}\tilde\sigma$; in particular, Q_δ is essential, and moreover we have $E_G(Q_\delta)_{\tilde\varphi}\subset\tilde M$. At this point, we can make the choice $\tilde M = E_G(Q_\delta)_{\tilde\varphi}$, and from it, we get that the ℤ-module homomorphism 11.7.7 is injective too, so it is bijective.

推论 11.9. 假定 k 是代数闭的. 设 $\mathcal{E}$ 是 $\mathcal{O}Gb$ 上的关键点群的 G-共轭类的代表集合使得 $\mathcal{E}$ 的元素都包含在 P_γ 里. 对 $\mathcal{O}Gb$ 上的任意局部点群 Q_δ 与任意 $\tilde\varphi,\tilde\varphi'\in E_G(Q_\delta,P_\gamma)$ 有

Corollary 11.9. *Assume that k is algebraically closed. Let $\mathcal{E}$ be a set of representatives for the G-conjugacy classes of essential pointed groups on $\mathcal{O}G$ such that P_γ contains all the elements of $\mathcal{E}$. For any local pointed group Q_δ on $\mathcal{O}G$ and any $\tilde\varphi,\tilde\varphi'\in E_G(Q_\delta,P_\gamma)$ we have*

11.9.1 $$\tilde\varphi'-\tilde\varphi = (\tilde\sigma-\widetilde{\mathrm{id}}_P)\circ\tilde\varphi + \sum_{i\in I}\tilde\mu_i\circ(\tilde\sigma_i-\widetilde{\mathrm{id}}_{Q^i})\circ\tilde\nu_i$$

其中 $\tilde\sigma\in E_G(P_\gamma)$ 与 $\{Q^i_{\delta^i}\}_{i\in I}$ 是 $\mathcal{E}$ 的有限序列, 以及对任意 $i\in I$ 有 $\tilde\mu_i\in E_{N_G(P_\gamma)}(Q^i_{\delta^i},P_\gamma)$ 与 $\tilde\nu_i\in E_G(Q_\delta,Q^i_{\delta^i})$ 与 $\tilde\sigma_i$ 是 $E_G(Q^i_{\delta^i})-E_G(Q^i_{\delta^i})_{\tilde\mu_i}$ 的 p'-元素.

for a suitable $\tilde\sigma\in E_G(P_\gamma)$, some finite sequence $\{Q^i_{\delta^i}\}_{i\in I}$ of elements of $\mathcal{E}$ and, for any $i\in I$, suitable $\tilde\mu_i\in E_{N_G(P_\gamma)}(Q^i_{\delta^i},P_\gamma)$, $\tilde\nu_i\in E_G(Q_\delta,Q^i_{\delta^i})$ and some p'-element $\tilde\sigma_i$ of $E_G(Q^i_{\delta^i})-E_G(Q^i_{\delta^i})_{\tilde\mu_i}$.

证明: 首先, 我们用 $\mathcal{E}\cup\{P_\gamma\}$ 代替 $\mathcal{E}$, 并对任意 $\tilde\mu\in E_G(P_\gamma)$ 记 $E_G(P_\gamma)_{\tilde\mu}=\{1\}$; 而且, 如果 Q_δ 是 $\mathcal{O}Gb$ 上的局部点群与 $\tilde\sigma$ 属于 $E_G(Q_\delta)$ 那么令

Proof: First of all, in the proof we replace $\mathcal{E}$ by $\mathcal{E}\cup\{P_\gamma\}$, setting $E_G(P_\gamma)_{\tilde\mu}=\{1\}$ for any $\tilde\mu\in E_G(P_\gamma)$; moreover, for any local pointed group Q_δ on $\mathcal{O}Gb$ and any element $\tilde\sigma$ of $E_G(Q_\delta)$, we set

11.9.2 $$\partial_{\tilde\sigma} = \tilde\sigma - \widetilde{\mathrm{id}}_Q\,.$$

对长度 $\ell(\tilde\varphi'-\tilde\varphi)$ 用归纳法证明

Arguing by induction on $\ell(\tilde\varphi'-\tilde\varphi)$, we will prove

11.9.3 $$\tilde\varphi'-\tilde\varphi = \sum_{i\in I}\tilde\mu_i\circ\partial_{\tilde\sigma_i}\circ\tilde\nu_i\,;$$

显然能假定 $\tilde\varphi'-\tilde\varphi$ 不可约; 特别是, Q_δ 是关键的, 从而存在 $x\in G$ 使得 $(Q_\delta)^x\in\mathcal{E}$, 以

it is clear that we may assume that $\tilde\varphi'-\tilde\varphi$ is not reducible; in particular, Q_δ is essential and therefore there is $x\in G$ such that

及 $(Q_\delta)^x \subset P_\gamma$. 令 $\iota: Q^x \to P$ 记包含映射; 由定理 11.7, 考虑 $\mathfrak{W}_{\delta^x,\gamma}$ 的时候 $E_G\big((Q_\delta)^x\big)$ 在 $E_G\big((Q_\delta)^x, P_\gamma\big)$ 的像上可迁作用; 所以, 存在 $\tilde\tau, \tilde\tau' \in E_G\big((Q_\delta)^x\big)$ 使得下面的 $\mathfrak{D}_G\big((Q_\delta)^x, P_\gamma\big)$ 的元素

$(Q_\delta)^x \in \mathcal{E}$, thus $(Q_\delta)^x \subset P_\gamma$. Denote by $\iota: Q^x \to P$ the inclusion map; it follows from Theorem 11.7 that, considering $\mathfrak{W}_{\delta^x,\gamma}$, $E_G\big((Q_\delta)^x\big)$ acts transitively on the image of $E_G\big((Q_\delta)^x, P_\gamma\big)$; hence, there are $\tilde\tau, \tilde\tau' \in E_G\big((Q_\delta)^x\big)$ such that the following elements of $\mathfrak{D}_G\big((Q_\delta)^x, P_\gamma\big)$

11.9.4 $$\tilde\rho = \tilde\varphi \circ (\tilde\kappa_x^Q)^{-1} - \tilde\iota \circ \tilde\tau\,, \quad \tilde\rho' = \tilde\varphi' \circ (\tilde\kappa_x^Q)^{-1} - \tilde\iota \circ \tilde\tau'$$

是可约的. 另一方面, 显然得到

are reducible. On the other hand, clearly

11.9.5 $$\tilde\varphi' - \tilde\varphi = \tilde\rho' \circ \tilde\kappa_x^Q + \tilde\iota \circ \partial_{\tilde\tau'} \circ \tilde\kappa_x^Q + \tilde\iota \circ \partial_{\tilde\tau^{-1}} \circ (\tilde\tau \circ \tilde\kappa_x^Q) + \tilde\rho \circ \tilde\kappa_x^Q\,.$$

令 $\tilde H$ 记被 $E_G((Q_\delta)^x)_{\tilde\iota}$ 的 p'-元素都生成的 $E_G((Q_\delta)^x)_{\tilde\iota}$ 的子群; 既然 $E_G((Q_\delta)^x)_{\tilde\iota}$ 包含一个 $E_G((Q_\delta)^x)$ 的 Sylow p-子群, 有

Denote by $\tilde H$ the subgroup generated by the set of all the p'-elements of $E_G((Q_\delta)^x)_{\tilde\iota}$; since $E_G((Q_\delta)^x)_{\tilde\iota}$ contains a Sylow p-subgroup of $E_G((Q_\delta)^x)$, we have

11.9.6 $$E_G\big((Q_\delta)^x\big) = \tilde H \cdot E_G\big((Q_\delta)^x\big)_{\tilde\iota}\,;$$

进一步, 如果 $\tilde\sigma, \tilde\sigma' \in E_G\big((Q_\delta)^x\big)$ 那么显然有

furthermore, if $\tilde\sigma, \tilde\sigma' \in E_G\big((Q_\delta)^x\big)$ then we clearly get

11.9.7 $$\partial_{\tilde\sigma \circ \tilde\sigma'} = \partial_{\tilde\sigma} \circ \tilde\sigma' + \partial_{\tilde\sigma'}\,.$$

此时, 不难完成等式 11.9.3 的证明.

At this point, it is not difficult to complete the proof of equality 11.9.3.

在 $\tilde\varphi' - \tilde\varphi$ 的所有如 11.9.3 的表达式中, 令 $n(\tilde\varphi' - \tilde\varphi)$ 记极小的表达式的和项数 $|I|$; 最后, 对 $n(\tilde\varphi' - \tilde\varphi)$ 用归纳法证明式子 11.9.1, 而显然能假定 $n(\tilde\varphi' - \tilde\varphi) \neq 0$; 那么, 只要应用下面的引理 11.10, 就可以验证对适当的 $\tilde\varphi'' \in E_G(Q_\delta, P_\gamma)$ 得到

Considering all the expressions of $\tilde\varphi' - \tilde\varphi$ as in equality 11.9.3, denote by $n(\tilde\varphi' - \tilde\varphi)$ the minimal number $|I|$ of terms in those sums; finally, we prove equality 11.9.1 arguing by induction on $n(\tilde\varphi' - \tilde\varphi)$, and clearly may assume that $n(\tilde\varphi' - \tilde\varphi) \neq 0$; then, it follows from Lemma 11.10 below that, for a suitable $\tilde\varphi'' \in E_G(Q_\delta, P_\gamma$, we get

11.9.8 $$n(\tilde\varphi'' - \tilde\varphi) = n(\tilde\varphi' - \tilde\varphi) - 1\,, \quad \tilde\varphi' - \tilde\varphi'' = \tilde\mu'' \circ \partial_{\tilde\sigma''} \circ \tilde\nu''\,,$$

其中 $\tilde\mu'' \in E_{N_G(P_\gamma)}(Q''_{\delta''}, P_\gamma)$, $Q''_{\delta''} \in \mathcal{E} \cup \{P_\gamma\}$, $\tilde\sigma''$ 是一个 $E_G(Q''_{\delta''}) - E_G(Q''_{\delta''})_{\tilde\mu''}$ 的 p'-元素, $\tilde\nu'' \in E_G(Q_\delta, Q''_{\delta''})$. 所以, $\tilde\varphi'' - \tilde\varphi$ 已经有下面的表达式

where $\tilde\mu'' \in E_{N_G(P_\gamma)}(Q''_{\delta''}, P_\gamma)$, $Q''_{\delta''} \in \mathcal{E} \cup \{P_\gamma\}$, $\tilde\sigma''$ is a p'-element of $E_G(Q''_{\delta''}) - E_G(Q''_{\delta''})_{\tilde\mu''}$ and $\tilde\nu'' \in E_G(Q_\delta, Q''_{\delta''})$. Consequently, the difference $\tilde\varphi'' - \tilde\varphi$ already has the following expression

11.9.9 $$\tilde\varphi'' - \tilde\varphi = \partial_{\tilde\sigma} \circ \tilde\varphi + \sum_{i \in I} \tilde\mu_i \circ \partial_{\tilde\sigma_i} \circ \tilde\nu_i\,.$$

如果 $Q''_{\delta''} \in \mathcal{E}$ 那么 $\tilde{\varphi}' - \tilde{\varphi}$ 也有形如 11.9.1 的表达式. 如果 $Q''_{\delta''} = P_\gamma$ 那么 $\tilde{\mu}''$ 是 P 的一个外自同构, 以及 $(\tilde{\mu}'')^{-1}$ 属于 $E_G(P_\gamma)$;可是,由等式11.9.8,有

If $Q''_{\delta''} \in \mathcal{E}$ then $\tilde{\varphi}' - \tilde{\varphi}$ already has the pattern of expression 11.9.1. If $Q''_{\delta''} = P_\gamma$ then $\tilde{\mu}''$ is an outer automorphism of P and we consider its inverse $(\tilde{\mu}'')^{-1}$ in $E_G(P_\gamma)$; but, according to equality 11.9.8, we have

11.9.10 $$\tilde{\varphi}' = \tilde{\mu}'' \circ \tilde{\sigma}'' \circ \tilde{\nu}'' , \quad \tilde{\varphi}'' = \tilde{\mu}'' \circ \tilde{\nu}'' ;$$

所以只要记 $\tilde{\sigma}' = \tilde{\mu}'' \circ \tilde{\sigma}'' \circ (\tilde{\mu}'')^{-1}$ 就得到

hence, setting $\tilde{\sigma}' = \tilde{\mu}'' \circ \tilde{\sigma}'' \circ (\tilde{\mu}'')^{-1}$, finally we get

11.9.11 $$\begin{aligned} \tilde{\varphi}' - \tilde{\varphi} &= \tilde{\sigma}' \circ \tilde{\varphi}'' - \tilde{\varphi} \\ &= (\tilde{\sigma}' \circ \tilde{\sigma}) \circ \tilde{\varphi} + \sum_{i\in I} (\tilde{\sigma}' \circ \tilde{\mu}_i) \circ \partial_{\tilde{\sigma}_i} \circ \tilde{\nu}_i - \tilde{\varphi} \\ &= \partial_{\tilde{\sigma}' \circ \tilde{\sigma}} \circ \tilde{\varphi} + \sum_{i\in I} (\tilde{\sigma}' \circ \tilde{\mu}_i) \circ \partial_{\tilde{\sigma}_i} \circ \tilde{\nu}_i . \end{aligned}$$

引理 11.10. 设 Q_δ, R_ε 是 $\mathcal{O}Gb$ 上的局部点群并设 $\tilde{\varphi}, \tilde{\varphi}'$ 是 $E_G(Q_\delta, R_\varepsilon)$ 的元素. 假定有

Lemma 11.10. *Let Q_δ and R_ε be local pointed groups on $\mathcal{O}Gb$ and consider elements $\tilde{\varphi}$ and $\tilde{\varphi}'$ of $E_G(Q_\delta, R_\varepsilon)$. Assume that*

11.10.1 $$0 \neq \tilde{\varphi}' - \tilde{\varphi} = \sum_{i\in I} \tilde{\nu}_i \circ (\tilde{\varphi}'_i - \tilde{\varphi}_i) \circ \tilde{\mu}_i ,$$

其中, 对适当的 $\mathcal{O}Gb$ 上的局部点群的有限序列 $\{Q^i_{\delta^i}\}_{i\in I}$ 与 $\{R^i_{\varepsilon^i}\}_{i\in I}$ 而对任意 $i \in I$, $\tilde{\varphi}_i$ 与 $\tilde{\varphi}'_i$ 属于 $E_G(Q^i_{\delta^i}, R^i_{\varepsilon^i})$, $\tilde{\mu}_i$ 属于 $E_G(Q_\delta, Q^i_{\delta^i})$ 与 $\tilde{\nu}_i$ 属于 $E_G(R^i_{\varepsilon^i}, R_\varepsilon)$. 那么, 存在 $n \in \mathbb{N}$ 与一个单射 $\sigma: [0, n] \to I$ 使得对任意 $\ell \in [1, n]$ 有

where, for suitable finite sequences $\{Q^i_{\delta^i}\}_{i\in I}$ and $\{R^i_{\varepsilon^i}\}_{i\in I}$ of local pointed groups on $\mathcal{O}G$ and for any $i \in I$, $\tilde{\varphi}_i$ and $\tilde{\varphi}'_i$ belong to $E_G(Q^i_{\delta^i}, R^i_{\varepsilon^i})$, $\tilde{\mu}_i$ belongs to $E_G(Q_\delta, Q^i_{\delta^i})$ and $\tilde{\nu}_i$ belongs to $E_G(R^i_{\varepsilon^i}, R_\varepsilon)$. Then, there are $n \in \mathbb{N}$ and an injective map $\sigma: [0, n] \to I$ such that, for any $\ell \in [1, n]$, we have

11.10.2 $$\begin{aligned} \tilde{\varphi} &= \tilde{\nu}_{\sigma(0)} \circ \tilde{\varphi}_{\sigma(0)} \circ \tilde{\mu}_{\sigma(0)} \\ \tilde{\nu}_{\sigma(\ell-1)} \circ \tilde{\varphi}'_{\sigma(\ell-1)} \circ \tilde{\mu}_{\sigma(\ell-1)} &= \tilde{\nu}_{\sigma(\ell)} \circ \tilde{\varphi}_{\sigma(\ell)} \circ \tilde{\mu}_{\sigma(\ell)} \\ \tilde{\nu}_{\sigma(n)} \circ \tilde{\varphi}'_{\sigma(n)} \circ \tilde{\mu}_{\sigma(n)} &= \tilde{\varphi}' . \end{aligned}$$

证明: 等式 11.10.1 显然等价于

Proof: Equality 11.10.1 is equivalent to

11.10.3 $$\tilde{\varphi}' + \sum_{i\in I} \tilde{\nu}_i \circ \tilde{\varphi}_i \circ \tilde{\mu}_i = \tilde{\varphi} + \sum_{i\in I} \tilde{\nu}_i \circ \tilde{\varphi}'_i \circ \tilde{\mu}_i ;$$

因为$\tilde{\varphi}' \neq \tilde{\varphi}$所以存在$i', i'' \in I$与双射$\pi: I - \{i''\} \longrightarrow I - \{i'\}$使得

since $\tilde{\varphi}' \neq \tilde{\varphi}$, there exist $i', i'' \in I$ and a bijective map $\pi: I - \{i''\} \longrightarrow I - \{i'\}$ such that

11.10.4
$$\begin{aligned}\tilde{\nu}_{i'} \circ \tilde{\varphi}_{i'} \circ \tilde{\mu}_{i'} &= \tilde{\varphi}\\ \tilde{\nu}_i \circ \tilde{\varphi}'_i \circ \tilde{\mu}_i &= \tilde{\nu}_{\pi(i)} \circ \tilde{\varphi}_{\pi(i)} \circ \tilde{\mu}_{\pi(i)}\\ \tilde{\varphi}' &= \tilde{\nu}_{i''} \circ \tilde{\varphi}'_{i''} \circ \tilde{\mu}_{i''}\end{aligned}$$

其中$i \in I - \{i'\}$.用归纳法得到n与$\sigma: [0, n] \to I$的定义如下:令$\sigma(0) = i'$;如果对$\ell \in \mathbb{N}$已经得到$\sigma(\ell)$的定义并$\sigma(\ell) \neq i''$,那么再令$\sigma(\ell+1) = \pi\big(\sigma(\ell)\big)$;特别是,有$\sigma(n) = i''$.

for any $i \in I - \{i'\}$. W alsoe get n and the definition of $\sigma: [0, n] \longrightarrow I$ in the following inductive way: set $\sigma(0) = i'$; if for $\ell \in \mathbb{N}$ we already know the definition of $\sigma(\ell)$ and $\sigma(\ell) \neq i''$ then we just set $\sigma(\ell+1) = \pi\big(\sigma(\ell)\big)$; in particular, we have $\sigma(n) = i''$.

最后,如果$\sigma(\ell) = \sigma(\ell')$,其中$\ell \neq \ell'$,那么有$\pi^{\ell}(i') = \pi^{\ell'}(i')$,从而还有$i' = \pi^{\ell'-\ell}(i')$,矛盾;所以,$\sigma$是单射.

Finally, if $\sigma(\ell) = \sigma(\ell')$ for $\ell \neq \ell'$ then we have $\pi^{\ell}(i') = \pi^{\ell'}(i')$ and therefore we still have $i' = \pi^{\ell'-\ell}(i')$, a contradiction; hence, σ is injective.

注意 11.11. 假定k是代数闭的.一般来说,不需要推论 11.9 那么高的精确度.假定T_θ与R_ε是$\mathcal{O}Gb$上的局部点群使得$|T| = |R|$与$T_\theta \subset P_\gamma \supset R_\varepsilon$,而令$\iota_T: T \to P$与$\iota_R: R \to P$记包含映射;不难验证,对任意$\tilde{\varphi} \in E_G(R_\varepsilon, T_\theta)$使得$\tilde{\iota}_T \circ \tilde{\varphi} \neq \tilde{\iota}_R$,从推论 11.9 可得到

Remark 11.11. Assume that k is algebraically closed. In general, when dealing with Corollary 11.9, we need not such a high precision. Assume that T_θ and R_ε are local pointed groups on $\mathcal{O}Gb$ such that $|T| = |R|$ and $T_\theta \subset P_\gamma \supset R_\varepsilon$, and denote by $\iota_T: T \to P$ and $\iota_R: R \to P$ the inclusion maps; it is easy to check that, for any $\tilde{\varphi} \in E_G(R_\varepsilon, T_\theta)$ such that $\tilde{\iota}_T \circ \tilde{\varphi} \neq \tilde{\iota}_R$, from Corollary 11.9 we get

11.11.1
$$\tilde{\iota}_T \circ \tilde{\varphi} - \tilde{\iota}_R = \sum_{h=0}^{n} \tilde{\iota}_h \circ (\tilde{\sigma}_h - \widetilde{\mathrm{id}}_{Q^h}) \circ \tilde{\nu}_h$$

其中$n \in \mathbb{N}$而任意$Q^h_{\delta^h}$是$\mathcal{O}Gb$上的自中心化的点群满足$Q^h_{\delta^h} \subset P_\gamma$与$\iota_h: Q^h \to P$是包含映射,$\tilde{\sigma}_h$是$E_G(Q^h_{\delta^h})$的$p'$-元素与$\tilde{\nu}_h \in E_G(R_\varepsilon, Q^h_{\delta^h})$;对后面的应用,令$\|\tilde{\varphi}\|$记满足等式 11.11.1 的极小的$n$.

where $n \in \mathbb{N}$ and any $Q^h_{\delta^h}$ is a self-centralizing local pointed group on $\mathcal{O}Gb$ such that $Q^h_{\delta^h} \subset P_\gamma$, $\iota_h: Q^h \to P$ is the inclusion map, $\tilde{\sigma}_h$ is a p'-element of $E_G(Q^h_{\delta^h})$ and $\tilde{\nu}_h \in E_G(R_\varepsilon, Q^h_{\delta^h})$; in order to be employed later, denote by $\|\tilde{\varphi}\|$ the smallest n appearing in equalities 11.11.1.

更精确地，由引理11.10，能把表达式11.11.1的形势选择和项使得有 $R_\varepsilon \subset Q^0_{\delta^0}$，$\nu_0 : R \to U^0$ 就是包含映射，而

More precisely, it follows from Lemma 11.10 that, in the expression 11.11.1, we can do our choice in such a way that $Q^h_{\delta^h} \subset P_\gamma$, that $\nu_0 : R \to U^0$ is the inclusion map, and that we have

11.11.2
$$\tilde\iota_{h-1} \circ \tilde\sigma_{h-1} \circ \tilde\nu_{h-1} = \tilde\iota_h \circ \tilde\nu_h$$
$$\tilde\iota_n \circ \tilde\sigma_n \circ \tilde\nu_n = \tilde\iota_T \circ \tilde\varphi$$

其中$1 \le h \le n$.最后，对任意$0 \le h \le n$ 显然可以提升 $\tilde\sigma_h$ 到一个融$_{\mathcal{O}G}(Q^h_{\delta^h})$的 p'-元素 σ_h，还可以一步步地提升 $\tilde\nu_h$ 到被 G 的 p'-元素决定的元素 $\nu_h \in$ 融$_{\mathcal{O}G}(R_\varepsilon, Q^h_{\delta^h})$（见定理10.19）使得对任意$1 \le h \le n$ 与适当的$u \in P$ 满足

for any $1 \le h \le n$. Finally, for any $0 \le h \le n$, we clearly can lift $\tilde\sigma_h$ to a p'-element σ_h of 融$_{\mathcal{O}G}(Q^h_{\delta^h})$ and, step by step, we can lift $\tilde\nu_h$ to an element $\nu_h \in$ 融$_{\mathcal{O}G}(R_\varepsilon, Q^h_{\delta^h})$ induced by a p'-element of G (see Theorem 10.19), in such a way that, for any $1 \le h \le n$ and a suitable $u \in P$, we get

11.11.3
$$\iota_{h-1} \circ \sigma_{h-1} \circ \nu_{h-1} = \iota_h \circ \nu_h$$
$$\iota_n \circ \sigma_n \circ \nu_n = \kappa^P_u \circ \iota_T \circ \varphi .$$

12. G-块的源代数

12. Source Algebras of Blocks

12.1. 设 G 是一个有限群并设 b 是一个 G-块; 令 $\alpha = \{b\}$ 记对应的 $\mathcal{O}G$ 上的 G-点, 并选择 G_α 的一个亏点群 P_γ. 我们这里考虑 $\mathcal{O}Gb$ 的源代数 $(\mathcal{O}G)_\gamma$; 这个 P-内代数是与 G-块 b 相关的最重要的结构. 已经知道 $\mathcal{O}Gb$ 和 $(\mathcal{O}G)_\gamma$ 是 Morita 等价的 (见 6.10); 事实上, 源代数还决定与块通常相关的其它结构. 我们只就融合的情况作个说明.

12.1. Let G be a finite group and b a block of G; denote by $\alpha = \{b\}$ the corresponding point of G on $\mathcal{O}G$, and choose a defect pointed group P_γ of G_α. In this section, we consider the source algebra $(\mathcal{O}G)_\gamma$ of $\mathcal{O}Gb$; this P-interior algebra is the most important structure associated with the block b of G. We already know that $\mathcal{O}Gb$ and $(\mathcal{O}G)_\gamma$ are Morita equivalent (see 6.10); actually, the source algebra determines all the current invariants associated with the block. We only explain it for the fusions.

命题 12.2. **对 $\mathcal{O}G$ 上的任意局部点群 Q_δ 与 R_ε, 都包含在 P_γ 里, 有**

Proposition 12.2. *For any local pointed groups Q_δ and R_ε on $\mathcal{O}G$ which are contained in P_γ, we have*

12.2.1 $$E_G(R_\varepsilon, Q_\delta) = F_{(\mathcal{O}G)_\gamma}(R_\varepsilon, Q_\delta).$$

证明: 只要应用推论 8.19 和定理 10.19 就完成证明.

Proof: It suffices to apply Corollary 8.19 and Theorem 10.19 to get the proof.

12.3. 选择 $i \in \gamma$ 并假定 $(\mathcal{O}G)_\gamma = i(\mathcal{O}G)i = B$; 从命题 12.2 和定理 8.20, 不难得到下面的正合序列

12.3. Choose $i \in \gamma$ and assume that $(\mathcal{O}G)_\gamma = i(\mathcal{O}G)i = B$; from Proposition 12.2 and Theorem 8.20, it is not difficult to obtain the following exact sequence

12.3.1 $$1 \longrightarrow P{\cdot}(B^P)^* \longrightarrow N_{B^*}(Pi) \longrightarrow E_G(P_\gamma) \longrightarrow 1.$$

如果 k 是代数闭的, 那么定理 10.14 推出 $E_G(P_\gamma)$ 是 p'-群; 也知道 $B^P/J(B^P) \cong k$, 从而

If k is algebraically closed, then Theorem 10.14 implies that $E_G(P_\gamma)$ is a p'-group; we also know that $B^P/J(B^P) \cong k$ and therefore

12.3.2 $$(B^P)^*/J(B^P)^* \cong k^*$$

其中 $J(B^P)^* = i + J(B^P)$ (见 3.2). 请注意, $P{\cdot}B^P$ 是 B 的子代数 并且, 由引理 7.3, 幂等元 i 在 $P{\cdot}B^P$ 中也是本原的, 而

where we set $J(B^P)^* = i + J(B^P)$ (see 3.2). Note that $P{\cdot}B^P$ is a subalgebra of B and, according to Lemma 7.3, the idempotent i is also primitive in $P{\cdot}B^P$ and we have

12.3.3 $$B^P/J(B^P) \cong P{\cdot}B^P/J(P{\cdot}B^P);$$

此时，不难验证

then, it is easily checked that

12.3.4 $$J(P{\cdot}B^P) = P{\cdot}J(B^P)\,, \quad J(P{\cdot}B^P)^* = P{\cdot}J(B^P)^*\,.$$

12.4. 令 $\bar{B} = k \otimes_{\mathcal{O}} B$ 并设 $\bar{M}$ 是一个非零 $\bar{B}$-模；已经知道，从 $\mathrm{End}_k(\bar{M})$ 到 k 的行列式映射决定下面的群同态并 B^* 的子群

12.4. Set $\bar{B} = k \otimes_{\mathcal{O}} B$ and let $\bar{M}$ be a nonzero $\bar{B}$-module; it is well-known that the *determinant map* from $\mathrm{End}_k(\bar{M})$ to k determines the following group homomorphism and the following subgroup of B^*

12.4.1 $$\det_{\bar{M}} : B^* \longrightarrow k^*\,, \quad K = \mathrm{Ker}(\det_{\bar{M}})\,.$$

请注意，因为对适当的 $n \in \mathbb{N}$ 有 $J(B^P)^{p^n} \subset J(\mathcal{O}){\cdot}B^P$，所以还有

Note that, since $J(B^P)^{p^n} \subset J(\mathcal{O}){\cdot}B^P$ for a suitable $n \in \mathbb{N}$, we have

12.4.2 $$P{\cdot}J(B^P)^* \subset K\,.$$

令 U 记 k^* 的唯一一个阶为 $\dim_k(\bar{M})_{p'}$ 的子群；因为显然 $\det_{\bar{M}}(\lambda{\cdot}i) = \bar{\lambda}^{\dim_k(\bar{M})}$，其中 $\lambda \in \mathcal{O}$ 与 $\bar{\lambda}$ 是 λ 在 k 中的像，所以，如果 k 是代数闭的，那么不难验证 $\det_{\bar{M}}$ 决定正合序列如下

Denote by U the unique subgroup of k^* of order $\dim_k(\bar{M})_{p'}$; since we evidently have $\det_{\bar{M}}(\lambda{\cdot}i) = \bar{\lambda}^{\dim_k(\bar{M})}$, where $\lambda \in \mathcal{O}$ and $\bar{\lambda}$ is the image of λ in k, if k is algebraically closed then it is not difficult to see that $\det_{\bar{M}}$ determines an exact sequence as follows

12.4.3 $$1 \longrightarrow \widehat{E_G(P_\gamma)} \longrightarrow N_{B^*}(Pi)/P{\cdot}J(B^P)^* \longrightarrow k^* \longrightarrow 1\,,$$

其中

where

12.4.4 $$\widehat{E_G(P_\gamma)} = N_K(Pi)/P{\cdot}J(B^P)^*$$

是 $E_G(P_\gamma)$ 上 U 的中心扩张；事实上，由引理如下，这个商可分裂.

is a central extension of $E_G(P_\gamma)$ by U; actually, according to the next lemma, this quotient splits.

引理 12.5. 设 B 是 $\mathcal{O}$-代数并设 D 与 E 是 B 的 $\mathcal{O}$-子代数；令 $N_{E^*}(D)$ 记 D 在 E^* 中的稳定化子，再令 $J^* = J(D)^* \cap E$. 对任意有限 p'-群 S 下面的自然映射是满射

Lemma 12.5. *Let B be an $\mathcal{O}$-algebra and consider two $\mathcal{O}$-subalgebras D and E of B; denote by $N_{E^*}(D)$ the stabilizer of D in E^*, and set $J^* = J(D)^* \cap E$. For any finite p'-group S, the following natural map is surjective*

12.5.1 $$\mathrm{Hom}\big(S, N_{E^*}(D)\big) \longrightarrow \mathrm{Hom}\big(S, N_{E^*}(D)/J^*\big)\,.$$

而且，对 $\mathrm{Hom}\big(S, N_{E^*}(D)/J^*\big)$ 的任意元素 σ，J^* 在 σ 的全原像上可迁地作用.

Moreover, for any σ of $\mathrm{Hom}\big(S, N_{E^*}(D)/J^*\big)$, J^* acts transitively on the converse image of σ.

证明: 设 $n \in \mathbb{N} - \{0\}$; 首先, 要证明: 如果 $\{a_s\}_{s\in S} \subset N_{E^*}(D)$ 满足对任意 $s, s' \in S$ 有

Proof: Fix $n \in \mathbb{N} - \{0\}$; first of all, we prove that if $\{a_s\}_{s\in S} \subset N_{E^*}(D)$ is such that, for any $s, s' \in S$, we have

12.5.2 $$a_s a_{s'} \in a_{ss'}(1 + J^n) = (1 + J^n)a_{ss'}$$

那么对任意 $s \in S$ 存在一个 $a_s(1 + J^n)$ 的元素 a'_s 使得对任意 $s, s' \in S$ 有

then, for any $s \in S$ there exists an element a'_s of $a_s(1 + J^n)$ such that, for any $s, s' \in S$, we have

12.5.3 $$a'_s a'_{s'} \in a'_{ss'}(1 + J^{n+1});$$

确实, 令

indeed, setting

12.5.4 $$z(s, s') = a_s a_{s'}(a_{ss'})^{-1} - 1, \quad \overline{z(s, s')} = z(s, s') + J^{n+1}$$

只要使用结合律就不难验证

and using the additive structure, we get

12.5.5 $$\overline{a_s z(s', s'')(a_s)^{-1}} - \overline{z(ss', s'')} + \overline{z(s, s's'')} - \overline{z(s, s')} = 0$$

其中 $s, s', s'' \in S$; 特别是, 如果 s'' 跑遍 S, 把所有这些等式加起来的和式产生下式

where $s, s', s'' \in S$; in particular, the sum of these equalities when s'' runs over S implies the following formula

12.5.6 $$\overline{z(s, s')} = \overline{w(s)} - \overline{w(ss')} + \overline{a_s w(s')(a_s)^{-1}}$$

其中, 对任意 $s \in S$ 记

where, for any $s \in S$, we set

12.5.7 $$w(s) = 1/|S| \cdot \sum_{s'' \in S} z(s, s'');$$

此时, 不难验证对任意 $s \in S$, $a'_s = \big(1 - w(s)\big)a_s$ 即为所求.

at this point, it is easy to check that, for any $s \in S$, $a'_s = \big(1 - w(s)\big)a_s$ fulfills the claim.

这样, 对 $N_{E^*}(D)$ 的任意子集合 $\{a_s\}_{s\in S}$ 使得 $a_s a_{s'}$ 属于 $a_{ss'}(1 + J)$, 其中 $s, s' \in S$, 可以得到 $a_{s,n} \in N_{E^*}(D)$ 使得

Thus, for any subset $\{a_s\}_{s\in S}$ of $N_{E^*}(D)$ such that $a_s a_{s'}$ belongs to $a_{ss'}(1 + J)$ for any $s, s' \in S$, we can find $a_{s,n} \in N_{E^*}(D)$ fulfilling

12.5.8 $$\begin{gathered} a_{s,0} = a_s, \quad a_{s,n+1} \in a_{s,n}(1 + J^{n+1}), \\ a_{s,n} a_{s',n} \in a_{ss',n}(1 + J^{n+1}), \end{gathered}$$

其中 $s, s' \in S$ 与 $n \in \mathbb{N}$; 那么, 由完备性, 只要考虑 $D \cap E$ 的序列 $\{(a_s)^{-1} a_{s,n}\}_{n\in\mathbb{N}}$, 就得到 $b_s \in N_{E^*}(D)$ 满足

for any $s, s' \in S$ and any $n \in \mathbb{N}$; then, by completeness, it suffices to consider the sequence $\{(a_s)^{-1} a_{s,n}\}_{n\in\mathbb{N}}$ in $D \cap E$, to obtain $b_s \in N_{E^*}(D)$ fulfilling

12.5.9 $$b_s \in a_{s,n}(1 + J^{n+1}), \quad b_s b_{s'} \in b_{ss'}(1 + J^{n+1})$$

其中 $s, s' \in S$ 与 $n \in \mathbb{N}$，从而有 $b_s b_{s'} = b_{ss'}$.

而且，设 $\{b'_s\}_{s\in S}$ 是对任意 $s, s' \in S$ 满足 $b'_s \in a_s(1+J)$ 与 $b'_s b'_{s'} = b'_{ss'}$ 的 $N_{E^*}(D)$ 的子集合；类似地，设 $n\in\mathbb{N}-\{0\}$；首先，要证明：如果 $d \in J$ 满足对任意 $s \in S$ 有

for any $s, s' \in S$ and any $n \in \mathbb{N}$ and therefore we get $b_s b_{s'} = b_{ss'}$.

Moreover, let $\{b'_s\}_{s\in S}$ be a subset of $N_{E^*}(D)$ such that we have $b'_s \in a_s(1+J)$ and $b'_s b'_{s'} = b'_{ss'}$ for any $s, s' \in S$; similarly, fix $n \in \mathbb{N}-\{0\}$; first of all, we prove that if $d \in J$ is such that, for any $s \in S$, we have

12.5.10 $$b'_s \in (b_s)^{1+d}(1+J^n)$$

那么存在 $d' \in d + J^n$ 使得

then there exists $d' \in d + j^n$ fulfilling

12.5.11 $$b'_s \in (b_s)^{1+d'}(1+J^{n+1})$$

其中 $s \in S$；确实，令

for any $s \in S$; indeed, setting

12.5.12 $$y(s) = (b_s)^{1+d}(b'_s)^{-1} - 1\,, \quad \overline{y(s)} = y(s) + J^{n+1}\,,$$

只要使用等式 $b_s b_{s'} = b_{ss'}$ 就不难验证

and employing the equalities $b_s b_{s'} = b_{ss'}$, we easily get

12.5.13 $$\overline{b'_s y(s')(b'_s)^{-1}} - \overline{y(ss')} + \overline{y(s)} = 0$$

其中 $s, s' \in S$；特别是，如果 s' 跑遍 S，把所有这些式加起来的和式产生下式

for any $s, s' \in S$; in particular, the sum of these equalities when s' runs over S produces the following formula

12.5.14 $$\overline{y(s)} = \overline{c} - \overline{b'_s c\,(b'_s)^{-1}}$$

其中

where we set

12.5.15 $$c = 1/|S|\cdot \sum_{s'\in S} y(s')\,;$$

此时，不难验证 $d' = d + c$ 即为所求.

at this point, it is easy to check that the element $d' = d + c$ fulfills the claim.

这样，对任意 $d \in J$ 使得 $b'_s \in (b_s)^{1+d}(1+J)$，其中 $s\in S$ 我们可以得到 $d_n \in J$ 使得

Consequently, for any $d \in J$ fulfilling $b'_s \in (b_s)^{1+d}(1+J)$ for any $s \in S$, we can find $d_n \in J$ such that

12.5.16 $$d_0 = d\,, \quad d_{n+1} \in d_n + J^{n+1}\,, \qquad b'_s \in (b_s)^{1+d_n}(1+J^{n+1})$$

其中 $s \in S$ 与 $n \in \mathbb{N}$；那么，由完备性，只要考虑 $D \cap E$ 的序列 $\{d_n\}_{n\in\mathbb{N}}$，就得到 $e \in D \cap E$ 使得 $e \in d_n + J^{n+1}$，其中 $n \in \mathbb{N}$;

for any $s \in S$ and any $n \in \mathbb{N}$; then, by completeness, it suffices to consider the sequence $\{d_n\}_{n\in\mathbb{N}}$ in $D \cap E$, to obtain $e \in D \cap E$ fulfilling $e \in d_n + J^{n+1}$ for any $n \in \mathbb{N}$;

所以, 也得到

hence, we also obtain

12.5.17 $$b'_s \in (b_s)^{1+e}(1+J^{n+1})$$

其中 $s \in S$ 与 $n \in \mathbb{N}$, 从而有 $b'_s = (b_s)^{1+e}$.

for any $s \in S$ and any $n \in \mathbb{N}$ and therefore we get $b'_s = (b_s)^{1+e}$.

12.6. 假定 k 是代数闭的. 那么有 幂$^{p^{\aleph}}(k) = k$ (见定义 2.13.1); 这样, 由定理 2.13, 自然的同态 $\mathcal{O} \to k$ 决定一个群同构

12.6. Assume that k is algebraically closed. Then 幂$^{p^{\aleph}}(k) = k$ (see definition 2.13.1); thus by Theorem 2.13, the natural homomorphism $\mathcal{O} \to k$ induces a group isomorphism

12.6.1 $$幂^{p^{\aleph}}(\mathcal{O}) - \{0\} \cong k^*;$$

也就是说, 存在一个自然的群同态 $k^* \to \mathcal{O}^*$, 这样我们也可以把任意 $\mathcal{O}$-代数看作 k^*-内代数 (当然这里 k^* 是无限群). 另一方面, 由命题 12.2, 商

that is to say, there is a natural group homomorphism $k^* \to \mathcal{O}^*$, thus any $\mathcal{O}$-algebra can be considered as a k^*-interior algebra (obviously, here k^* is an infinite group). On the other hand, by Proposition 12.2, the quotient

12.6.2 $$\hat{E}_G(P_\gamma) = N_{B^*}(Pi)/P{\cdot}J(B^P)^*$$

是 $E_G(P_\gamma)$ 上 k^* 的中心扩张 (见 12.3); 请注意, $\hat{E}_G(P_\gamma)$ 包含商 $\widehat{E_G(P_\gamma)}$ (见 12.4.4) 并且不难验证

is a central extension of $E_G(P_\gamma)$ by k^* (see 12.3); note that $\hat{E}_G(P_\gamma)$ contains the quotient $\widehat{E_G(P_\gamma)}$ (see 12.4.4) and it is easily checked that

12.6.3 $$\hat{E}_G(P_\gamma) = k^*{\cdot}\widehat{E_G(P_\gamma)}, \quad k^* \cap \widehat{E_G(P_\gamma)} = U.$$

12.7. 进一步, 因为 $E_G(P_\gamma)$ 是一个 p'-群 (见定理 10.14) 所以, 由后面的引理 14.10, 下面的正合序列可分裂

12.7. Furthermore, since $E_G(P_\gamma)$ is a p'-group (see Theorem 10.14), it follows from Lemma 14.10 in section 14 that the following exact sequence splits

12.7.1 $$1 \longrightarrow P/Z(P) \longrightarrow N_G(P_\gamma)/C_G(P) \longrightarrow E_G(P_\gamma) \longrightarrow 1$$

并且提升同态都共轭. 选择一个从 $E_G(P_\gamma)$ 到 $\mathrm{Aut}(P)$ 的提升同态 σ 并且考虑 $\mathcal{O}P$ 的 k^*-内 $\hat{E}_G(P_\gamma)$-代数的对应的结构; 那么,下面的 P-内代数 (见等式 9.1.1)的同构类不依赖 σ 的选择

and that all the sections are conjugate. Choose a section σ from $E_G(P_\gamma)$ to $\mathrm{Aut}(P)$ and consider the corresponding k^*-interior $\hat{E}_G(P_\gamma)$-algebra structure of $\mathcal{O}P$; then, the isomorphism class of the following P-interior algebra does not depend on the choice of σ

12.7.2 $$\mathcal{O}P \otimes_{k^*} \hat{E}_G(P_\gamma) = \mathcal{O}P \otimes_U \widehat{E_G(P_\gamma)}.$$

定理 12.8. 假定 k 是代数闭的. 存在 B 的幺子 $\mathcal{O}$-代数 D 的唯一一个 $(B^P)^*$-共轭类使得既 D 包含 P 的像也有下面的 P-内代数同构

Theorem 12.8. *Assume that k is algebraically closed. There exists a unique $(B^P)^*$-conjugation class of unitary $\mathcal{O}$-subalgebras D of B such that D contains the image of P and we have the P-interior algebra isomorphism*

12.8.1 $$D \cong \mathcal{O}P \otimes_{k^*} \hat{E}_G(P_\gamma)\,.$$

而且 B 有一个 D 的 $D\otimes_{\mathcal{O}}D^\circ$-模补 M, 并 $p|P|$ 整除 B/D 的任意 $\mathcal{O}(P\times P)$-模直和项 L 的 $\mathcal{O}$-秩. 特别是

Moreover, D has a complement M in B as $D\otimes_{\mathcal{O}} D^\circ$-modules and $p|P|$ divises the $\mathcal{O}$-rank of any direct summand L of B/D as $\mathcal{O}(P\times P)$-module. In particular

12.8.2 $$\frac{\mathrm{rank}_{\mathcal{O}}(B)}{|P|} \equiv |E_G(P_\gamma)| \pmod{p}\,.$$

证明: 由引理 10.22, 存在一个 B 的 $\mathcal{O}$-基 T 使得 $P{\cdot}T{\cdot}P = T$ 并对任意 $t\in T$ 有 $|Pt| = |P|$, 并且有 $|PtP| = |P|$ 当且仅当存在 $x\in N_G(P)$ 使得

Proof: According to Lemma 10.22, there is an $\mathcal{O}$-basis T of B such that $P{\cdot}T{\cdot}P = T$ and $|Pt| = |P|$ for any $t\in T$, and that we have $|PtP| = |P|$ if and only if there exists $x\in N_G(P)$ fulfilling

12.8.3 $$(\mathcal{O}P){\cdot}t = \mathcal{O}(PtP) \cong \mathrm{Ind}_{\Delta_x(P)}^{P\times P}(\mathcal{O})\,;$$

如果 $t\in T$ 满足这个条件, 那么 $(\mathcal{O}P){\cdot}tx^{-1}$ 在 $\mathcal{O}G$ 里是一个满足 $(\mathcal{O}P){\cdot}tx^{-1}\cong \mathcal{O}P$ 的 $\mathcal{O}(P\times P)$-模直和项; 这样, 因为 $(\mathcal{O}P)(P)\cong kZ(P)$, 所以对适当的 $u\in P$ 有

if an element $t\in T$ fulfills this condition, then $(\mathcal{O}P){\cdot}tx^{-1}$ is a direct summand of $\mathcal{O}G$ as $\mathcal{O}(P\times P)$-modules, which fulfills $(\mathcal{O}P){\cdot}tx^{-1}\cong \mathcal{O}P$; thus, since we have $(\mathcal{O}P)(P)\cong kZ(P)$, for a suitable $u\in P$, we get

12.8.4 $$tux^{-1}\in(\mathcal{O}G)^P\,,\quad \mathrm{Br}_P(tux^{-1})\neq 0\,;$$

既然 $itux^{-1}i^{x^{-1}} = tux^{-1}$, 存在一个 $C_G(P)$-块 e 使得

as far as we have $itux^{-1}i^{x^{-1}} = tux^{-1}$, there exists a block e of $C_G(P)$ fulfilling

12.8.5 $$e\mathrm{Br}_P(i) = \mathrm{Br}_P(i)\,,\quad e\mathrm{Br}_P(i^{x^{-1}}) = \mathrm{Br}_P(i^{x^{-1}})\,.$$

现在, 只要使用同构 10.14.3 就得到 x 属于 $N_G(P_\gamma)$, 从而 x 决定一个 $E_G(P_\gamma)$ 的元素.

Now, it follows from isomorphism 10.14.3 that x belongs to $N_G(P_\gamma)$ and therefore x determines an element of $E_G(P_\gamma)$.

另一方面, 只要应用引理 12.5 于 B, 于其子代数 $P{\cdot}B^P$,

On the other hand, by applying Lemma 12.5 to B, to its subalgebra $P{\cdot}B^P$ and to

于有限 p'-群 $\widehat{E_G(P_\gamma)}$，就得到一个群同态

the finite p'-group $\widehat{E_G(P_\gamma)}$, we get a group homomorphism

12.8.6 $$\widehat{\sigma}\colon \widehat{E_G(P_\gamma)} \longrightarrow N_{B^*}(P{\cdot}i)$$

使得 $\widehat{\sigma}$ 提升从 $\widehat{E_G(P_\gamma)}$ 到 $E_G(P_\gamma)$ 的自然同态 (见 12.3.1)；特别是，$\widehat{\sigma}(U) \subset \mathcal{O}^*{\cdot}P{\cdot}J(B^P)^*$，从而不难验证 $\widehat{\sigma}(U) \subset 幂^{p^{\aleph}}(\mathcal{O})$；也就是说，从式 12.6.3 可得到一个群同态

which lifts the natural homomorphism from $\widehat{E_G(P_\gamma)}$ to $E_G(P_\gamma)$ (see 12.3.1); in particular, we have $\widehat{\sigma}(U) \subset \mathcal{O}^*{\cdot}P{\cdot}J(B^P)^*$ and then it is not difficult to check that $\widehat{\sigma}(U) \subset 幂^{p^{\aleph}}(\mathcal{O})$; that is to say, from equalities 12.6.3, we obtain a group homomorphism

12.8.7 $$\hat{\sigma}\colon \hat{E}_G(P_\gamma) \longrightarrow N_{B^*}(P{\cdot}i) \subset B^*$$

使得包含映射 $\mathcal{O}P \to B$ 成为 k^*-内 $\hat{E}_G(P_\gamma)$-代数同态；这样，$\hat{\sigma}$ 决定一个 P-内代数同态如下

such that the inclusion map $\mathcal{O}P \to B$ becomes a k^*-interior $\hat{E}_G(P_\gamma)$-algebra homomorphism; thus, $\hat{\sigma}$ induces a P-interior algebra homomorphism

12.8.8 $$s : \mathcal{O}P \otimes_{k^*} \hat{E}_G(P_\gamma) \longrightarrow B$$

但是，由引理10.24，$(\mathcal{O}P){\cdot}i$ ($\cong \mathcal{O}P$) 是 B 的 $\mathcal{O}(P\times P)$-模直和项；事实上，能假定 $i \in T$；而且，因为 (见同构 10.14.3)

But, by Lemma 10.24, $(\mathcal{O}P){\cdot}i$ ($\cong \mathcal{O}P$) is a direct summand of B as $\mathcal{O}(P\times P)$-modules; actually, we may assume that $i \in T$; moreover, since (see isomorphism 10.14.3)

12.8.9 $$k \otimes_{kZ(P)} kC_G(P)e \cong (\mathcal{O}G)(P_\gamma)$$

并且 $\mathrm{Br}_P(i)$ 是本原的，所以有

and $\mathrm{Br}_P(i)$ is primitive, we have

12.8.10 $$k \otimes_{kZ(P)} B(P) \cong k \cong k \otimes_{kZ(P)} (\mathcal{O}P)(P);$$

也就是说，$\mathcal{O}(P\times P)$-模 $\mathcal{O}P$ 在 B 里的重数就是 1. 又，对任意 $\hat{\varepsilon} \in \hat{E}_G(P_\gamma)$，$\mathcal{O}(P\times P)$-模 $(\mathcal{O}P){\cdot}\hat{\sigma}(\hat{\varepsilon})$ 显然是 B 的直和项；特别是，存在 $t_{\hat{\varepsilon}} \in T$ 使得

that is to say, the $\mathcal{O}(P\times P)$-module $\mathcal{O}P$ has multiplicity 1 in B. Furthermore, for any $\hat{\varepsilon} \in \hat{E}_G(P_\gamma)$, the $\mathcal{O}(P\times P)$-module $(\mathcal{O}P){\cdot}\hat{\sigma}(\hat{\varepsilon})$ certainly is a direct summand of B too; in particular, exists $t_{\hat{\varepsilon}} \in T$ fulfilling

12.8.11 $$(\mathcal{O}P){\cdot}t_{\hat{\varepsilon}} = \mathcal{O}(Pt_{\hat{\varepsilon}}P) \cong (\mathcal{O}P){\cdot}\hat{\sigma}(\hat{\varepsilon}),$$

从而 $\mathcal{O}(P\times P)$-模 $(\mathcal{O}P){\cdot}t_{\hat{\varepsilon}}\hat{\sigma}(\hat{\varepsilon})^{-1}$ 既是 B 的直和项也是同构于 $\mathcal{O}P$；所以 $\mathcal{O}(P\times P)$-模 $(\mathcal{O}P){\cdot}\hat{\sigma}(\hat{\varepsilon})$ 在 B 里的重数也是 1.

and thus the $\mathcal{O}(P\times P)$-module $(\mathcal{O}P){\cdot}t_{\hat{\varepsilon}}\hat{\sigma}(\hat{\varepsilon})^{-1}$ is a direct summand of B and is isomorphic to $\mathcal{O}P$; consequently, the $\mathcal{O}(P\times P)$-module $(\mathcal{O}P){\cdot}\hat{\sigma}(\hat{\varepsilon})$ also has multiplicity 1 in B.

最后, 令 $\hat{\mathcal{E}} \subset \hat{E}_G(P_\gamma)$ 记一个 $E_G(P_\gamma)$ 的代表集合, 再令 $D = \mathrm{Im}(s)$; 不难验证 $\{(\mathcal{O}P)\cdot\hat\sigma(\hat\varepsilon)\}_{\hat\varepsilon\in\hat{\mathcal{E}}}$ 是相互不同构的 $\mathcal{O}(P\times P)$-模; 所以, $\hat{\mathcal{E}}$ 与 $\{t_{\hat\varepsilon}\}_{\hat\varepsilon\in\hat{\mathcal{E}}}$ 的基数一样. 更精确地, 对任意 $\hat\varepsilon\in\hat{\mathcal{E}}$, $k^*\cdot t_{\hat\varepsilon}$ 与 $k^*\cdot\hat\sigma(\hat\varepsilon)$ 在 $\tilde B = B/J\big(\mathcal{O}(P\times P)\big)\cdot B$ 中的像一样; 确实, 我们使用反证法; 假定存在 $\hat\varepsilon\in\hat{\mathcal{E}}$ 使得 $k^*\cdot t_{\hat\varepsilon}$ 与 $k^*\cdot\hat\sigma(\hat\varepsilon)$ 在 $\tilde B$ 中的像不一样; 那么存在 $t\in T$ 满足

Finally, denote by $\hat{\mathcal{E}} \subset \hat{E}_G(P_\gamma)$ a set of representatives for $E_G(P_\gamma)$, and set $D = \mathrm{Im}(s)$; it is not difficult to check that the $\mathcal{O}(P\times P)$-modules $\{(\mathcal{O}P)\cdot\hat\sigma(\hat\varepsilon)\}_{\hat\varepsilon\in\hat{\mathcal{E}}}$ are mutually not isomorphic; hence, the cardinals of $\hat{\mathcal{E}}$ and $\{t_{\hat\varepsilon}\}_{\hat\varepsilon\in\hat{\mathcal{E}}}$ coincide. More precisely, for any $\hat\varepsilon\in\hat{\mathcal{E}}$, the images of $k^*\cdot t_{\hat\varepsilon}$ and $k^*\cdot\hat\sigma(\hat\varepsilon)$ in $\tilde B = B/J\big(\mathcal{O}(P\times P)\big)\cdot B$ coincide; indeed, we argue by contradiction; assume that there is $\hat\varepsilon\subset\hat{\mathcal{E}}$ such that the image of $k^*\cdot t_{\hat\varepsilon}$ and $k^*\cdot\hat\sigma(\hat\varepsilon)$ in $\tilde B$ are not the same; then, there exists $t\in T$ fulfilling

12.8.12 $$\tilde B = k\cdot\tilde t_{\hat\varepsilon} \oplus k\cdot\widetilde{\hat\sigma(\hat\varepsilon)} \oplus \Big(\bigoplus_{\tilde t'} k\cdot\tilde t'\Big)$$

其中 $\tilde t'$ 跑遍 $T-\{t,t_{\hat\varepsilon}\}$ 在 $\tilde B$ 中的像; 此时, 由 Nakayama 引理, 不难得到矛盾.

where $\tilde t'$ runs over the image of $T-\{t,t_{\hat\varepsilon}\}$ in $\tilde B$; at this point, it is not difficult to get a contradiction from the Nakayama Lemma.

那么, 从 Nakayama 引理, 我们能假定 $\hat\sigma(\hat{\mathcal{E}})\subset T$, 从而有

Then, from the Nakayama Lemma, we may assume that $\hat\sigma(\hat{\mathcal{E}})\subset T$, and therefore

12.8.13 $$D \cong \mathcal{O}P\otimes_{k^*}\hat{E}_G(P_\gamma)\,,\quad B\cong D\oplus M\,,$$

这里 $M = \bigoplus_t \mathcal{O}t$, 其中 t 跑遍 $T-\hat\sigma(\hat{\mathcal{E}})$; 这样, M 是 B 的子 $\mathcal{O}(P\times P)$-模, 而 $p|P|$ 整除 M 的任意 $\mathcal{O}(P\times P)$-模直和项 L 的 $\mathcal{O}$-秩; 特别是, $P\times P$ 稳定 $\mathrm{End}_\mathcal{O}(B)$ 的幂等元 f 使得 $f(B)=D$ 与 $(\mathrm{id}_B - f)(B)=M$. 现在, 考虑

for $M = \bigoplus_t \mathcal{O}t$ where t runs over $T-\hat\sigma(\hat{\mathcal{E}})$; thus, M is an $\mathcal{O}(P\times P)$-submodule of B, and $p|P|$ divises the $\mathcal{O}$-rank of any direct summand L of M as $\mathcal{O}(P\times P)$-modules; in particular, $P\times P$ stabilizes an idempotent f of $\mathrm{End}_\mathcal{O}(B)$ such that $f(B)=D$ and $(\mathrm{id}_B - f)(B) = M$. Now, consider

12.8.14 $$g = \frac{1}{|\hat{\mathcal{E}}|^2}\sum_{\hat\varepsilon,\hat\varepsilon'\in\hat{\mathcal{E}}} h_{\hat\sigma(\hat\varepsilon),\hat\sigma(\hat\varepsilon')}\circ f\circ\big(h_{\hat\sigma(\hat\varepsilon),\hat\sigma(\hat\varepsilon')}\big)^{-1}$$

其中 $h_{\hat\sigma(\hat\varepsilon),\hat\sigma(\hat\varepsilon')}$ 是 $\mathrm{End}_\mathcal{O}(B)$ 的元素使得对任意 $a\in B$ 有

where $h_{\hat\sigma(\hat\varepsilon),\hat\sigma(\hat\varepsilon')}$ is an element of $\mathrm{End}_\mathcal{O}(B)$ which, for any $a\in B$, fulfills

12.8.15 $$h_{\hat\sigma(\hat\varepsilon),\hat\sigma(\hat\varepsilon')}(a) = \hat\sigma(\hat\varepsilon)\,a\hat\sigma(\hat\varepsilon')^{-1}\,;$$

不难验证 g 是 $D\otimes_\mathcal{O} D^\circ$-模自同态; 而且, 对任意 $d\in D$ 有 $g(d)=d$ 并对任意 $a\in B$ 还

it is easy to check that g is a $D\otimes_\mathcal{O} D^\circ$-module endomorphism; moreover, for any $d\in D$, we have $g(d)=d$ and, for any $a\in B$, we get

有 $g(a+D)=D$, 从而 $\mathrm{Ker}(g)$ 是 D 在 B 里的 $D\otimes_{\mathcal{O}}D^{\circ}$-模补.

$g(a+D)=D$, so that $\mathrm{Ker}(g)$ is complement of D in B as $D\otimes_{\mathcal{O}}D^{\circ}$-modules.

如果一个 B 的幺子 P-内代数 D' 满足同构式 12.8.1, 那么这个同构决定另一个群同态

If a unitary P-interior subalgebra D' of B fulfills isomorphism 12.8.1, then this isomorphism induces a group homomorphism

12.8.16 $$\hat{\sigma}'\colon \hat{E}_G(P_\gamma)\longrightarrow N_{B^*}(P{\cdot}i)$$

使得 $\hat{\sigma}$ 提升从 $\hat{E}_G(P_\gamma)$ 到 $E_G(P_\gamma)$ 的自然同态 (见 12.3.1); 只要考虑 $\hat{\sigma}'$ 在 $\widehat{E_G(P_\gamma)}$ 上的限制就能应用引理 12.5; 所以存在 $u\in P$ 与 $c\in J(B^P)^*$ 使得对任意 $\hat{\varepsilon}\in\hat{E}_G(P_\gamma)$ 有 $\hat{\sigma}'(\hat{\varepsilon})=\hat{\sigma}(\hat{\varepsilon})^{uc}$, 从而还有 $D'=D^c$.

which lifts the natural homomorphism from $\hat{E}_G(P_\gamma)$ to $E_G(P_\gamma)$ (see 12.3.1); thus, we can apply Lemma 12.5 to the restriction of $\hat{\sigma}'$ to $\widehat{E_G(P_\gamma)}$; consequently, there exist $u\in P$ and $c\in J(B^P)^*$ such that, for any $\hat{\varepsilon}\in\hat{E}_G(P_\gamma)$, we have $\hat{\sigma}'(\hat{\varepsilon})=\hat{\sigma}(\hat{\varepsilon})^{uc}$, and therefore we still have $D'=D^c$.

推论 12.9. 假定 k 是代数闭的. 如果 $P\lhd G$ 那么

Corollary 12.9. *Assume that k is algebraically closed. If $P\lhd G$ then we have*

12.9.1 $$B\cong\mathcal{O}P\otimes_{k^*}\hat{E}_G(P_\gamma)\,.$$

证明: 如果 $P\lhd G$ 那么 $\mathcal{O}G$ 的不可分解 $\mathcal{O}(P\times P)$-模直和项都是秩 $|P|$ 的; 所以 $D\otimes_{\mathcal{O}}D^{\circ}$-模 B/D 是零.

Proof: If $P\lhd G$ then all the indecomposable direct summands of $\mathcal{O}G$ as $\mathcal{O}(P\times P)$-module have $\mathcal{O}$-rank $|P|$; consequently, the $D\otimes_{\mathcal{O}}D^{\circ}$-module B/D is zero.

推论 12.10. 假定 k 是代数闭的. 对任意 P 的子群 Q 存在 $\mathcal{O}Gb$ 上的局部 Q-点 δ 使得 p 不整除 $\mathrm{m}_\delta^\gamma=\big(\mathrm{res}_Q^P(\gamma)\big)(\delta)$.

Corollary 12.10. *Assume that k is algebraically closed. For any subgroup Q of P there exists a local point δ of Q on $\mathcal{O}Gb$ such that p does not divide $\mathrm{m}_\delta^\gamma=\big(\mathrm{res}_Q^P(\gamma)\big)(\delta)$.*

证明: 设 J 是一个 B 的相互正交本原幂等元的集合使得 $\sum_{j\in J}j=i$; 那么, 不难验证

Proof: Let J be a set of pairwise orthogonal primitive idempotents of B fulfilling $\sum_{j\in J}j=i$; then, it is easy to check

12.10.1 $$\mathrm{rank}_{\mathcal{O}}(B)=\sum_{j\in J}\mathrm{rank}_{\mathcal{O}}(Bj)=\sum_{\rho\in\mathcal{P}_B(1)}\mathrm{m}_\rho^\gamma\,\mathrm{rank}_{\mathcal{O}}(Bj_\rho)\,,$$

其中 $j_\rho\in\rho\cap J$ (见 3.11.1); 可是, 由引理 10.22, $|P|$ 整除 $\mathrm{rank}_{\mathcal{O}}(Bj)$, 其中 $j\in J$; 所以, 由式 12.8.2 和定理 10.14, 存在 $\rho\in\mathcal{P}_B(1)$ 使得 p 不整除 m_ρ^γ. 而且, 如果 Q 是 P 的子群, 那

where we choose $j_\rho\in\rho\cap J$ (see 3.11.1); but, it follows from Lemma 10.22 that $|P|$ divides $\mathrm{rank}_{\mathcal{O}}(Bj)$ for any $j\in J$; hence, according to equality 12.8.2 and Theorem 10.14, there exists $\rho\in\mathcal{P}_B(1)$ such that p does not divide m_ρ^γ. Moreover, if Q is a subgroup of P,

么不难验证 (见 3.11.1)

it is easily checked that (see 3.11.1)

12.10.2 $$\mathrm{m}_\rho^\gamma = \sum_{\delta\in\mathcal{P}_B(Q)} \mathrm{m}_\rho^\delta \mathrm{m}_\delta^\gamma ;$$

可是, 如果$\delta\in\mathcal{P}_B(Q)$不是局部的, 那么只要选择$Q_\delta$的亏点群$R_\varepsilon$就得到一个有正交$Q/R$-迹幂等元$\ell\in\varepsilon$使得$\mathrm{Tr}_R^Q(\ell)\in\delta$ (见定理 7.2); 此时, 由 Mackey 公式,在$\mathcal{D}(B)$里有

however, if $\delta \in \mathcal{P}_B(Q)$ is not local then, choosing a defect pointed group R_ε of Q_δ, we get an idempotent $\ell \in \varepsilon$ having an orthogonal Q/R-trace and fulfilling $\mathrm{Tr}_R^Q(\ell) \in \delta$ (see Theorem 7.2); in this situation, by the Mackey formula, in $\mathcal{D}(B)$ we get

12.10.3 $$\mathrm{res}_1^Q(\delta) = \sum_u \mathrm{res}_1^{R^u}(\varepsilon^u) = |Q/R|\,\mathrm{res}_1^R(\varepsilon)$$

其中$u\in Q$跑遍一个Q/R的代表集合; 所以存在$\delta\in\mathcal{LP}_B(Q)$使得$p$不整除$\mathrm{m}_\rho^\delta\mathrm{m}_\delta^\gamma$.

where $u \in Q$ runs over a set of representatives for Q/R; hence, there is $\delta \in \mathcal{LP}_B(Q)$ such that p does not divide $\mathrm{m}_\rho^\delta \mathrm{m}_\delta^\gamma$.

命题 12.11. **如果 $f{:}\,B\cong B$ 是满足 $k\otimes f=\mathrm{id}_{k\otimes_{\mathcal{O}}B}$ 的 B 的 P-内代数自同构那么对适当的 $n\in J(\mathcal{O})\cdot B^P$ 有 $f(a)=a^{i+n}$, 其中 $a\in B$.**

Proposition 12.11. *If $f{:}\,B \cong B$ is a P-interior algebra automorphism which fulfills $k \otimes f = \mathrm{id}_{k\otimes_{\mathcal{O}}B}$ then, for a suitable $n \in J(\mathcal{O})\cdot B^P$, we have $f(a) = a^{i+n}$ for any $a \in B$.*

证明: 由 G-内代数 $\mathrm{Ind}_P^G(B)$ 的定义 (见注意 5.8), f 显然决定一个 G-内代数自同构如下

Proof: By the definition of the G-interior algebra $\mathrm{Ind}_P^G(B)$ (see Remark 5.8), f clearly induces a G-interior algebra automorphism

12.11.1 $$\mathrm{Ind}_P^G(f) : \mathrm{Ind}_P^G(B) \cong \mathrm{Ind}_P^G(B)$$

使得

fulfilling

12.11.2 $$\mathrm{Ind}_P^G(f)(x\otimes a\otimes y) = x\otimes f(a)\otimes y$$

其中$x,y\in G$ 与 $a\in B$; 特别是, 显然有

for any $x, y \in G$ and any $a \in B$; in particular, we clearly have

12.11.3 $$k\otimes \mathrm{Ind}_P^G(f) = \mathrm{id}_{\mathrm{Ind}_P^G(k\otimes_{\mathcal{O}}B)} .$$

一方面, 由定理 5.11, 存在一个完备诱导的 G-内代数 A 并 $\omega\in\mathcal{D}_A(G)$ 使得 $\mathcal{O}G\cong A_\omega$; 此时, 由同构式 5.3.2 和 5.8.6, 得到

On the one hand, by Theorem 5.11, there exist an inductively complete G-interior algebra A and $\omega \in \mathcal{D}_A(G)$ fulfilling $\mathcal{O}G \cong A_\omega$; then, from isomorphisms 5.3.2 and 5.8.6, we get

12.11.4 $$\mathcal{O}Gb \cong A_\alpha , \quad B\cong A_\gamma , \quad \mathrm{Ind}_P^G(B)\cong A_{\mathrm{ind}_P^G(\gamma)} ;$$

特别是, 因为 $\alpha \subset \mathrm{ind}_P^G(\gamma)$ 所以也得到

in particular, since we have $\alpha \subset \mathrm{ind}_P^G(\gamma)$, we also get

12.11.5 $$\mathcal{O}Gb \cong \mathrm{Ind}_P^G(B)_\alpha\,.$$

现在, 由等式 12.11.3, α 与 $\mathrm{Ind}_P^G(f)(\alpha)$ 在 $\mathrm{Ind}_P^G(k \otimes_\mathcal{O} B)$ 里的像是一样的.

Now, from equality 12.11.3, the images of α and $\mathrm{Ind}_P^G(f)(\alpha)$ in $\mathrm{Ind}_P^G(k \otimes_\mathcal{O} B)$ coincide.

另一方面, 由等式 5.8.3 并引理 10.22, 不难验证存在满足 $G{\cdot}T{\cdot}G = T$ 的 $\mathrm{Ind}_P^G(B)$ 的 $\mathcal{O}$-基 T; 特别是, 从 $\mathrm{Ind}_P^G(B)^G$ 到 $\mathrm{Ind}_P^G(k \otimes_\mathcal{O} B)^G$ 的自然同态是满射. 所以, 由推论 2.14, 有 $\mathrm{Ind}_P^G(f)(\alpha) = \alpha$ (请注意, 可能 $\mathrm{Ind}_P^G(f)$ 并不稳定 α 的元素), 从而 $\mathrm{Ind}_P^G(f)$ 决定一个 $\mathcal{O}Gb$ 的 G-内代数自同构 $\mathrm{Ind}_P^G(f)_\alpha$ (见 4.12.2); 可是, 恒等同构是 $\mathcal{O}Gb$ 的唯一 G-内代数自同构; 这样, $\mathrm{Ind}_P^G(f)_\alpha$ 在 B 上的限制也是恒等的.

On the other hand, from equality 5.8.3 and Lemma 10.22 it is not difficult to prove the existence of an $\mathcal{O}$-basis T of $\mathrm{Ind}_P^G(k \otimes_\mathcal{O} B)$ fulfilling $G{\cdot}T{\cdot}G = T$; in particular, the canonical homomorphism from $\mathrm{Ind}_P^G(B)^G$ to $\mathrm{Ind}_P^G(k \otimes_\mathcal{O} B)^G$ is surjective. So, by Corollary 2.14, we get $\mathrm{Ind}_P^G(f)(\alpha) = \alpha$ (note that $\mathrm{Ind}_P^G(f)$ need not fix any element of α), and therefore $\mathrm{Ind}_P^G(f)$ induces a G-interior algebra automorphism $\mathrm{Ind}_P^G(f)_\alpha$ of $\mathcal{O}Gb$ (see 4.12.2); however, the unique G-interior algebra automorphism of $\mathcal{O}Gb$ is the identity; thus, the restriction of $\mathrm{Ind}_P^G(f)_\alpha$ on B is the identity too.

进一步, 因为 $1 \otimes i \otimes 1$ 属于 γ 所以, 由等式 12.11.2, $\mathrm{Ind}_P^G(f)$ 在 B 上的限制就是 f. 最后, 只要应用 $\mathrm{Ind}_P^G(f)_\alpha$ 的定义, 就得到 f 是可嵌入自同构, 从而对适当的 $c \in (B^P)^*$ 有 $f(a) = a^c$, 其中 $a \in B$ (见 4.6). 此时, c 在 $k \otimes_\mathcal{O} B$ 中的像属于 $Z(k \otimes_\mathcal{O} B)^*$; 可是, 由后面的引理 14.6, 自然的同态

Moreover, since $1 \otimes i \otimes 1$ belongs to γ, it follows from equality 12.11.2 that the restriction of $\mathrm{Ind}_P^G(f)$ to B coincides with f. Finally, according to the very definition of $\mathrm{Ind}_P^G(f)_\alpha$, we get that f is an *embeddable* automorphism and therefore, for a suitable $c \in (B^P)^*$, we have $f(a) = a^c$ for any $a \in B$ (see 4.6). Then, the image of c in $k \otimes_\mathcal{O} B$ belongs to $Z(k \otimes_\mathcal{O} B)^*$; but, by Lemma 14.6 in section 14, the natural homomorphism

12.11.6 $$Z(B)^* \longrightarrow Z(k \otimes_\mathcal{O} B)^*$$

是满射; 这样, 存在 $z \in Z(B)^*$ 使得 $z^{-1}c$ 属于 $i + J(\mathcal{O}){\cdot}B^P$.

is surjective; thus, there exists $z \in Z(B)^*$ such that $z^{-1}c$ belongs to $i + J(\mathcal{O}){\cdot}B^P$.

12.12. 在下面的例子里我表示一个这种偶对 (G, b) 的无限集合使得对应的代数 $\mathcal{O}Gb$ 相互不是同构的, 但对应的亏群和对应的源代数都是相互同

12.12. In the example below, we exhibit an infinite set of those pairs (G, b), where the corresponding algebras $\mathcal{O}Gb$ are mutually not isomorphic, whereas all the corresponding defect groups and source algebras

构的. 更一般来说, 我猜想是: **固定一个有限 p-群 P, 考虑这种偶对 (G,b) 使得 P 是 $G_{\{b\}}$ 的亏群, 那么 $\mathcal{O}Gb$ 的源代数作为 P-内代数的同构类的集合是有限的.**

are mutually ismorphic. More generally, our conjecture is: *fix a finite p-group P and consider all the pairs (G,b) where P is a defect group of $G_{\{b\}}$; then, the set of isomorhism classes, as P-interior algebras, of the source algebras of those $\mathcal{O}Gb$ is finite.*

例子 12.13. 设 q 是一个素数使得 p 整除 $q-1$, 并且对 $m\in\mathbb{N}-p\mathbb{N}$, 设 K 是 q^m 元域. 固定 $n\in\mathbb{N}$ 使得 $0<n<p$, 并设 V 是一个维 n K-向量空间. 考虑 V 的 K-线性自同构的群 $GL(n,q^m)$ 和它的主块 $b_\circ(n,q^m)$ (见 10.2). 别忘了, 有

Example 12.13. Let q be a prime number such that p divides $q-1$, and for $m\in\mathbb{N}-p\mathbb{N}$, let K be the field of cardinal q^m. Fix $n\in\mathbb{N}$ fulfilling $0<n<p$, and let V be a K-linear vector space of dimension n. Consider the group $GL(n,q^m)$ of K-linear automorphisms of V and its principal block $b_\circ(n,q^m)$ (see 10.2). Recall that we have

12.13.1 $$|GL(n,q^m)| = q^{m\binom{n}{2}}\prod_{h=1}^{n}(q^{mh}-1)$$

并且, 因为显然有

and moreover, since clearly we have

12.13.2 $$(q^\ell-1)/(q-1)=\sum_{h=1}^{\ell-1}q^h\equiv\ell \pmod p\,,$$

所以得到

we get

12.13.3 $$|GL(n,q^m)|_p = (|K^*|_p)^n\,.$$

这样, 考虑 V 的一个基 X, 并令 T 记在 $GL(n,q^m)$ 中的分解式 $V=\oplus_{v\in X}K{\cdot}v$ 的稳定化子; 那么, 显然 $T\cong(K^*)^n$ 并且 T 的一个 Sylow p-子群 P 也是 $GL(n,q^m)$ 的 Sylow p-子群; 事实上, 显然 $T\cong P\times W$, 其中 W 是 p'-群. 特别是, P 也是 $GL(n,q^m)_{\{b_\circ(n,q^m)\}}$ 的亏群.

Thus, consider a basis X of V and denote by T the stabilizer in $GL(n,q^m)$ of the decomposition $V=\oplus_{v\in X}K{\cdot}v$; then, it is clear that $T\cong(K^*)^n$ and a Sylow p-subgroup of T is a Sylow p-subgroup of $GL(n,q^m)$ too; actually, it is quite clear that $T\cong P\times W$ for a suitable p'-group W. In particular, P is also a defect group of $GL(n,q^m)_{\{b_\circ(n,q^m)\}}$.

另一方面, 固定 X 的一个全序; 令 U 记满足 $f(v)-v$ 属于 $\oplus_{v'<v}K{\cdot}v'$, 其中 $v\in X$, 的元素 $f\in GL(n,q^m)$ 的子群, 再令 $H=U{\cdot}T$, 即所谓 Borel 子群; 最后, 令 $S_n\subset GL(n,q^m)$

On the other hand, fix a total order in X; denote by U the subgroup of all the elements $f\in GL(n,q^m)$ such that $f(v)-v$ belongs to $\oplus_{v'<v}K{\cdot}v'$ for any $v\in X$, and set $H=U{\cdot}T$, i. e. the so-called *Borel subgroup*; finally, denote by $S_n\subset GL(n,q^m)$ the group

记 X 的置换群. 那么, 已经知道而不难验证有

of permutations of X. Then, it is well-known and easy to check that

12.13.4 $$GL(n,q^m) = \bigsqcup_{s\in S_n} HsH\,, \quad N_{GL(n,q^m)}(P) = T{\cdot}S_n\,.$$

而且, 令 $A = \mathcal{O}GL(n,q^m)$ 与 $L = U{\cdot}W$; 既然 p 不整除 L 的阶, 在 A 中能考虑 A^P 的幂等元

Moreover, set $A = \mathcal{O}GL(n,q^m)$ and $L = U{\cdot}W$; as far as p does not divide the order of L, in A^P we can consider the idempotent

12.13.5 $$\ell = \frac{1}{|L|}\cdot\sum_{x\in L} x\,;$$

因为 $\varepsilon_G(\ell) = 1$, 所以得到 $\mathrm{Br}_P(\ell b_\circ(n,q^m)) \neq 0$ (见 10.2); 这样, 存在 A 上的唯一局部 P-点 γ 并 $i\in\gamma$ 使得

since we have $\varepsilon_G(\ell) = 1$, we get $0 \neq \mathrm{Br}_P(\ell b_\circ(n,q^m))$ (see 10.2); thus, there are a unique point γ of P on A and $i\in\gamma$ fulfilling

12.13.6 $$P_\gamma \subset GL(n,q^m)_{\{b_\circ(n,q^m)\}}\,, \quad i\ell = i = \ell i\,, \quad \varepsilon_G(i) = 1\,;$$

特别是, 也得到 (见 12.13.4)

in particular, we still get

12.13.7 $$N_{GL(n,q^m)}(P_\gamma) = T{\cdot}S_n\,, \quad E_{GL(n,q^m)}(P_\gamma) \cong S_n\,.$$

假定 k 是代数闭的, 并令 $B = iAi$; 因为 ε_G 决定 $\mathcal{O}$ 上的 B-模结构 (见式子 12.13.6), 所以有

Assume that k is algebraically closed and set $B = iAi$; since ε_G determines a B-module struture on $\mathcal{O}$ (see 12.13.6), we have

12.13.8 $$\hat{E}_{GL(n,q^m)}(P_\gamma) \cong k^* \times S_n\,;$$

此时, 由定理 12.8, 存在 B 的一个幺子 P-内代数 D 并 B 的子 $D\otimes_{\mathcal{O}} D^\circ$-模 M 使得

then, it follows from Theorem 12.8 that there are a unitary P-interior algebra D and a $D\otimes_{\mathcal{O}} D^\circ$-module M such that

12.13.9 $$D \cong \mathcal{O}P \rtimes S_n\,, \quad B = D \oplus M\,;$$

可是 B 是 $\ell A\ell$ 的直和项, 并且有

but B is a direct summand of $\ell A\ell$, so that

12.13.10 $$\ell A\ell = \bigoplus_{s\in S_n} \ell\,\mathcal{O}(HsH)\,\ell = \bigoplus_{s\in S_n}\sum_{u\in P} \mathcal{O}{\cdot}\ell us\ell\,;$$

这样, 只要把它们的$\mathcal{O}$-秩比一比就得到 $M = \{0\}$ 与 $i = \ell$; 也就是说, 我们得到下面的不依赖于 m 的 P-内代数结构

thus, it suffices to compare both $\mathcal{O}$-ranks, to get $M = \{0\}$ and $i = \ell$; that is to say, we obtain the following P-interior algebra structure, which does not depend on m.

12.13.11 $$B \cong \mathcal{O}(P \rtimes S_n)\,.$$

13. 超聚焦子代数的局部结构

13. Local Structure of the Hyperfocal Subalgebra

13.1. 设 G 是一个有限群并设 S 是 G 的 Sylow p-群; 令 $\mathfrak{f}_G(S)$ 记所谓 S 的聚焦子群, 也就是说由元素 $u^{-1}u^x$, 其中 $x \in G$ 与 $u, u^x \in S$, 生成的子群. 已经知道, 存在唯一 G 的规正子群 H 使得 $G = H{\cdot}S$ 与 $H \cap S = \mathfrak{f}_G(S)$, 并且由所谓 Alperin 融合定理, 能得到

13.1. Let G be a finite group and S a Sylow p-subgroup of G; denote by $\mathfrak{f}_G(S)$ the so-called *focal subgroup* of S, that is to say, the subgroup of S generated by the elements $u^{-1}u^x$, where $x \in G$ and $u, u^x \in S$. It is well-known that there exists a unique normal subgroup H of G such that $G = H{\cdot}S$ and $H \cap S = \mathfrak{f}_G(S)$, and from the so-called *Alperin Fusion Theorem* we may obtain

13.1.1 $$\mathfrak{f}_G(S) = \big\langle\, [N_G(T), T] \mid T \in \mathfrak{S}(S) \big\rangle .$$

13.2. 进一步, 对任意 S 的子群 T 只考虑 $N_G(T)$ 的 p'-子群 K 的集合, 记 $\mathfrak{S}_{p'}\big(N_G(T)\big)$; 那么, 令

13.2. Going further, for any subgroup T of S, consider only the set of all the p'-subgroups of $N_G(T)$, noted $\mathfrak{S}_{p'}\big(N_G(T)\big)$; then, we set

13.2.1 $$\mathfrak{h}_G(S) = \big\langle\, [K, T] \mid T \in \mathfrak{S}(S)\,,\ K \in \mathfrak{S}_{p'}\big(N_G(T)\big) \big\rangle ,$$

称为 S 的超聚焦子群; 只要使用归纳法就不难证明存在唯一 G 的规正子群 L 使得 $G = L{\cdot}S$ 与 $L \cap S = \mathfrak{h}_G(S)$.

called *hyperfocal subgroup* of S; arguing by induction, it is not difficult to prove that there exists a unique normal subgroup L of G such that $G = L{\cdot}S$ and $L \cap S = \mathfrak{h}_G(S)$.

13.3. 假定 k 是代数闭的. 事实上, 对任意 G-块 b 有类似结果如下. 设 $P_{\hat\gamma}$ 是一个 $G_{\{b\}}$ 的亏点群, 并考虑包含在 $P_{\hat\gamma}$ 里的在 $\mathcal{O}G$ 上的局部点群的集合, 记 $\mathfrak{S}_{\mathcal{L}}(P_{\hat\gamma})$; 那么, 令

13.3. Assume that k is algebraically closed. Actually, we have the similar result below for any block b of G. Let P_γ be a defect pointed group of $G_{\{b\}}$ and consider the set of local pointed groups on $\mathcal{O}G$ contained in $P_{\hat\gamma}$, noted $\mathfrak{S}_{\mathcal{L}}(P_{\hat\gamma})$; then, we set

13.3.1 $$\mathfrak{h}_G(P_{\hat\gamma}) = \big\langle\, [K, Q] \mid Q_{\hat\delta} \in \mathfrak{S}_{\mathcal{L}}(P_{\hat\gamma})\,,\ K \in \mathfrak{S}_{p'}\big(N_G(Q_\delta)\big) \big\rangle ,$$

称为 $P_{\hat\gamma}$ 的超聚焦子群. 选择 $i \in \hat\gamma$ 并且记

called *hyperfocal subgroup* of $P_{\hat\gamma}$. Choose $i \in \hat\gamma$ and set

13.3.2 $$B = i(\mathcal{O}G)i \cong (\mathcal{O}G)_{\hat\gamma}\,, \quad Q = \mathfrak{h}_G(P_{\hat\gamma});$$

后面的很重要定理 15.10 说: **存在唯一一个 B 的 P-稳定幺子代数 D 的 $(B^P)^*$-共轭类使得**

in §15, the significant Theorem 15.10 affirms: *there is a unique $(B^P)^*$-conjugacy class of P-stable unitary subalgebras D of B fulfilling*

13.3.3 $$D \cap Pi = Qi\,, \quad B = D \otimes_Q P\,;$$

D 称为块 b 的超聚焦子代数. 首先我们决定 D 的所谓局部结构, 也就是说在 D 上的局部点群并它们之间的 D-融合.

we call D the *hyperfocal subalgebra* of the block b. First of all, we study the so-called *local structure* of D; in other terms, the local pointed groups and their fusions in D.

13.4. 更一般来说, 设 Q 是 P 的规正子群并设 D 是 B 的 P-稳定幺子代数使得

13.4. More generally, let Q be a normal subgroup of P and D a P-stable unitary subalgebra of B such that

13.4.1 $$D \cap Pi = Qi\,, \quad B = D \otimes_Q P\,;$$

这样, D 成为一个 Q-内 P-代数, 以及 B 就是对应的 P-内代数 (见 9.1); 特别是, D 在 B 里是 $\mathcal{O}(Q\times Q)P$-模直和项, 从而只要应用引理 10.22 和推论 3.7, 就不难证明 D 有一个 P-稳定 $\mathcal{O}$-基 T 使得 $Q{\cdot}T{\cdot}Q = T$. 请注意, 由引理 10.24, 还有 $P \cong P{\cdot}i$, 并且 $\gamma = \{i\}$ 是 D 上的局部 P-点.

thus, D becomes a Q-interior P-algebra and B is just the corresponding P-interior algebra (see 9.1); in particular, D is a direct summand of B as $\mathcal{O}(Q\times Q)P$-modules and then, from Lemma 10.22 and Corollary 3.7, it is not diffficult to prove that D has a P-stable $\mathcal{O}$-basis T fulfilling $Q{\cdot}T{\cdot}Q = T$. Note that, according to Lemma 10.24, we have $P \cong P{\cdot}i$, and that $\gamma = \{i\}$ is a local point of P on D.

命题 13.5. **假定 k 是代数闭的. 对 P 的任意子群 R 有 $J(D^R) = D^R \cap J(B^R)$ 并 D 上的任意 R-点 ε 包含在 B 上的唯一 R-点 $\hat{\varepsilon}$ 中. 而且, ε 是局部的当且仅当 $\hat{\varepsilon}$ 是局部的, 以及那么有**

Proposition 13.5. *Assume that k is algebraically closed. For any subgroup R of P, we have $J(D^R) = D^R \cap J(B^R)$ and any point ε of R on D is contained in a unique point $\hat{\varepsilon}$ of R on B. Moroever, ε is local if and only if $\hat{\varepsilon}$ is local, and then we have*

13.5.1 $$F_D(R_\varepsilon, P_\gamma) = F_B(R_{\hat{\varepsilon}}, P_{\hat{\gamma}})\,.$$

证明: 只要使用定理 9.5 就得到第一个论断. 假定 ε 是局部的; 由定理 9.2, 对任意 $\varphi \in 融_B(R_{\hat{\varepsilon}}, P_{\hat{\gamma}})$ 存在 $u \in P$ 与 $\psi \in 融_D\big((R_\varepsilon)^u, P_\gamma\big)$ 使得 $\varphi = \psi \circ \kappa_u^R$; 这样, 因为显然 $\tilde{\kappa}_u^P = \widetilde{\mathrm{id}}_P$, 所以得到

Proof: It suffices to apply Theorem 9.5 to get the first statement. Assume that ε is local; it follows from Theorem 9.2 that, for any $\varphi \in 融_B(R_{\hat{\varepsilon}}, P_{\hat{\gamma}})$, there exist $u \in P$ and $\psi \in 融_D\big((R_\varepsilon)^u, P_\gamma\big)$ fulfilling $\varphi = \psi \circ \kappa_u^R$; thus, since obviously $\varphi = \psi \circ \kappa_u^R$, we get

13.5.2 $$\tilde{\varphi} = (\tilde{\kappa}_u^P)^{-1} \circ \varphi = (\tilde{\kappa}_u^P)^{-1} \circ \tilde{\psi} \circ \tilde{\kappa}_u^P\,,$$

从而 $\tilde{\varphi}$ 属于 $F_D(R_\varepsilon, P_\gamma)$ (见等式 8.8.1).

and therefore $\tilde{\varphi}$ belongs to $F_D(R_\varepsilon, P_\gamma)$ (see equality 8.8.1).

推论 13.6. 假定 k 是代数闭的. 那么有 $\mathfrak{h}_G(P_{\hat{\gamma}}) \subset Q$.

证明: 设 $R_{\hat{\varepsilon}}$ 是 B 上的局部点群使得 $R_{\hat{\varepsilon}} \subset P_{\hat{\gamma}}$ 并设 K 是 $N_G(R_{\hat{\varepsilon}})$ 的 p'-子群; 由推论 9.3, 融$_B(R_{\hat{\varepsilon}})$/融$_D(R_\varepsilon)$ 是 p-群, 从而 K 在 $\widetilde{\mathrm{Aut}}(R)$ 里的像包含在 $F_D(R_\varepsilon)$ 里 (见命题 12.2); 特别是, 由融合的定义 (见 8.8), 有 $[K,R] \subset Q$.

13.7. 设 R 是 P 的子群; 由定理 10.7, 存在唯一 $\tilde{C}_G(R)$-块 g 使得 $\mathrm{Br}_R(\hat{\gamma})g = \mathrm{Br}_R(\hat{\gamma})$; 特别是, 对任意 B 上的局部 R-点 $\hat{\varepsilon}$ 有 $\mathrm{Br}_R(\hat{\varepsilon})g = \mathrm{Br}_R(\hat{\varepsilon})$, 从而还有 $R_{\hat{\varepsilon}} \subset \tilde{C}_G(R)_{\hat{\zeta}}$, 其中 $\hat{\zeta}$ 是 $\mathcal{O}G$ 上的 $\tilde{C}_G(R)$-点使得 $\mathrm{Br}_R(\hat{\zeta}) = \{g\}$ (见 10.12); 这样, 由定理 6.8, 任意 $\tilde{C}_G(R)_{\hat{\zeta}}$ 的亏点群 $U_{\hat{\nu}}$ 包含这些 $R_{\hat{\varepsilon}}$, 并且存在 $x \in G$ 使得 $(U_{\hat{\nu}})^x \subset P_{\hat{\gamma}}$.

13.8. 也就是说, 存在

Corollary 13.6. *Assume that k is algebraically closed. Then we have* $\mathfrak{h}_G(P_{\hat{\gamma}}) \subset Q$.

Proof: Let R_ε be a local pointed group on B fulfilling $R_{\hat{\varepsilon}} \subset P_{\hat{\gamma}}$ and K a p'-subgroup of $N_G(R_{\hat{\varepsilon}})$; it follows from Corollary 9.3 that 融$_B(R_{\hat{\varepsilon}})$/融$_D(R_\varepsilon)$ is a p-group and therefore the image of K in $\widetilde{\mathrm{Aut}}(R)$ is contained in $F_D(R_\varepsilon)$ (see Proposition 12.2); in particular, by the very definition of the fusions (see 8.8), we have $[K,R] \subset Q$.

13.7. Let R be a subgroup of P; according to Theorem 10.7, there is a unique block g of $\tilde{C}_G(R)$ such that $\mathrm{Br}_R(\hat{\gamma})g = \mathrm{Br}_R(\hat{\gamma})$; in particular, for any local point $\hat{\varepsilon}$ of R on B, we have $\mathrm{Br}_R(\hat{\varepsilon})g = \mathrm{Br}_R(\hat{\varepsilon})$ and therefore we still have $R_{\hat{\varepsilon}} \subset \tilde{C}_G(R)_{\hat{\zeta}}$, where $\hat{\zeta}$ is the point of $\tilde{C}_G(R)$ on $\mathcal{O}G$ fulfilling $\mathrm{Br}_R(\hat{\zeta}) = \{g\}$ (see 10.12); thus, according to Theorem 6.8, any defect pointed group $U_{\hat{\nu}}$ of $\tilde{C}_G(R)_{\hat{\zeta}}$ contains all these $R_{\hat{\varepsilon}}$, and there exists $x \in G$ such that $(U_{\hat{\nu}})^x \subset P_{\hat{\gamma}}$.

13.8. That is to say, there exist

13.8.1 $$\varphi \in \bigcap_{\hat{\varepsilon} \in \mathcal{LP}_B(R)} 融_B(R_{\hat{\varepsilon}}, P_{\hat{\gamma}}), \quad \Phi: \mathcal{LP}_B(R) \longrightarrow \mathcal{LP}_B(T),$$

其中 $T = \varphi(R)$, 使得被 φ 决定的 $\psi \in \mathrm{Hom}(R,T)$ 属于融$_B(R_{\hat{\varepsilon}}, T_{\Phi(\hat{\varepsilon})})$, 其中 $\hat{\varepsilon}$ 跑遍 $\mathcal{LP}_B(R)$, 并 Φ 是单射; 特别是, 有

where $T = \varphi(R)$, such that the group homomorphism $\psi \in \mathrm{Hom}(R,T)$ determined by φ belongs to 融$_B(R_{\hat{\varepsilon}}, T_{\Phi(\hat{\varepsilon})})$ where $\hat{\varepsilon}$ runs over $\mathcal{LP}_B(R)$, and Φ is injective; in particular, we have

13.8.2 $$\mathrm{res}_1^R(\hat{\varepsilon}) = \mathrm{res}_1^T(\Phi(\hat{\varepsilon})) = \omega_{\hat{\varepsilon}};$$

此时, 由命题 8.13, 存在一个 R-内代数同构

then, by Proposition 8.13, there exists a R-interior algebra isomorphism

13.8.3 $$f_\psi : B_{\hat{\varepsilon}} \cong \mathrm{Res}_\psi(B_{\Phi(\hat{\varepsilon})})$$

使得下面的 $\mathcal{O}$-代数同构是可嵌入的 (见 4.6)

such that the following $\mathcal{O}$-algebra isomorphism is *embeddable* (see 4.6)

13.8.4 $$\mathrm{Res}_1^P(B)_{\omega_{\hat{\varepsilon}}} \cong \mathrm{Res}_1^R(B_{\hat{\varepsilon}}) \overset{f_\psi}{\cong} \mathrm{Res}_1^T(B_{\Phi(\hat{\varepsilon})}) \cong \mathrm{Res}_1^P(B)_{\omega_{\hat{\varepsilon}}}.$$

而且$P_{\hat{\gamma}}$包含$\tilde{C}_G(T)_{\hat{\xi}}$的亏点群，其中$\hat{\xi}$是满足$T_{\Phi(\hat{\varepsilon})}\subset\tilde{C}_G(T)_{\hat{\xi}}$的$\mathcal{O}G$上的$\tilde{C}_G(T)$-点. 请注意，如果$\hat{\theta}$是一个$\mathcal{O}G$上的局部$\tilde{C}_P(T)$-点使得$\tilde{C}_P(T)_{\hat{\theta}}\subset P_{\hat{\gamma}}$ (见论断 10.5.1)，那么$\tilde{C}_G(T)_{\hat{\xi}}$包含$\tilde{C}_P(T)_{\hat{\theta}}$ (见 10.12)，从而$\tilde{C}_P(T)_{\hat{\theta}}$是$\tilde{C}_G(T)_{\xi}$的亏点群.

Moreover, P_γ contains a defect pointed group of $\tilde{C}_G(T)_{\hat{\xi}}$, where $\hat{\xi}$ is the point of $\tilde{C}_G(T)$ on $\mathcal{O}G$ fulfilling $T_{\Phi(\hat{\varepsilon})}\subset\tilde{C}_G(T)_{\hat{\xi}}$. Note that, if $\hat{\theta}$ is a local point of $\tilde{C}_P(T)$ on $\mathcal{O}G$ such that $\tilde{C}_P(T)_{\hat{\theta}}\subset P_{\hat{\gamma}}$ (see statement 10.5.1), then $\tilde{C}_G(T)_{\hat{\xi}}$ contains $\tilde{C}_P(T)_{\hat{\theta}}$ (see 10.12), and therefore $\tilde{C}_P(T)_{\hat{\theta}}$ is a defect pointed group of $\tilde{C}_G(T)_{\hat{\xi}}$.

定理 13.9. **假定 k 是代数闭的. 设 $R_{\hat{\varepsilon}}$ 是 B 上的局部点群并令 $\hat{\zeta}$ 记 $\mathcal{O}G$ 上的 $\tilde{C}_G(R)$-点使得 $R_{\hat{\varepsilon}}\subset\tilde{C}_G(R)_{\hat{\zeta}}$. 假定 $P_{\hat{\gamma}}$ 包含一个 $\tilde{C}_G(R)_{\hat{\zeta}}$ 的亏点群. 那么, $C_P(R)$ 在满足 $\varepsilon\subset\hat{\varepsilon}$ 的 $\varepsilon\in\mathcal{LP}_D(R)$ 的集合上可迁地作用. 特别是, 如果 $R_{\hat{\varepsilon}}$ 是自中心化子的, 那么存在唯一 D 上的局部 R-点 ε 使得 $\varepsilon\subset\hat{\varepsilon}$.**

Theorem 13.9. *Assume that k is algebraically closed. Let R_ε be a local pointed group on B and denote by $\hat{\zeta}$ the point of $\tilde{C}_G(R)$ on $\mathcal{O}G$ such that $R_{\hat{\varepsilon}}\subset\tilde{C}_G(R)_{\hat{\zeta}}$. Assume that P_γ contains a defect pointed group of $\tilde{C}_G(R)_{\hat{\zeta}}$. Then, $C_P(R)$ acts transitively on the set of $\varepsilon\in\mathcal{LP}_D(R)$ fulfilling $\varepsilon\subset\hat{\varepsilon}$. In particular, if R_ε is self-centralizing, there exists a unique local point ε of R on D such that $\varepsilon\subset\hat{\varepsilon}$.*

证明: 设 ε 与 δ 是 D 上的局部 R-点使得 $\varepsilon\subset\hat{\varepsilon}$ 与 $\delta\subset\hat{\varepsilon}$ 并取 $\ell\in\varepsilon$ 与 $j\in\delta$; 因为 ℓ 与 j 在 B^R 中是相互共轭的, 所以 DR-模 $B\ell$ 与 Bj 是相互同构的, 其中 DR-模结构的定义是左边乘以 D 并 R-共轭.

可是, 从式子 13.4.1 可得到下面的 DR-模直和分解

Proof: Let ε and δ be local points of R on D such that $\varepsilon\subset\hat{\varepsilon}$ and $\delta\subset\hat{\varepsilon}$, and pick $\ell\in\varepsilon$ and $j\in\delta$; since ℓ and j are mutually conjugate in B^R, the DR-modules $B\ell$ and Bj are mutually isomorphic, where the DR-module structures are defined by the multiplication on the left by D and by the conjugation by R.

But, from 13.4.1, we get the following DR-module direct sum decompositions.

13.9.1 $$B\ell=\bigoplus_{u\in X}DR\otimes_{DR_u}Du\ell\,,\quad Bj=\bigoplus_{u\in X}DR\otimes_{DR_u}Duj\,,$$

其中 $X\subset P$ 是在 P/Q 上的 R-共轭作用的轨道的代表集合, 而 R_u 是稳定 Du 在 R 中的子群, 其中 $u\in X$; 而且, 如果 $R_u=R$ 那么不难验证

where $X\subset P$ is a set of representatives for the orbits of R acting by conjugation in P/Q, and, for any $u\in X$, R_u is the stabilizer of Du in R; moreover, if $R_u=R$ then it is easily checked that

13.9.2 $$\mathrm{End}_{DR}(Du\ell)^\circ\cong\ell D^R\ell\,,\quad \mathrm{End}_{DR}(Duj)^\circ\cong jD^Rj\,,$$

从而 $Du\ell$ 与 Duj 是不可分解 DR-模.

and therefore the DR-modules $Du\ell$ and Duj are indecomposable.

这样, 因为 (见推论 6.6),

Thus, since we have (see Corollary 6.6),

13.9.3 $$(D\ell)(R) = D(R)\mathrm{Br}_R(\ell) \neq \{0\}$$

所以存在 $u \in P$ 使得既 R 稳定 Du, 也 DR-模 $D\ell$ 与 Duj 是相互同构的 (见推论 3.7 和命题 6.3). 更精确地, 对任意 $v \in R$ 有 $D[v, u^{-1}] = D$, 从而还有$[R, u^{-1}] \subset Q$. 特别是, u 稳定DR; 令 $\kappa_u: DR \cong DR$ 记对应的自同构; 那么, 因为 $R^{u^{-1}}$稳定 $j^{u^{-1}}$, 所以 $Dj^{u^{-1}}$ 成为 DR-模并且右边乘以 u 决定一个 DR-模同构

there exists $u \in P$ such that R stabilizes Du and that the DR-modules $D\ell$ and Duj are isomorphic (see Corollary 3.7 and Proposition 6.3). Explicitly, for any $v \in R$, we have $D[v, u^{-1}] = D$, and therefore we still have $[R, u^{-1}] \subset Q$. In particular, u stabilizes DR; denote by $\kappa_u: DR \cong DR$ the corresponding automorphism; then, since $R^{u^{-1}}$ stabilizes $j^{u^{-1}}$, $Dj^{u^{-1}}$ becomes a DR-module and the multiplication on the right by u determines a DR-module isomorphism

13.9.4 $$\mathrm{Res}_{\kappa_u}(Dj^{u^{-1}}) \cong Duj \cong D\ell\,.$$

而且, 既然 u-共轭也稳定D,不难验证u-共轭决定一个DR-模同构$\mathrm{Res}_{\kappa_u}(D) \cong D$. 这样, 由推论 3.7, 也存在一个$DR$-模同构

Moreover, since the conjugation by u stabilizes D too, it is easy to check that this conjugation induces a DR-module isomorphism $\mathrm{Res}_{\kappa_u}(D) \cong D$. Thus, by Corollary 3.7, we get a DR-module isomorphism

13.9.5 $$Du(i-j) \cong \mathrm{Res}_{\kappa_u}\big(D(i - j^{u^{-1}})\big) \cong D(i-\ell)\,.$$

也就是说, 存在一个 DR-模同构 $f: D \cong Du$ 使得

That is to say, there exists a DR-module isomorphism $f: D \cong Du$ fulfilling

13.9.6 $$f(D\ell) = Duj\,, \quad f\big(D(i-\ell)\big) = Du(i-j)\,;$$

令 $f(i) = du$, 其中 $d \in D$; 这样, du 属于 $(Du)^R \cap B^*$ 与 ℓdu 属于 Duj 与$(i-\ell)du$ 属于 $Du(i-j)$, 从而 $\ell^d = uju^{-1}$.

set $f(i) = du$ for a suitable $d \in D$; thus, du belongs to $(Du)^R \cap B^*$, ℓdu belongs to $(i-\ell)du$ and $(i-\ell)du$ belongs to $Du(i-j)$, and therefore we have $\ell^d = uju^{-1}$.

最后, 令 S 记 $v \in P$ 使得 R 稳定 Dv 的集合, 再令 T 记 $v \in S$ 使得 $(Dv)^R \cap B^* \neq \emptyset$的集合; 因为$(Dv)\cdot(Dw) = Dvw$ 其中$v, w \in P$, 所以 S 与T是 P 的子群; 进一步, 从式子 13.4.1 可得到下面的 $kC_Q(R)$-模直和分解

Finally, denote by S the set of $v \in P$ such that R stabilizes Dv, and by T the set of $v \in S$ fulfilling $(Dv)^R \cap B^* \neq \emptyset$; since $(Dv)\cdot(Dw) = Dvw$ for any $v, w \in P$, S and T are subgroups of P; furthermore, from 13.4.1, we may obtain the following direct sum decomposition of $kC_Q(R)$-modules

13.9.7 $$B(R) = \bigoplus_{w \in X \cap S} (Dw)(R)$$

并如果 $v \in T$, $a \in (Dv)^R \cap B^*$ 那么右边乘以 $\mathrm{Br}_R(a)$ 决定一个 $kC_Q(R)$-模同构

and if $v \in T$ and $a \in (Dv)^R \cap B^*$ then the multiplication on the right by $\mathrm{Br}_R(a)$ determines a $kC_Q(R)$-module isomorphism

13.9.8 $$(Dw)(R) \cong (Dwv)(R)$$

其中 $w \in S$.

for any $w \in S$.

另一方面, 因为 $B(R)$ 显然是 $kC_G(R)$ 的 $kC_Q(R)$-模直和项, 所以这些 $(Dw)(R)$ 是投射 $kC_Q(R)$-模, 从而 $|C_Q(R)|$ 整除 $(Dw)(R)$ 的维数 (见引理 10.20). 这样, 因为考虑 B 上的右边乘以 T 的元素的时候 Q 就是 Dw 的稳定化子, 所以 $|T/Q||C_Q(R)|$ 整除 $B(R)$ 的维数; 进一步, 因为 T 包含 $C_P(R)$, 所以有

On the other hand, since $B(R)$ clearly is a direct summand of the $kC_Q(R)$-module $kC_G(R)$, those terms $(Dw)(R)$ are projective $kC_Q(R)$-modules and therefore $|C_Q(R)|$ divides the dimension of $(Dw)(R)$ (see Lemma 10.20). Thus, since Q is the stabilizer of Dw in T by the multiplication on the right on B, $|T/Q||C_Q(R)|$ divides the dimension of $B(R)$; furthermore, since T contains $C_P(R)$, we have

13.9.9 $$|T/Q||C_Q(R)| = |T/Q{\cdot}C_P(R)||C_P(R)|\,;$$

此时, 由以下引理 13.10, 得到 $T = Q.C_P(R)$; 所以有 $u = vz$, 其中 $v \in Q$, $z \in C_P(R)$, 还有 $\ell^{dv} = zjz^{-1}$; 也就是说, $\delta = \varepsilon^z$.

at this point, by Lemma 13.10 below, we get $T = Q.C_P(R)$; consequently, we have $u = vz$ for suitable $v \in Q$ and $z \in C_P(R)$, and thus we still have $\ell^{dv} = zjz^{-1}$; hence, $\delta = \varepsilon^z$.

引理 13.10. 假定 k 是代数闭的. 设 R_ε 是 $\mathcal{O}G$ 上的局部点群并令 α 与 ζ 分别记 $\mathcal{O}G$ 上的 G-点与 $\tilde{C}_G(R)$-点使得 $R_\varepsilon \subset \tilde{C}_G(R)_\zeta \subset G_\alpha$; 取一个 G_α 的亏点群 P_γ 使得它包含一个 $\tilde{C}_G(R)_\zeta$ 的亏点群. 那么 $|C_P(R)|$ 整除 $(\mathcal{O}G)_\gamma(R)$ 的维数, 并且有

Lemma 13.10. *Assume that k is algebraically closed. Let R_ε be a local pointed group on $\mathcal{O}G$ and denote by α and ζ the respective points of G and $\tilde{C}_G(R)$ on $\mathcal{O}G$ such that $R_\varepsilon \subset \tilde{C}_G(R)_\zeta \subset G_\alpha$; pick a defect pointed group P_γ of G_α which contains a defect group of $\tilde{C}_G(R)_\zeta$. Then, $|C_P(R)|$ divides the dimension of $(\mathcal{O}G)_\gamma(R)$, and we have*

13.10.1 $$\frac{\dim_k\big((\mathcal{O}G)_\gamma(R)\big)}{|C_P(R)|} \not\equiv 0 \pmod p\,;$$

证明: 令 $B = (\mathcal{O}G)_\gamma$; 因为 $B(R)$ 是 $kC_G(R)$ 的 $kC_P(R)$-模直和项, 所以它是投射的, 从而 $|C_P(R)|$ 整除 $B(R)$ 的维数 (见引理 10.20). 设 T_θ 是 $\tilde{C}_G(R)_\zeta$

Proof: Set $B = (\mathcal{O}G)_\gamma$; since $B(R)$ is a direct summand of $kC_G(R)$ as $kC_P(R)$-modules, it is projective and therefore $|C_P(R)|$ divides the dimension of $B(R)$ (see Lemma 10.20). Let T_θ be a defect pointed group

的亏点群使得 $T_\theta \subset P_\gamma$；请注意，由 13.8，显然 $T = \tilde{C}_P(R)$．从等式 10.2.2 不难验证对任意 $C_G(R)$ 的子群 L 有

of $\tilde{C}_G(R)_\zeta$ such that $T_\theta \subset P_\gamma$; note that, by 13.8, we clearly have $T = \tilde{C}_P(R)$. From equality 10.2.2, it is easily checked that, for any subgroup L of $C_G(R)$, we have

13.10.2 $$\mathrm{Br}_R\big((\mathcal{O}G)^{R\cdot L}\big) = kC_G(R)^L .$$

令 $\bar{X}$ 记 $X \subset (\mathcal{O}G)^R$ 在 $(\mathcal{O}G)(R)$ 中的像；特别是，$\bar{\zeta}$ 与 $\bar{\theta}$ 分别是 $kC_G(R)$ 上的 $C_G(R)$-点与 $C_P(R)$-点（见命题 3.23）；而且，显然 $C_P(R)_{\bar{\theta}} \subset C_G(R)_{\bar{\zeta}}$；进一步，因为

Denote by $\bar{X}$ the image of $X \subset (\mathcal{O}G)^R$ in $(\mathcal{O}G)(R)$; in particular, $\bar{\zeta}$ and $\bar{\theta}$ respectively are points of $C_G(R)$ and $C_P(R)$ on $kC_G(R)$ (see Proposition 3.23); moreover, we clearly have $C_P(R)_{\bar{\theta}} \subset C_G(R)_{\bar{\zeta}}$; furthermore, since we have

13.10.3 $$\overline{\mathrm{Tr}_T^{\tilde{C}_G(R)}(a)} = \mathrm{Tr}_{C_P(R)}^{C_G(R)}(\bar{a}), \quad \mathrm{Br}_{C_P(R)}^{kC_G(R)}(\bar{\theta}) = \mathrm{Br}_T(\theta) \neq \{0\}$$

其中 $a \in (\mathcal{O}G)^T$，所以 $\bar{\zeta}$ 包含在 $\big(kC_G(R)\big)_{C_P(R)}^{C_G(R)}$ 里，而 $\bar{\theta}$ 是局部的（见推论 6.6）．这样，由推论 6.12，$C_P(R)_{\bar{\theta}}$ 是 $C_G(R)_{\bar{\zeta}}$ 的亏点群；特别是，由式 13.10.2 和定理 10.14，得到

for any $a \in (\mathcal{O}G)^T$, $\big(kC_G(R)\big)_{C_P(R)}^{C_G(R)}$ contains $\bar{\zeta}$ and $\bar{\theta}$ is local (see Corollary 6.6). Thus, according to Corollary 6.12, $C_P(R)_{\bar{\theta}}$ is a defect pointed group of $C_G(R)_{\bar{\zeta}}$; in particular, from equality 13.10.2 and Theorem 10.14, we get

13.10.4 $$\frac{\dim_k\big(kC_G(R)_{\bar{\theta}}\big)}{|C_P(R)|} \equiv |E_{C_G(R)}\big(C_P(R)_{\bar{\theta}}\big)| \not\equiv 0 \pmod p .$$

另一方面，显然 $kC_G(R)_{\bar{\theta}}$ 与 $B_\theta(R)$ 是同构的；进一步，由定理 10.14，T_θ 是自中心化的，从而，由推论 10.16，其它 B 上的 T-点不是局部的；这样，对任意 $\eta \in \mathcal{P}_B(T) - \{\theta\}$ 有

On the other hand, $kC_G(R)_{\bar{\theta}}$ and $B_\theta(R)$ are clearly isomorphic; moreover, by Theorem 10.14, T_θ is self-centralizing and therefore, according to Corollary 10.16, all the other points of T on B are not local; thus, for any $\eta \in \mathcal{P}_B(T) - \{\theta\}$, we have

13.10.5 $$\mathrm{Br}_{C_P(R)}^{kC_G(R)}(\bar{\eta}) = \mathrm{Br}_T(\eta) = \{0\} ,$$

从而，或者 $\bar{\eta} = \{0\}$，或者 $\bar{\eta}$ 是 $B(R)$ 上的非局部 $C_P(R)$-点．在第二情况里存在一个 $C_P(R)$ 的真子群 U，以及一个 $B(R)$ 的有正交 $C_P(R)/U$-迹幂等元 $\bar{\ell}$ 使得 $\mathrm{Tr}_U^{C_P(R)}(\bar{\ell})$ 属于 $\bar{\eta}$．而且，对任意 $B(R)^{C_P(R)}$ 的幂等元 $\bar{\jmath}$ 与 $u \in C_P(R)$，

and therefore either $\bar{\eta} = \{0\}$, or $\bar{\eta}$ is a nonlocal point of $C_P(R)$ on $B(R)$. In the second case, there exist a proper subgroup U of $C_P(R)$ and an idempotent $\bar{\ell}$ of $B(R)$ having an orthogonal $C_P(R)/U$-trace such that $\mathrm{Tr}_U^{C_P(R)}(\bar{\ell})$ belongs to $\bar{\eta}$. Moreover, for any idempotent $\bar{\jmath}$ in $B(R)^{C_P(R)}$ and any $u \in C_P(R)$, $\bar{\jmath}B(R)\bar{\ell}^u$ and $\bar{\ell}^u B(R)\bar{\jmath}$

$\bar{\jmath}B(R)\bar{\ell}^u$ 与 $\bar{\ell}^u B(R)\bar{\jmath}$ 分别是投射左与右$kC_P(R)$-模, 所以, 由引理 10.20, $p|C_P(R)|$ 整除戏面的维数

respectively are projective left and right $kC_P(R)$-modules; consequently, it follows from Lemma 10.20 that $p|C_P(R)|$ divides the following dimensions

13.10.6 $$\dim_k\big(\bar{\jmath}B(R)\mathrm{Tr}_U^{C_P(R)}(\bar{\ell})\big)\,, \quad \dim_k\big(\mathrm{Tr}_U^{C_P(R)}(\bar{\ell})B(R)\bar{\jmath}\big)\,.$$

最后, 只要考虑一个 B^T 的相互正交本原幂等元的集合 J 使得 $\sum_{j\in J} j \in \gamma$ 就得到

Finally, considering a set J of pairwise orthogonal primitive idempotents of B^T such that $\sum_{j\in J} j \in \gamma$, we get

13.10.7 $$\dim_k\big(B(R)\big) = \sum_{\eta,\eta'\in\mathcal{P}_B(T)} \mathrm{m}_\eta^\gamma\, \mathrm{m}_{\eta'}^\gamma \dim_k\big(\bar{\jmath}_\eta\, B(R)\bar{\jmath}_{\eta'}\big)\,,$$

其中 $\bar{\jmath}_\eta \in \bar{\eta}$; 所以上面的研究推出

where we choose $\bar{\jmath}_\eta \in \bar{\eta}$; hence, all the considerations above imply that

13.10.8 $$\frac{\dim_k\big(B(R)\big)}{|C_P(R)|} \equiv (\mathrm{m}_\theta^\gamma)^2\,\frac{\dim_k\big(kC_G(R)_{\bar{\theta}}\big)}{|C_P(R)|} \pmod p$$

并, 由推论 12.10, p 不整除 m_θ^γ. 引理证毕.

and, according to Corollary 12.10, p does not divide m_θ^γ. We are done.

注意 13.11. 事实上, 可以证明对 B 上的自中心化的点群 T_θ 有 $(\mathrm{m}_\theta^\gamma)^2 \equiv 1 \pmod p$, 从而还有

Remark 13.11. Actually, it is possible to prove that, for any self-centralizing pointed group T_θ on B, we have $(\mathrm{m}_\theta^\gamma)^2 \equiv 1 \pmod p$ and therefore we get

13.11.1 $$\frac{\dim_k\big(B(R)\big)}{|C_P(R)|} \equiv |E_{C_G(R)}\big(C_P(R)_{\bar{\theta}}\big)| \pmod p\,.$$

定理 13.12. 假定 k 是代数闭的. 那么 $k\otimes_{kZ(Q)} D(Q)$ 是单代数. 特别是, Q 在 D 上有唯一局部点 δ.

Theorem 13.12. *Assume that k is algebraically closed. Then, $k\otimes_{kZ(Q)}D(Q)$ is a simple algebra. In particular, Q has a unique local point δ on D.*

证明: 因为 $[P,Q]\subset Q$, 所以任意 $u\in P$ 满足 Q 稳定 Du, 从而有

Proof: Since we have $[P,Q]\subset Q$, Q stabilizes Du for any $u\in P$ and therefore we have

13.12.1 $$B(Q) = \bigoplus_{u\in X}(Du)(Q) = \sum_{u\in P}(Du)(Q)$$

其中 X 是 P/Q 的一个代表集合; 此时, 只要应用引理 7.3, 就显然得到

where X is a set of representatives for P/Q; at this point, it follows from Lemma 7.3 that we have

13.12.2 $$J\big(D(Q)\big) = D(Q) \cap J\big(B(Q)\big)$$

从而对任意 $\hat{\delta} \in \mathcal{LP}_B(Q)$ 有

and therefore, for any $\hat{\delta} \in \mathcal{LP}_B(Q)$, we get

13.12.3 $$B(Q_{\hat{\delta}}) = \sum_{u \in P} s_{\hat{\delta}}\big((Du)^Q\big), \quad s_{\hat{\delta}}(D) \cong \prod_{\delta} D(Q_\delta).$$

其中 δ 跑遍包含在 $\hat{\delta}$ 中的 $\mathcal{LP}_D(Q)$ 的元素.

where δ runs over the set of elements $\mathcal{LP}_D(Q)$ contained in $\hat{\delta}$.

而且, 令 $\hat{\zeta}$ 与 $\hat{\xi}$ 分别记 $\mathcal{O}G$ 上的 $\tilde{C}_G(Q)$-点与 $P{\cdot}C_G(Q)$-点使得

Moreover, denote by $\hat{\zeta}$ and $\hat{\xi}$ the respective points of $\tilde{C}_G(Q)$ and $P{\cdot}C_G(Q)$ on $\mathcal{O}G$ fulfilling

13.12.4 $$Q_{\hat{\delta}} \subset \tilde{C}_G(Q)_{\hat{\zeta}} \subset P{\cdot}C_G(Q)_{\hat{\xi}};$$

因为 $P_{\hat{\gamma}} \subset P{\cdot}C_G(Q)_{\hat{\xi}}$ (见 10.12) 所以 $P_{\hat{\gamma}}$ 是 $P{\cdot}C_G(Q)_{\hat{\xi}}$ 的亏点群; 特别是, 如果 $T_{\hat{\theta}}$ 是一个 $\tilde{C}_G(Q)_{\hat{\zeta}}$ 的亏点群, 那么存在 $z \in C_G(Q)$ 使得 $(T_{\hat{\theta}})^z \subset P_{\hat{\gamma}}$. 此时, 我们能应用定理 13.9, 从而 $C_P(Q)$ 在满足 $\delta \subset \hat{\delta}$ 的 $\delta \in \mathcal{LP}_D(Q)$ 的集合上可迁地作用. 所以从式子 13.12.3, 得到

since we have $P_{\hat{\gamma}} \subset P{\cdot}C_G(Q)_{\hat{\xi}}$ (see 10.12), $P_{\hat{\gamma}}$ is a defect pointed group of $P{\cdot}C_G(Q)_{\hat{\xi}}$; in particular, if $T_{\hat{\theta}}$ is a defect pointed group of $\tilde{C}_G(Q)_{\hat{\zeta}}$ then there exists $z \in C_G(Q)$ such that $(T_{\hat{\theta}})^z \subset P_{\hat{\gamma}}$. At this point, we can apply Theorem 13.9 and therefore $C_P(Q)$ acts transitively on the set of $\delta \in \mathcal{LP}_D(Q)$ fulfilling $\delta \subset \hat{\delta}$. Consequently, from the isomorphism in 13.12.3, we get

13.12.5 $$\mathrm{m}_{\hat{\delta}}^{\hat{\gamma}} = |C_P(Q)/N_{C_P(Q)}(Q_\delta)|\mathrm{m}_{\delta}^{\gamma}$$

其中 $\delta \in \mathcal{P}_D(Q)$ 包含在 $\hat{\delta}$ 中.

for some $\delta \in \mathcal{P}_D(Q)$ contained in $\hat{\delta}$.

另一方面, 从式 10.2.2 不难验证 $\mathrm{Br}_Q\big((\mathcal{O}G)^P\big) = kC_G(Q)^P$ 从而 $\tilde{\imath} = \mathrm{Br}_Q(i)$ 在 $kC_G(Q)^P$ 里是本原幂等元; 这样, 由定理 9.5, $\tilde{\imath}$ 在 $k\big(P{\cdot}C_G(Q)\big)^P$ 里也是本原的. 进一步, 显然

On the other hand, from 10.2.2, it is easy to check that $\mathrm{Br}_Q\big((\mathcal{O}G)^P\big) = kC_G(Q)^P$ and therefore $\tilde{\imath} = \mathrm{Br}_Q(i)$ is a primitive idempotent in $kC_G(Q)^P$; thus, by Theorem 9.5, $\tilde{\imath}$ is also primitive in $k\big(P{\cdot}C_G(Q)\big)^P$. Furthermore, we clearly have

13.12.6 $$\mathrm{Br}_P^{k(P\cdot C_G(Q))}(\tilde{\imath}) = \mathrm{Br}_P(i) \neq \{0\}$$

从而 $\tilde{\imath}$ 包含在一个 $k\big(P{\cdot}C_G(Q)\big)$ 上的局部 P-点 $\tilde{\gamma}$ 中; 而且, 因为对任意 $P{\cdot}C_G(Q)$ 的真包含 P

and therefore $\tilde{\imath}$ is contained in a local point $\tilde{\gamma}$ of P on $k\big(P{\cdot}C_G(Q)\big)$; moreover, since we have $\mathrm{Br}_R(i) = \{0\}$ for any p-subgroup R

的 p-子群 R 有 $\mathrm{Br}_R(i)=\{0\}$，所以 $P_{\tilde\gamma}$ 在 $k\big(P{\cdot}C_G(Q)\big)$ 上是极大的局部点群; 特别是, 下面的 P-内代数

of $P{\cdot}C_G(Q)$ strictly containing P, the local pointed group $P_{\tilde\gamma}$ on $k\big(P{\cdot}C_G(Q)\big)$ is maximal; in particular, the following P-interior algebra

13.12.7 $$\tilde C=\tilde\imath k\big(P{\cdot}C_G(Q)\big)\tilde\imath=B(Q)\otimes_{C_P(Q)}P$$

是一个 $P{\cdot}C_G(Q)$-块的源代数. 此时, 与 13.12.5 类似, 对任意 $\tilde\delta\in\mathcal{P}(\tilde C)$ 我们能证明

is the source algebra of a block of $P{\cdot}C_G(Q)$. At this point, as in 13.12.5 above, for any $\tilde\delta\in\mathcal{P}(\tilde C)$ we can prove that

13.12.8 $$\mathrm{m}_{\tilde\delta}^{\tilde\gamma}=|P/N_P(Q_{\hat\delta})|\mathrm{m}_{\hat\delta}^{\hat\gamma}$$

其中 $\hat\delta$ 是满足 $\tilde\delta$ 包含 $\mathrm{Br}_Q(\hat\delta)$ 的 $\mathcal{LP}_B(Q)$ 的某个元素.

where $\hat\delta$ is some element of $\mathcal{LP}_B(Q)$ such that $\mathrm{Br}_Q(\hat\delta)$ is contained in $\tilde\delta$.

可是, 由推论 12.10, 存在 $\tilde\delta\in\mathcal{P}(\tilde C)$ 使得 p 不整除 $\mathrm{m}_{\tilde\delta}^{\tilde\gamma}$; 所以也存在唯一 $\hat\delta\in\mathcal{P}_B(Q)$ 使得 $\mathrm{Br}_Q(\hat\delta)\subset\tilde\delta$ 并且 P 稳定它; 类似, 那么存在唯一 $\delta\in\mathcal{LP}_D(Q)$ 使得 $\delta\subset\hat\delta$ 并且 $C_P(Q)$ 稳定它.

But, according to Corollary 12.10, there is $\tilde\delta\in\mathcal{P}(\tilde C)$ such that p does not divide $\mathrm{m}_{\tilde\delta}^{\tilde\gamma}$; hence, there exists a unique $\hat\delta\in\mathcal{P}_B(Q)$ such that $\mathrm{Br}_Q(\hat\delta)\subset\tilde\delta$ and thus P stabilizes it; similarly, then there exists a unique $\delta\in\mathcal{LP}_D(Q)$ such that $\delta\subset\hat\delta$ and thus $C_P(Q)$ stabilizes it.

特别是, 有

In particular, we have

13.12.9 $$S=D(Q_\delta)\cong B(Q_{\hat\delta})$$

而 p 不整除 S 的维数; 这样, 我们有 $S=Z(S)\oplus\mathrm{Ker}(\mathrm{tr})$，其中 $\mathrm{tr}\colon S\to K$ 记迹映射. 令

and p does not divide the dimension of S; thus, we have $S=Z(S)\oplus\mathrm{Ker}(\mathrm{tr})$, where $\mathrm{tr}\colon S\to K$ denotes the trace map. Set

13.12.10 $$\overline{B(Q)}=k\otimes_{kZ(Q)}B(Q)\,,\quad\overline{(Du)(Q)}=k\otimes_{kZ(Q)}(Du)(Q)$$

再令 $\bar s\colon\overline{D(Q)}\to S$ 记自然同态; 我们要考虑下面的商

and denote by $\bar s\colon\overline{D(Q)}\to S$ the natural map; we will consider the following quotient

13.12.11 $$\begin{aligned}\overline{\overline{B(Q)}}&=\overline{B(Q)}\Big/\overline{B(Q)}{\cdot}\mathrm{Ker}(\bar s){\cdot}\overline{B(Q)}\\&=\bigoplus_{u\in X}\overline{(Du)(Q)}\Big/\sum_{v\in P}\overline{(Dv)(Q)}{\cdot}\mathrm{Ker}(\bar s){\cdot}\overline{(Dv^{-1}u)(Q)}\,.\end{aligned}$$

请注意, 对任意 $u\in X\cap\tilde C_P(Q)$ 能假定 $u\in C_P(Q)$；此湿

Note that, for any $u\in X\cap\tilde C_P(Q)$, we may assume that $u\in C_P(Q)$; then, we have

13.12.12 $$\overline{(Du)(Q)}=\overline{D(Q)}\bar u$$

其中 $\bar{u}$ 是 u 在 $\overline{B(Q)}$ 中的像; 进一步, 这个选择决定一个双射

where $\bar{u}$ is the image of u in $\overline{B(Q)}$; moreover, this choice determines a bijective map

13.12.13 $$X \cap \tilde{C}_P(Q) \longrightarrow \bar{C}_P(Q)\,.$$

这个研究推出 $S\bar{C}_P(Q)$ 是 $\overline{\overline{B(Q)}}$ 的幺子代数, 也是 $\overline{\overline{B(Q)}}$ 作为 $S\bar{C}_P(Q)$-双模时的直和项. 它也推出, 因为有

These remarks imply that $S\bar{C}_P(Q)$ is a unitary subalgebra of $\overline{\overline{B(Q)}}$, and that, as $S\bar{C}_P(Q)$-bimodules, it is a direct summand of $\overline{\overline{B(Q)}}$. They also imply that, since we have

13.12.14 $$S\bar{C}_P(Q) = k\bar{C}_p(Q) \oplus \sum_{\bar{u}\in\bar{C}_P(Q)} \mathrm{Ker}(\mathrm{tr})\bar{u}$$

并且 $\bar{C}_P(Q)$ 稳定 Ker(tr), 所以 $k\bar{C}_P(Q)$ 是 $S\bar{C}_P(Q)$ 作为 $k\bar{C}_P(Q)$-双模时的直和项. 特别是, 投射 $\overline{\overline{B(Q)}}$-模限制为 $k\bar{C}_P(Q)$-模时也是投射的; 也就是说, 如果 J 是一个 $B(Q)$ 的相互正交本原幂等元使得 $\sum_{j\in J} j = \tilde{\imath}$, 那么对任意 $j \in J$, $\overline{B(Q)}$-模 $\overline{\overline{B(Q)}j}$ 满足 (见例子 4.15 与同构式 5.8.6 与定理 5.11 并 5.12)

and $\bar{C}_P(Q)$ stabilizes Ker(tr), $k\bar{C}_P(Q)$ is a direct summand of $S\bar{C}_P(Q)$ considered as a $k\bar{C}_P(Q)$-bimodule. In particular, the restriction of any projective $\overline{\overline{B(Q)}}$-module to $k\bar{C}_P(Q)$ remains projective; that is to say, if J is a set of pairwise orthogonal primitive idempotents of $B(Q)$ fulfilling $\sum_{j\in J} j = \tilde{\imath}$, then, for any $j \in J$, the $\overline{B(Q)}$-module $\overline{\overline{B(Q)}j}$ fulfills (see Example 4.15, isomorphism 5.8.6 and Theorems 5.11 and 5.12)

13.12.15 $$\mathrm{End}_k\Big(\overline{\overline{B(Q)j}}\Big)^{\bar{C}_P(Q)} = \mathrm{End}_k\Big(\overline{\overline{B(Q)j}}\Big)_1^{\bar{C}_P(Q)}\,.$$

另一方面, 记 $\mathrm{Br}_Q(\hat{\zeta}) = \{\tilde{e}\}$; 因为在对应的完备诱导的 G-内代数里有 $\hat{\zeta} \subset \mathrm{ind}_{\tilde{C}_P(Q)}^{\tilde{C}_G(Q)}(\hat{\theta})$ (见定理 5.11 并 6.8), 所以还有 (见定理 5.12)

On the other hand, set $\mathrm{Br}_Q(\hat{\zeta}) = \{\tilde{e}\}$; since we have $\hat{\zeta} \subset \mathrm{ind}_{\tilde{C}_P(Q)}^{\tilde{C}_G(Q)}(\hat{\theta})$ in the corresponding inductively complete G-interior algebra (see Theorems 5.11 and 6.8), we still have (see Theorem 5.12)

13.12.16 $$\tilde{e} = \mathrm{Br}_Q\big(\mathrm{Tr}_{C_P(Q)}^{C_G(Q)}(a\ell a')\big) = \mathrm{Tr}_{C_P(Q)}^{C_G(Q)}\big(\mathrm{Br}_Q(a\ell a')\big)\,,$$

其中 $a, a' \in (\mathcal{O}G)^{\tilde{C}_P(Q)}$ 与 $\ell \in \hat{\theta}$, 从而 $\bar{e} = \mathrm{Tr}_{\bar{C}_P(Q)}^{\bar{C}_G(Q)}(\bar{a}\bar{\ell}\bar{a}')$, 其中 $\bar{e}, \bar{a}, \bar{a}', \bar{\ell}$ 分别是 $\tilde{e}, a, a', \ell$ 在 $k\bar{C}_G(Q)$ 中的像. 这样, 令

for suitable $a, a' \in (\mathcal{O}G)^{\tilde{C}_P(Q)}$ and $\ell \in \hat{\theta}$, and therefore $\bar{e} = \mathrm{Tr}_{\bar{C}_P(Q)}^{\bar{C}_G(Q)}(\bar{a}\bar{\ell}\bar{a}')$, where $\bar{e}$, $\bar{a}$, $\bar{a}'$ and $\bar{\ell}$ respectively are the images of $\tilde{e}$, a, a' and ℓ in $k\bar{C}_G(Q)$. Thus, setting

13.12.17 $$\bar{\zeta} = \mu_{Z(k\bar{C}_G(Q))}^{\bar{e}}\,, \qquad \bar{\theta} = \mu_{k\bar{C}_G(Q)^{\bar{C}_P(Q)}}^{\bar{\ell}}\,,$$

再令$\bar\gamma$记γ在$\mathcal{D}_{k\bar C_G(Q)}(\bar C_P(Q))$里的像; 那么在对应的完备诱导的$\bar C_G(Q)$-内代数里有

and denoting by $\bar\gamma$ the image of γ in $\mathcal{D}_{k\bar C_G(Q)}(\bar C_P(Q))$, in the corresponding inductively complete $\bar C_G(Q)$-interior algebra we have

13.12.18 $$\bar\zeta \subset \mathrm{ind}_{\bar C_P(Q)}^{\bar C_G(Q)}(\bar\theta) \subset \mathrm{ind}_{\bar C_P(Q)}^{\bar C_G(Q)}(\bar\gamma)\,.$$

特别是, 不难证明对任意$k\bar C_G(Q)_{\bar\zeta}$ $(=k\bar C_G(Q)\bar e)$上的局部点群$\bar R_{\bar\varepsilon}$存在$\bar z\in\bar C_G(Q)$使得

In particular, it is easy to prove that, for any local pointed group $\bar R_{\bar\varepsilon}$ on $k\bar C_G(Q)_{\bar\zeta}$ $(=k\bar C_G(Q)\bar e)$, there is $\bar z\in\bar C_G(Q)$ fulfilling

13.12.19 $$\bar R^{\bar z}\subset \bar C_P(Q)\,,\quad \bar\varepsilon^{\bar z}\subset \mathrm{res}_{\bar R}^{\bar C_P(Q)}(\bar\gamma)\,,$$

从而$(\bar R_{\bar\varepsilon})^{\bar z}$也是一个$k\bar C_G(Q)_{\bar\gamma}$ $(=\overline{B(Q)})$上的局部点群. 这样, 只要把$\bar R=1$应用推论 6.4, 就得到

and therefore $(\bar R_{\bar\varepsilon})^{\bar z}$ is also a local pointed group on $k\bar C_G(Q)_{\bar\gamma}$ $(=\overline{B(Q)})$. Thus, by applying Corollary 6.4 to $\bar R=1$, we get

13.12.20 $$\mathcal{P}\big(k\bar C_G(Q)\bar e\big)=\mathcal{P}\big(\overline{B(Q)}\big)\,,$$

从而$k\bar C_G(Q)\bar e$是 Morita 等价于$\overline{B(Q)}$ (见 3.13).

and therefore $k\bar C_G(Q)\bar e$ is Morita equivalent to $\overline{B(Q)}$ (see 3.15).

取$j\in J$, 令$\overline{\overline{M}}$记对应于$\overline{B(Q)}$-模$\overline{\overline{B(Q)j}}$的$k\bar C_G(Q)$-模再令$\bar{\bar\zeta}$与$\bar{\bar\gamma}$分别记$\bar\zeta$与$\bar\gamma$在$\mathrm{End}_k(\overline{\overline{M}})$上的对应的$\bar C_G(Q)$-除子与$\bar C_P(Q)$-除子(见4.12.1); 类似地, 因为$\bar{\bar\zeta}\subset\mathrm{ind}_{\bar C_P(Q)}^{\bar C_G(Q)}(\bar{\bar\gamma})$, 所以对任意$\mathrm{End}_k(\overline{\overline{M}})$上的局部点群$\bar R_{\bar{\bar\varepsilon}}$存在$\bar z\in\bar C_G(Q)$使得

Pick $j\in J$ and denote by $\overline{\overline{M}}$ the $k\bar C_G(Q)$-module corresponding to the $\overline{B(Q)}$-module $\overline{\overline{B(Q)j}}$, and by $\bar{\bar\zeta}$ and $\bar{\bar\gamma}$ the respective divisors of $\bar C_G(Q)$ and $\bar C_P(Q)$ on $\mathrm{End}_k(\overline{\overline{M}})$ corresponding to $\bar\zeta$ and $\bar\gamma$ (see 4.12.1); similarly, since we have $\bar{\bar\zeta}\subset\mathrm{ind}_{\bar C_P(Q)}^{\bar C_G(Q)}(\bar{\bar\gamma})$, for any local pointed group $\bar R_{\bar{\bar\varepsilon}}$ on $\mathrm{End}_k(\overline{\overline{M}})$ there exists $\bar z\in\bar C_G(Q)$ such that

13.12.21 $$\bar R^{\bar z}\subset\bar C_P(Q)\,,\quad \tilde\varepsilon^{\bar z}\subset\mathrm{res}_{\bar R}^{\bar C_P(Q)}(\tilde\gamma)\,,$$

从而$(\bar R_{\bar{\bar\varepsilon}})^{\bar z}$也是$\mathrm{End}_k(\overline{\overline{B(Q)j}})$上的局部点群.

and therefore $(\bar R_{\bar{\bar\varepsilon}})^{\bar z}$ is also a local pointed group on $\mathrm{End}_k(\overline{\overline{B(Q)j}})$.

此时, 从式 13.12.15 可推出$\bar R=1$, 也就是说$\overline{\overline{M}}$是投射的; 所以$\overline{\overline{B(Q)j}}$也是一个投射$\overline{B(Q)}$-模, 从而有$\overline{B(Q)j}\cong\overline{\overline{B(Q)j}}$; 特别是, 对任意满足$\bar{\bar\jmath}'=0$的$j'\in J$有$\overline{j'B(Q)j}=\{0\}$并且, 既然

At this point, equality 13.12.15 implies that $\bar R=1$ and, in other terms, that $\overline{\overline{M}}$ is projective; consequently, $\overline{\overline{B(Q)j}}$ is a projective $\overline{B(Q)}$-module too, and therefore we have $\overline{B(Q)j}\cong\overline{\overline{B(Q)j}}$; in particular, for any $j'\in J$ such that $\bar{\bar\jmath}'=0$, we have $\overline{j'B(Q)j}=\{0\}$ and, as far as the algebra

$\overline{B(Q)}$ 是对称的 (见同构式 10.1.1), 还有 $\overline{jB(Q)j'} = \{0\}$; 现在, 由 Nakayama 引理, 有

$\overline{B(Q)}$ is symmetric (see 10.1.1), we still have $\overline{jB(Q)j'} = \{0\}$; now, from the Nakayama Lemma, we get

13.12.22 $$j'B(Q)j = \{0\} = jB(Q)j' .$$

最后, 令 $\tilde{f}$ 记满足 $\bar{\jmath} \neq 0$ 的 $j \in J$ 的和; 因为

Finally, denote by $\tilde{f}$ the sum of all the elements $j \in J$ such that $\bar{\jmath} \neq 0$; since

13.12.23 $$(\tilde{\imath} - \tilde{f})B(Q)\tilde{f} = \{0\} = \tilde{f}B(Q)(\tilde{\imath} - \tilde{f}) ,$$

所以 $\tilde{f}$ 属于 $Z\big(B(Q)\big)$; 此时, 只要应用定理 10.7, 就得到 $\tilde{f} = \tilde{\imath}$; 这样, $\overline{B(Q)} = \overline{\overline{B(Q)}}$, 从而得到 $\mathrm{Ker}(\bar{s}) = \{0\}$ 与 $\overline{D(Q)} \cong S$. 特别是, 由推论 6.6, 有 $\mathcal{LP}_D(Q) = \{\delta\}$.

$\tilde{f}$ belongs to $Z\big(B(Q)\big)$; then, it follows from Theorem 10.7 that $\tilde{f} = \tilde{\imath}$; thus, we obtain $\overline{B(Q)} = \overline{\overline{B(Q)}}$ and therefore $\mathrm{Ker}(\bar{s}) = \{0\}$ and $\overline{D(Q)} \cong S$. In particular, according to Corollary 6.6, we get $\mathcal{LP}_D(Q) = \{\delta\}$.

推论 13.13. **假定 k 是代数闭的. 如果 $Q = 1$ 那么 D 是一个 $\mathcal{O}$ 上的矩阵代数.**

Corollary 13.13. *Assume that k is algebraically closed. If $Q = 1$ then D is a matrix algebra over $\mathcal{O}$.*

注意 13.14. 由推论 13.6, 如果 $Q = 1$ 那么 $\mathfrak{h}_G(P_{\hat{\gamma}}) = 1$; 请注意 $\mathfrak{h}_G(P_{\hat{\gamma}}) = 1$ 等价于对任意 $P_{\hat{\gamma}}$ 的局部子点群 $R_{\hat{\varepsilon}}$ 有 $E_G(R_{\hat{\varepsilon}})$ 是 p-群; 也就是说, b 是所谓幂零块.

Remark 13.14. According to Corollary 13.6, if $Q = 1$ then $\mathfrak{h}_G(P_{\hat{\gamma}}) = 1$; note that $\mathfrak{h}_G(P_{\hat{\gamma}}) = 1$ is equivalent to say that, for any local pointed subgroup $R_{\hat{\varepsilon}}$ of P_γ, $E_G(R_{\hat{\varepsilon}})$ is a p-group; that is to say, b is a so-called *nilpotent block.*

证明: 如果 $Q = 1$ 那么 $D(Q) = k \otimes_{\mathcal{O}} D$; 设 J 是 D 的相互正交本原幂等元的集合使得 $\sum_{j\in J} j = i$; 由定理 13.12, 对任意 $j, j' \in J$ 有 $k \otimes_{\mathcal{O}} jDj'$ 与 k 是同构的, 从而还有 jDj' 与 $\mathcal{O}$ 也是同构的; 此时, 不难验证 D 就是一个 $\mathcal{O}$ 上的矩阵代数.

Proof: If $Q = 1$ then $D(Q) = k \otimes_{\mathcal{O}} D$; let J be a set of pairwise orthogonal primitive idempotents of D fulfilling $\sum_{j\in J} j = i$; according to Theorem 13.12, for any $j, j' \in J$, $k \otimes_{\mathcal{O}} jDj'$ is isomorphic to k and therefore jDj' is isomorphic to $\mathcal{O}$ too; at this point, it is not difficult to prove that D is a matrix algebra over $\mathcal{O}$.

14. 超聚焦子代数的唯一性

14. Uniqueness of the Hyperfocal Subalgebra

14.1. 这里，我们假定 $\mathcal{O}$ 的分式域 $\mathcal{K}$ 是特征为零的域. 如同在§12里一样，G 是有限群并 b 是一个 G-块；取 $\mathcal{O}Gb$ 上的极大的局部点群 P_γ 与 $i\in\gamma$，而记

14.1. In this section, we assume that the quotient field $\mathcal{K}$ of $\mathcal{O}$ has characteristic zero. As in section 12, G is a finite group and b a block of G; choose a maximal local pointed group P_γ on $\mathcal{O}Gb$ and $i \in \gamma$, and set

14.1.1 $$B = i(\mathcal{O}G)i \cong (\mathcal{O}G)_\gamma\,.$$

设 Q 是 P 的规正子群并设 D 是 B 的 P-稳定幺子代数使得

Let Q be a normal subgroup of P and D a P-stable unitary subalgebra of B fulfilling

14.1.2 $$D \cap Pi = Qi\,, \quad B = D \otimes_Q P\,;$$

这样 D 成为一个 Q-内 P-代数，以及 B 就是对应的 P-内代数 (见 9.1).

thus D becomes a Q-interior P-algebra, and B is the corresponding P-interior algebra (see 9.1).

14.2. 假定k是代数闭的；这里，要证明满足条件 14.1.2 的 B 的 P-稳定幺子代数都是相互 $(B^P)^*$-共轭的；事实上，下面的结果是更精确的，我们在后面的存在性的证明里需要这个更精确的体式.

14.2. Assume that k is algebraically closed; here, we will prove that all the P-stable unitary subalgebras of B fulfilling condition 14.1.2 are mutually $(B^P)^*$-conjugate; actually, the result below is more precise since we after need it in the proof of the existence.

命题 14.3. 假定 k 是代数闭的. 那么$C_Q(P)$-内代数 $D(P)$ 与 $kC_Q(P)$ 是同构的. 特别是，有

Proposition 14.3. *Assume that k is algebraically closed. Then $D(P)$ and $kC_Q(P)$ are isomorphic $C_P(Q)$-interior algebras. In particular, we have*

14.3.1 $$D^P = \mathcal{O}C_Q(P) + \mathrm{Ker}(\mathrm{Br}_P^D)\,,\qquad \Big(\sum_{u\in P-Q\cdot Z(P)} D\otimes u\Big)^P \subset \mathrm{Ker}(\mathrm{Br}_P^B)\,.$$

证明: 由引理 10.22，存在一个 B 的 $\mathcal{O}$-基 T 使得 $P{\cdot}T{\cdot}P = T$ 并且 $|P{\cdot}t| = |P| = |t{\cdot}P|$，其中$t\in T$，从而不难验证下面的集合

Proof: By Lemma 10.22, there exists an $\mathcal{O}$-basis T of B such that $P{\cdot}T{\cdot}P = T$ and that $|P{\cdot}t| = |P| = |t{\cdot}P|$ for any $t \in T$, and then it is easily checked that the set

14.3.2 $$\overline{T^P} = \{\mathrm{Br}_P(t)\}_{t\in T\cap B^P}$$

是满足 $Z(P)\cdot\overline{T^P}=\overline{T^P}$ 的 $B(P)$ 的 k-基，并且对任意 $\bar{t}\in\overline{T^P}$ 有 $|Z(P)\cdot\bar{t}|=|Z(P)|$. 而且，因为 k 是代数闭的，所以只要使用同构 10.14.3 , 就得到

is an $\mathcal{O}$-basis of $B(P)$ fulfilling $Z(P)\cdot\overline{T^P} = \overline{T^P}$ and $|Z(P)\cdot\bar{t}| = |Z(P)|$ for any $\bar{t} \in \overline{T^P}$. Moreover, since k is algebraically closed, from isomorphism 10.14.3, we get

14.3.3
$$\begin{aligned} k\otimes_{kZ(P)} B(P) &\cong k\otimes_{kZ(P)} \big(\mathrm{Br}_P(i)kC_G(P)\mathrm{Br}_P(i)\big) \\ &\cong s_\gamma(i)(\mathcal{O}G)(P_\gamma)s_\gamma(i) \cong k \end{aligned}$$

这样，有 $\dim_k\big(B(P)\big)=|Z(P)|$; 可是，由引理 10.24, $kZ(P)$-模 $kZ(P)\mathrm{Br}_P(i)$ 是 $B(P)$ 的直和项; 所以有 $B(P)\cong kZ(P)$.

and thus, we have $\dim_k\big(B(P)\big) = |Z(P)|$; but, according to Lemma 10.24, the $kZ(P)$-module $kZ(P)\mathrm{Br}_P(i)$ is a direct summand of $B(P)$; hence, we have $B(P)\cong kZ(P)$.

另一方面，令 $U\subset P$ 记一个 P/Q 的代表集合; 因为 P 稳定分解如下

On the other hand, denote by $U\subset P$ a set of representatives for P/Q; since P stabilizes the decomposition

14.3.4
$$B = \bigoplus_{u\in U} D\otimes u\,,$$

所以不难验证有

it is easy to check that

14.3.5
$$B(P)\cong \bigoplus_{u\in U_P} (D\otimes u)(P)$$

其中 U_P 是满足 $[P,u]\subset Q$ 的元素 $u\in U$ 的集合; 特别是，$D(P)$ 在 $kC_Q(P)$-模 $B(P)$ 里是直和项，并且对任意 $z\in Z(P)$ 有 $(D\otimes z)(P)\cong D(P)$. 这样，从同构 $B(P)\cong kZ(P)$ 可推出

where U_P is the set of $u\in U$ such that $[P,u]\subset Q$; in particular, $D(P)$ is a direct summand in the $kC_Q(P)$-module $B(P)$, and we get $(D\otimes z)(P)\cong D(P)$ for any $z\in Z(P)$. Thus, the isomorphism $B(P)\cong kZ(P)$ implies that

14.3.6
$$D(P)\cong kC_Q(P)\,,\quad (D\otimes u)(P)=\{0\}$$

其中 $u\in U_P-Q\cdot Z(P)$.

for any $u\in U_P-Q\cdot Z(P)$.

推论 14.4. 假定 k 是代数闭的. 那么有

Corollary 14.4. *Assume that k is algebraically closed. Then we have*

14.4.1
$$N_{J(B^P)^*}(D) = J\big(Z(B)\big)^*\cdot J(D^P)^*\,.$$

证明: 令 $U\subset P$ 记一个 P/Q 的代表集合并考虑分解如下

Proof: Denote by $U\subset P$ a set of representatives for P/Q and consider the decomposition

14.4.2
$$B = \bigoplus_{u\in U} D\otimes u\,;$$

这样, 对任意 $a \in N_{J(B^P)^*}(D)$ 有 $a = \sum_{u\in U} a_u$, 其中 a_u 属于 $D\otimes u$; 由式 14.3.1, 有

thus, for any $a \in N_{J(B^P)^*}(D)$, we have $a = \sum_{u\in U} a_u$ for suitable $a_u \in D\otimes u$; by the inclusion in 14.3.1, we have

$$14.4.3 \qquad \sum_{z\in U\cap Q\cdot Z(P)} \mathrm{Br}_P(a_z) \in J\big(B(P)\big)^* \cong J\big(kZ(P)\big)^*$$

从而对适当的 $z \in U\cap Q\cdot Z(P)$ 有 $\mathrm{Br}_P(a_z) \notin J\big(B(P)\big)$; 事实上, 能假定 $z\in Z(P)$, 并对适当的 $\lambda\in\mathcal{O}^*$, $\lambda\cdot a_z\cdot z^{-1}$ 属于 $J(D^P)^*$.

and therefore $\mathrm{Br}_P(a_z) \notin J\big(B(P)\big)$ for a suitable $z \in U\cap Q\cdot Z(P)$; actually, we may assume that $z\in Z(P)$, and then, for a suitable $\lambda\in\mathcal{O}^*$, $\lambda\cdot a_z\cdot z^{-1}$ belongs to $J(D^P)^*$.

最后令 $c=\lambda^{-1}z\cdot(a_z)^{-1}a$; 那么 c 也属于 $N_{J(B^P)^*}(D)$ 并且有

Finally, set $c = \lambda^{-1}z\cdot(a_z)^{-1}a$; then c still belongs to $N_{J(B^P)^*}(D)$ and we have

$$14.4.4 \qquad c = i + \sum_{u\in U-Q} c_u\,;$$

这样, 对任意 $d\in D$ 存在 $d'\in D$ 使得

thus, for any $d\in D$ there exists $d'\in D$ such that

$$14.4.5 \qquad (d\otimes 1)\big(i+\sum_{u\in U-Q} c_u\big) = \big(i+\sum_{u\in U-Q} c_u\big)(d'\otimes 1)\,,$$

从而 $d=d'$ 并对任意 $u\in U-Q$ 得到 $(d\otimes 1)c_u = c_u(d\otimes 1)$; 这样, c 既与 D 可交换也与 P 可交换, 从而它属于 $Z(B)$; 因为 $Z(B)\subset B^P$ 所以 c 属于 $J\big(Z(B)\big)^*$.

and therefore we get $d = d'$ and, for any $u\in U-Q$, $(d\otimes 1)c_u = c_u(d\otimes 1)$; thus, c centralizes both D and P, and therefore it belongs to $Z(B)$; since $Z(B)\subset B^P$, finally c belongs to $J\big(Z(B)\big)^*$.

命题 14.5. 使固定 $\pi\in J(\mathcal{O})$, 并记 $\bar{\mathcal{O}}=\mathcal{O}/\pi\mathcal{O}$ 与 $\bar{D}=D/\pi\cdot D$. 对任意满足 $\bar{\mathcal{O}}\otimes_{\mathcal{O}} f = \mathrm{id}_{\bar{D}}$ 的 Q-内 P-代数自同构 $f\colon D\cong D$ 存在 $d\in D^P$ 使得对任意 $a\in D$ 有 $f(a)=a^{i+\pi\cdot d}$. 特别是, 有

Proposition 14.5. *Fix $\pi\in J(\mathcal{O})$ and set $\bar{\mathcal{O}}=\mathcal{O}/\pi\mathcal{O}$ and $\bar{D}=D/\pi\cdot D$. For any Q-interior P-algebra automorphism $f\colon D\cong D$ fulfilling $\bar{\mathcal{O}}\otimes_{\mathcal{O}} f = \mathrm{id}_{\bar{D}}$, there exists $d\in D^P$ such that $f(a)=a^{i+\pi\cdot d}$ for any $a\in D$. In particular, we have*

$$14.5.1 \qquad N_{i+\pi\cdot B^P}(D) = \big(i+\pi\cdot Z(B)\big)\cdot\big(i+\pi\cdot D^P\big)\,.$$

证明: 对 $|P/Q|$ 用归纳法; 如果 $Q=P$ 那么 $D=B$ 与 $k\otimes f = \mathrm{id}_{k\otimes_{\mathcal{O}}B}$; 这样, 由命题 12.11, 存在 $n\in J(\mathcal{O})\cdot B^P$ 使得对任意 $a\in B$ 有 $f(a)=a^{i+n}$;

Proof: We argue by induction on $|P/Q|$; if $Q=P$ then $D=B$ and $k\otimes f = \mathrm{id}_{k\otimes_{\mathcal{O}}B}$; thus, according to Proposition 12.11, there exists $n\in J(\mathcal{O})\cdot B^P$ fulfilling $f(a)=a^{i+n}$ for any $a\in B$; in particular, since we

特别是, 因为 $\bar{\mathcal{O}}\otimes_{\mathcal{O}} f=\mathrm{id}_{\bar{B}}$ 所以 $i+n$ 在 $\bar{B}$ 中的像属于 $Z(\bar{B})^*$; 从而, 只要应用下面的引理 14.6, 就得到 $z\in Z(B)^*$ 使得 $z^{-1}(i+n)$ 属于 $i+\pi\cdot B^P$.

假定 $Q\neq P$ 并设 R 是包含 Q 与满足 $|R/Q|=p$ 的 P 的规正子群; 令 E 记 B 的由 D 和 Ri 生成的子代数; 也就是说, 有

assume that $\bar{\mathcal{O}}\otimes_{\mathcal{O}} f=\mathrm{id}_{\bar{B}}$, the image of $i+n$ in $\bar{B}$ belongs to $Z(\bar{B})^*$; hence, from the Lemma 14.6 below we find an element $z\in Z(B)^*$ such that $z^{-1}(i+n)$ belongs to $i+\pi\cdot B^P$.

Assume that $Q\neq P$ and let R be a normal subgroup of P containing Q and fulfilling $|R/Q|=p$; denote by E the subalgebra of B generated by D and Ri; that is to say, we have

14.5.2 $$E=D\otimes_Q R,\quad B=E\otimes_R P,\quad E\cap Pi=Ri.$$

当然, f 决定一个 R-内 P-代数自同构 $g\colon E\cong E$, 并且我们有 $\bar{\mathcal{O}}\otimes_{\mathcal{O}} g=\mathrm{id}_{\bar{E}}$; 所以, 存在 $m\in E^P$ 使得 $g(a)=a^{i+\pi\cdot m}$, 其中 $a\in E$.

另一方面, 因为 P 在 R/Q 上的作用是平凡的, 所以 P 稳定 $D\otimes v$ 其中 $v\in R$; 那么, 如果 $W\subset R$ 是 R/Q 的代表集合, 那么不难验证

Obviously, f induces an R-interior P-algebra automorphism $g\colon E\cong E$, and we still have $\bar{\mathcal{O}}\otimes_{\mathcal{O}} g=\mathrm{id}_{\bar{E}}$; hence, there exists $m\in E^P$ fulfilling $g(a)=a^{i+\pi\cdot m}$, for any $a\in E$.

On the other hand, since P acts trivially on R/Q, P stabilizes $D\otimes v$ for any $v\in R$; thus, choosing a set of representatives $W\subset R$ for R/Q, it is easy to check that we have

14.5.3 $$E^P=\bigoplus_{w\in W}(D\otimes w)^P,\quad C_E(D)=\bigoplus_{w\in W}C_E(D)\cap(D\otimes w),$$

其中 $C_E(D)$ 是对任意 $d\in D$ 满足 $e(d\otimes 1)=(d\otimes 1)e$ 的 $e\in E$ 的集合; 这样, 既然 $C_E(D)^P=Z(E)^P$, 就得到

where $C_E(D)$ denotes the set of $e\in E$ such that $e(d\otimes 1)=(d\otimes 1)e$ for any $d\in D$; in this way, since $C_E(D)^P=Z(E)^P$, we clearly get

14.5.4 $$Z(E)^P=\bigoplus_{w\in W}Z(E)^P\cap(D\otimes w).$$

对任意 $w\in W$ 不态难验证下面的商

For any $w\in W$, it is easy to prove that the quotient

14.5.5 $$(D\otimes w)^P/Z(E)\cap(D\otimes w)^P$$

是由自 $\mathcal{O}$-模; 所以在 $(D\otimes w)^P$ 里存在一个 $Z(E)^P\cap(D\otimes w)$ 的补 M_w; 这样, 下面的直和

is a free $\mathcal{O}$-module; hence, $Z(E)^P\cap(D\otimes w)$ has a complement M_w in $(D\otimes w)^P$; thus, the direct sum

14.5.6 $$M=\bigoplus_{w\in W}M_w$$

在$\mathcal{O}$-模E^P中就是一个$Z(E)^P$的补. 此时, 由后面的命题16.8, 能选择唯一$m \in M$使得$g(a) = a^{i+\pi \cdot m}$, 其中$a \in E$.

becomes a complement of $Z(E)^P$ in E^P as $\mathcal{O}$-modules. Then, by Proposition 16.8 in section 16, we can choose a unique $m \in M$ fulfilling $g(a) = a^{i+\pi \cdot m}$ for any $a \in E$.

最后, 令$\hat{\mathcal{K}}$记包含一个本原p-次单位根的$\mathcal{K}$的扩张, 再令$\hat{\mathcal{O}}$记$\hat{\mathcal{K}}$中的$\mathcal{O}$上的整元构成的环; 如果$\varphi: R \to \hat{\mathcal{O}}^*$是群同态使得$\mathrm{Ker}(\varphi) = Q$, 那么显然存在一个$\hat{\mathcal{O}}$-代数自同构

Finally, denote by $\hat{\mathcal{K}}$ an extension of $\mathcal{K}$ containing a primitive p-th root of unity, and by $\hat{\mathcal{O}}$ the integral closure of $\mathcal{O}$ in $\hat{K}$; if $\varphi: R \to \hat{\mathcal{O}}^*$ is a group homomorphism such that $\mathrm{Ker}(\varphi) = Q$, then it is clear that there exists an $\hat{\mathcal{O}}$-algebra isomorphism

14.5.7 $$t_\varphi: \hat{\mathcal{O}} \otimes_{\mathcal{O}} E \cong \hat{\mathcal{O}} \otimes_{\mathcal{O}} E = \hat{D} \otimes_Q R$$

其中$\hat{D} = \hat{\mathcal{O}} \otimes_{\mathcal{O}} D$, 使得

where $\hat{D} = \hat{\mathcal{O}} \otimes_{\mathcal{O}} D$, fulfilling

14.5.8 $$t_\varphi(d \otimes 1) = d \otimes 1, \quad t_\varphi(i \otimes v) = \varphi(v) \cdot i \otimes v$$

其中$d \in D$与$v \in R$.

for any $d \in D$ and any $v \in R$.

进一步, 不难验证g的扩张$\hat{\mathcal{O}} \otimes g$与$t_\varphi$可交换; 又, 既然$D \otimes w$上的$t_\varphi$-作用就是乘以$\varphi(w)$, t_φ稳定M_w,其中$w \in W$; 特别是, $t_\varphi(m)$属于M, 并且对任意$a \in E$还有

Moreover, it is easily checked that the extension $\hat{\mathcal{O}} \otimes g$ of g and t_φ commute; further, since the action of t_φ on $D \otimes w$ is just the multiplication by $\varphi(w)$, t_φ stabilizes M_w for any $w \in W$; in particular, $t_\varphi(m)$ belongs to M and, for any $a \in E$, we still have

14.5.9 $$\begin{aligned} g(a) &= t_\varphi\Big(g\big((t_\varphi)^{-1}(a)\big)\Big) \\ &= t_\varphi\big((t_\varphi)^{-1}(a)^{i+\pi \cdot m}\big) = a^{i+\pi \cdot t_\varphi(m)}, \end{aligned}$$

从而, 由m的唯一性 (见命题16.8), 得到$t_\varphi(m) = m$; 可是, 显然有

and therefore, from the uniqueness of m (see Proposition 16.8), we get $t_\varphi(m) = m$; but, we clearly have

14.5.10 $$(\hat{\mathcal{O}} \otimes_{\mathcal{O}} E)^{\langle t_\varphi \rangle} = \hat{D} \otimes_Q 1$$

所以$m \in D^P \otimes 1$.

and therefore $m \in D^P \otimes 1$.

引理 14.6. 使固定$\pi \in J(\mathcal{O})$, 令$\bar{\mathcal{O}} = \mathcal{O}/\pi\mathcal{O}$与$\bar{B} = B/\pi \cdot B$, 再令$\bar{b}$与$\bar{\imath}$分别记$b$与$i$在$\bar{\mathcal{O}}G$中的像. 那么有

Lemma 14.6. *Fix $\pi \in J(\mathcal{O})$, set $\bar{\mathcal{O}} = \mathcal{O}/\pi\mathcal{O}$ and $\bar{B} = B/\pi \cdot B$, and denote by $\bar{b}$ and $\bar{\imath}$ the respective images of b and i in $\bar{\mathcal{O}}G$. Then, we have*

14.6.1 $$Z(\bar{B}) = Z(\bar{\mathcal{O}}G)\bar{\imath} \cong Z(\bar{\mathcal{O}}G\bar{b}).$$

特别是,自然同态$Z(B) \to Z(\bar{B})$是满射.

In particular, the natural map $Z(B) \to Z(\bar{B})$ is surjective.

证明: 因为 $\bar{B} = \bar{\imath}(\bar{\mathcal{O}}G)\bar{\imath}$ 所以显然 $Z(\bar{\mathcal{O}}G)\bar{\imath} \subset Z(\bar{B})$. 另一方面, 由定理 5.11, 存在一个完备诱导的 G-内代数 A 和 $\omega \in \mathcal{D}_A(G)$ 使得 $\mathcal{O}G \cong A_\omega$; 此时, 由同构式 5.3.2 和 5.8.6, 得到下面的可嵌入同构

Proof: Since $\bar{B} = \bar{\imath}(\bar{\mathcal{O}}G)\bar{\imath}$, we clearly have $Z(\bar{\mathcal{O}}G)\bar{\imath} \subset Z(\bar{B})$. On the other hand, by Theorem 5.11, there exist an inductively complete G-interior algebra A and $\omega \in \mathcal{D}_A(G)$ such that $\mathcal{O}G \cong A_\omega$; then, from isomorphisms 5.3.2 and 5.8.6, we get the following *embeddable* isomorphisms

14.6.2 $$\mathcal{O}Gb \cong A_\alpha\,, \quad B \cong A_\gamma\,, \quad \mathrm{Ind}_P^G(B) \cong A_{\mathrm{ind}_P^G(\gamma)}\,,$$

其中 α 是被 b 决定的 A 上的 G-点; 特别是, 因为 $\alpha \subset \mathrm{ind}_P^G(\gamma)$ 所以存在 $\mathrm{Ind}_P^G(B)^G$ 的本原幂等元 ℓ 使得 $\mathcal{O}Gb \cong \ell\mathrm{Ind}_P^G(B)\ell$ (见 3.15 和 4.6.3). 令 $\bar{\ell}$ 记 ℓ 在 $\mathrm{Ind}_P^G(\bar{B})$ 中的像; 类似地, 右乘以 $\bar{\ell}$ 决定 $\bar{\mathcal{O}}$-代数同态如下

where α is the point of G on A determined by b; in particular, since $\alpha \subset \mathrm{ind}_P^G(\gamma)$, there is a primitive idempotent in $\mathrm{Ind}_P^G(B)^G$ such that $\mathcal{O}Gb \cong \ell\mathrm{Ind}_P^G(B)\ell$ (see 3.15 and 4.6.3). Denote by $\bar{\ell}$ the image of ℓ in $\mathrm{Ind}_P^G(\bar{B})$; similarly, the multiplication by $\bar{\ell}$ on the right induces an $\bar{\mathcal{O}}$-algebra homomorphism

14.6.3 $$Z(\bar{B}) \cong Z\big(\mathrm{Ind}_P^G(\bar{B})\big) \longrightarrow Z(\bar{\mathcal{O}}G\bar{b})\,.$$

而且, 下面的合成

Moreover, the composition

14.6.4 $$Z(\bar{B}) \longrightarrow Z(\bar{\mathcal{O}}G\bar{b}) \longrightarrow Z(\bar{B})$$

是被下面的可嵌入的 P-内代数自同构决定的 $\mathcal{O}$-代数同态

is the homomorphism determined by the *embeddable* P-interior algebra automorphism

14.6.5 $$B \cong \mathrm{Ind}_P^G(B)_\gamma \cong A_\gamma \cong (A_\alpha)_\gamma \cong (\mathcal{O}G)_\gamma = B\,;$$

可是, B 的可嵌入的 P-内代数自同构是被 $(B^P)^*$ 决定的 (见 4.6); 所以, 这个合成就是恒等同构, 从而有

but, all the *embeddable* P-interior algebra automorphisms of B are induced by $(B^P)^*$ (see 4.6); hence, that composition is the identity, and therefore we get

14.6.6 $$Z(\bar{B}) = Z(\bar{\mathcal{O}}G)\bar{\imath} \cong Z(\bar{\mathcal{O}}G\bar{b})\,.$$

另一方面, 在 $\bar{\mathcal{O}}G$ 里考虑 G 的任意 G-共轭类的所有元素的和; 它们的集合是 $Z(\bar{\mathcal{O}}G)$ 的 $\bar{\mathcal{O}}$-基而自然映射 $Z(\mathcal{O}G) \to Z(\bar{\mathcal{O}}G)$ 是满射; 所以, 自然的映射 $Z(B) \to Z(\bar{B})$ 也是满射.

On the other hand, in $\bar{\mathcal{O}}G$ consider the sum of all the elements of each G-conjugacy class of G; the set of these sums is an $\bar{\mathcal{O}}$-basis of $Z(\bar{G})$ and therefore the natural map $Z(\mathcal{O}G) \to Z(\bar{\mathcal{O}}G)$ is surjective, so that the natural map $Z(B) \to Z(\bar{B})$ is surjective too.

定理 14.7. 假定 k 是代数闭的并使固定 $\pi \in \mathcal{O}$. 如果 D' 是满足条件 14.1.2 的 B 的 P-稳定幺子代数使得

Theorem 14.7. *Assume that k is algebraically closed and fix $\pi \in \mathcal{O}$. If D' is a P-stable unitary subalgebra of B which fulfills condition* 14.1.2 *and*

14.7.1 $$D' + \pi{\cdot}B = D + \pi{\cdot}B$$

那么存在一个 $n \in J(B^P)$ 使得 $D' = D^{i+\pi.n}$.

then, we have $D' = D^{i+\pi.n}$ for a suitable $n \in J(B^P)$.

注意 14.8. 如果 $\pi \in \mathcal{O}^*$ 那么 $\pi{\cdot}B = B$, 从而得到超聚焦子代数的唯一性.

Remark 14.8. If $\pi \in \mathcal{O}^*$ then we have $\pi{\cdot}B = B$ and therefore we get the uniqueness of the *hyperfocal subalgebra*.

证明: 对 $|P/Q|$ 用归纳法, 显然能假定 $Q \neq P$; 设 R 是 P 的规正子群使得 $Q \subset R$ 与 $|R/Q| = p$; 令 E 与 E' 分别记 B 的被 D 和 Ri 生成的子代数与被 D' 和 Ri 生成的子代数; 因为显然有

Proof: We argue by induction on $|P/Q|$ and we clearly may assume that $Q \neq P$; let R be a normal subgroup of P such that $Q \subset R$ and $|R/Q| = p$; denote by E and E' the subalgebras of B respectively generated by D and Ri, and by D' and Ri; since we clearly have

14.7.2 $$E \otimes_R P = B = E' \otimes_R P\,, \quad E \cap Pi = Ri = E' \cap Pi\,,$$

所以存在 $m \in J(B^P)$ 使得 $E' = E^{i+\pi\cdot m}$; 也就是说, 只要用 $(D')^{(i+\pi\cdot m)^{-1}}$ 代替 D', 就能假定 $E' = E$, 从而有

there is $m \in J(B^P)$ fulfilling $E' = E^{i+\pi\cdot m}$; that is to say, replacing D' by $(D')^{(i+\pi\cdot m)^{-1}}$, we may assume that $E' = E$, and in this case we have

14.7.3 $$D \otimes_Q R = E = D' \otimes_Q R\,.$$

特别是, 包含映射 $D \to E$ 与 $D' \to E$ 决定下面的 P-内代数同态

In particular, the inclusion maps $D \to E$ and $D' \to E$ induce the following P-interior algebra homomorphisms

14.7.4 $$B = D \otimes_Q P \stackrel{t}{\longrightarrow} \mathrm{Res}^R_Q(E) \otimes_Q P \stackrel{t'}{\longleftarrow} D' \otimes_Q P = B\,.$$

这样, 我们令

Thus, we set

14.7.5 $$\tilde{B} = \mathrm{Res}^R_Q(E) \otimes_Q P\,, \quad \tilde{E} = \mathrm{Res}^R_Q(E) \otimes_Q R\,;$$

不难验证对任意 $v \in R$, $\tilde{B}$ 的元素 $v^{-1}i\otimes_Q v$ 属于 $Z(\tilde{B}) \cap \tilde{E}^*$, 并且这些元素的集合 Z 在这个群里就是阶为 p 的子群. 请注意, 因为 $Z \subset Z(\tilde{B}) \cap \tilde{E}^*$, 所以 $\tilde{B}$ 与 $\tilde{E}$ 也有一个自然的

for any $v \in R$, it is easy to check that the element $v^{-1}i \otimes_Q v$ of $\tilde{B}$ belongs to $Z(\tilde{B}) \cap \tilde{E}^*$ and moreover, the set Z of them is a subgroup of order p of this intersection. Note that, since $Z \subset Z(\tilde{B}) \cap \tilde{E}^*$, $\tilde{B}$ and $\tilde{E}$ also have an $\mathcal{O}Z$-algebra structure, and it is not

$\mathcal{O}Z$-代数结构, 并且不难验证从等式 14.7.2 可推出

difficult to see that equality 14.7.2 implies the following equalities

14.7.6 $$\mathcal{O}Z \otimes_{\mathcal{O}} B \cong \bigoplus_{z\in Z} t(B)z = \tilde{B} = \bigoplus_{z\in Z} t'(B)z \cong \mathcal{O}Z \otimes_{\mathcal{O}} B\,.$$

进一步, 考虑自然的 P-内代数同态 (见 9.7.1)

Further, consider the natural P-interior algebra homomorphism (see 9.7.1)

14.7.7 $$\tilde{B} = \mathrm{Res}^R_Q(E) \otimes_Q P \xrightarrow{\ f\ } E \otimes_R P = B \quad ;$$

显然有 $f(v^{-1}i \otimes_Q v) = 1_B$ 其中 $v \in R$, 并且从等式14.7.6可推出

we clearly have $f(v^{-1}i \otimes_Q v) = 1_B$ for any $v \in R$ and equality 14.7.6 implies that

14.7.8 $$\mathrm{Ker}(f) = \bigoplus_{z\in Z} t(B)(z - \tilde{\imath})$$

其中 $\tilde{\imath} = i \otimes_Q 1$. 而且, 有 $f \circ t = \mathrm{id}_B = f \circ t'$; 此时, 由下面的引理 14.9, 存在 $\tilde{c} \in (\tilde{B}^P)^*$ 使得对任意 $a \in B$ 有 $t'(a) = t(a)^{\tilde{c}}$; 特别是, 还有 $a = f\big(t'(a)\big) = a^{f(\tilde{c})}$, 从而 $f(\tilde{c})$ 属于 $Z(B)$.

where we set $\tilde{\imath} = i \otimes_Q 1$. Moreover, we have $f \circ t = \mathrm{id}_B = f \circ t'$; then, by Lemma 14.9 below, there exists $\tilde{c} \in (\tilde{B}^P)^*$ such that we have $t'(a) = t(a)^{\tilde{c}}$ for any $a \in B$; in particular, we still have $a = f\big(t'(a)\big) = a^{f(\tilde{c})}$, and therefore $f(\tilde{c})$ belongs to $Z(B)$.

现在, 记 $\bar{\mathcal{O}} = \mathcal{O}/\pi\mathcal{O}$, 再记

Now, set $\bar{\mathcal{O}} = \mathcal{O}/\pi\mathcal{O}$ and

14.7.9 $$\bar{B} = \bar{\mathcal{O}} \otimes_{\mathcal{O}} B\,, \quad \bar{E} = \bar{\mathcal{O}} \otimes_{\mathcal{O}} E\,,$$
$$\bar{D} = \bar{\mathcal{O}} \otimes_{\mathcal{O}} D\,, \quad \bar{D}' = \bar{\mathcal{O}} \otimes_{\mathcal{O}} D'\,;$$

类似地, 令

similarly, set

14.7.10 $$\bar{B} = \bar{D} \otimes_Q P \xrightarrow{\ \bar{t}\ } \mathrm{Res}^R_Q(\bar{E}) \otimes_Q P \xleftarrow{\ \bar{t}'\ } \bar{D}' \otimes_Q P = \bar{B}\,,$$

还令 $\bar{\tilde{c}}$ 记 $\tilde{c}$ 在 $\mathrm{Res}^R_Q(\bar{E}) \otimes_Q P$ 中的像; 因为假定 $\bar{D} = \bar{D}'$, 所以有 $\bar{t} = \bar{t}'$, 从而 $\bar{\tilde{c}}$ 属于 $Z(\mathrm{Res}^R_Q(\bar{E}) \otimes_Q P)$. 最后, 因为显然

and, in $\mathrm{Res}^R_Q(\bar{E}) \otimes_Q P$, denote by $\bar{\tilde{c}}$ the image of $\tilde{c}$; since we assume that $\bar{D} = \bar{D}'$, we have $\bar{t} = \bar{t}'$ and therefore $\bar{\tilde{c}}$ belongs to $Z(\mathrm{Res}^R_Q(\bar{E}) \otimes_Q P)$. Finally, since we clearly have

14.7.11 $$\mathrm{Ker}(f) \cap \pi{\cdot}\tilde{B} = \pi{\cdot}\mathrm{Ker}(f)\,,$$

所以 $\tilde{c}$ 在 $\tilde{B}/\pi{\cdot}\mathrm{Ker}(f)$ 中的像也属于 $Z\big(\tilde{B}/\pi{\cdot}\mathrm{Ker}(f)\big)$.

the image of $\tilde{c}$ in $\tilde{B}/\pi{\cdot}\mathrm{Ker}(f)$ belongs to $Z\big(\tilde{B}/\pi{\cdot}\mathrm{Ker}(f)\big)$ too.

设 $\hat{\mathcal{K}}$ 是被一个本原 p-次单位根生成的 $\mathcal{K}$ 的扩张并令 $\hat{\mathcal{O}}$ 记 $\hat{\mathcal{K}}$ 中的 $\mathcal{O}$ 上的整元构成的环 (见 2.12); 类似地, 记

Let $\hat{\mathcal{K}}$ be the extension of $\mathcal{K}$ generated by a primitive p-th root of unity and denote by $\hat{\mathcal{O}}$ the integral closure of $\mathcal{O}$ in $\hat{\mathcal{K}}$ (see 2.12); similarly, set

$$14.7.12 \qquad \hat{B} = \hat{\mathcal{O}} \otimes_{\mathcal{O}} B, \quad \hat{E} = \hat{\mathcal{O}} \otimes_{\mathcal{O}} E,$$
$$\hat{D} = \hat{\mathcal{O}} \otimes_{\mathcal{O}} D, \quad \hat{D}' = \hat{\mathcal{O}} \otimes_{\mathcal{O}} D'.$$

而且, 设 $\varphi: R \to \hat{\mathcal{O}}^*$ 是一个群同态使得 $\mathrm{Ker}(\varphi) = Q$, 而考虑对应的同态 $\mathcal{O}Z \to \hat{\mathcal{O}}$ (别忘了, 有 $R/Q \cong Z$); 令

Moreover, let $\varphi: R \to \hat{\mathcal{O}}^*$ be a group homomorphism such that $\mathrm{Ker}(\varphi) = Q$, and consider the corresponding homomorphism $\mathcal{O}Z \to \hat{\mathcal{O}}$ (recall that $R/Q \cong Z$); we set

$$14.7.13 \qquad \hat{\tilde{B}} = \hat{\mathcal{O}} \otimes_{\mathcal{O}Z} \tilde{B}, \quad \hat{\tilde{E}} = \hat{\mathcal{O}} \otimes_{\mathcal{O}Z} \tilde{E},$$

再令 $\hat{\tilde{c}}$ 记 $\tilde{c}$ 在 $\hat{\tilde{B}}$ 中的像.

and denote by $\hat{\tilde{c}}$ the image of $\tilde{c}$ in $\hat{\tilde{B}}$.

请注意, 因为有

Note that, since we have

$$14.7.14 \qquad (\xi^h - 1)\hat{\mathcal{O}} = (\xi - 1)\hat{\mathcal{O}}$$

其中 $1 \le h < p$, 所以从等式 14.7.8 可推出 $\mathrm{Ker}(f)$ 在 $\hat{\tilde{B}}$ 中的像是 $(\xi - 1)\cdot\hat{\tilde{B}}$. 进一步, t 与 t' 决定下面的 P-内代数同态

for any $1 \le h < p$, it follows from equality 14.7.8 that the image of $\mathrm{Ker}(f)$ in $\hat{\tilde{B}}$ is equal to $(\xi - 1)\cdot\hat{\tilde{B}}$. Further, t and t' induce the following P-interior algebra homomorphisms

$$14.7.15 \quad \hat{t}: \hat{B} \to \hat{\mathcal{O}} \otimes_{\mathcal{O}} \tilde{B} \to \hat{\tilde{B}}, \quad \hat{t}': \hat{B} \to \hat{\mathcal{O}} \otimes_{\mathcal{O}} \tilde{B} \to \hat{\tilde{B}},$$

并且, 不难验证它们是双射 (见 14.7.6) 和有 (见 14.7.2).

and it is easy to check that they are bijective (see 14.7.6) and that we have (see 14.7.2)

$$14.7.16 \qquad \hat{t}(\hat{E}) = \hat{\tilde{E}} = \hat{t}'(\hat{E}).$$

特别是, 因为 $t'(a) = t(a)^{\tilde{c}}$, 其中 $a \in B$, 所以 $\hat{\tilde{B}}$ 的被 $\hat{\tilde{c}}$ 决定的自同构稳定 $\hat{\tilde{E}}$; 进一步, 既然 $\tilde{c}$ 在 $\tilde{B}/\pi\cdot\mathrm{Ker}(f)$ 中的像属于 $Z(\tilde{B}/\pi\cdot\mathrm{Ker}(f))$, 这个 $\hat{\tilde{B}}$ 的自同构在 $\hat{\tilde{E}}/\pi(\xi-1)\cdot\hat{\tilde{E}}$ 上平凡作用; 这样, 由应用命题 14.5, 就存在 $\hat{\tilde{e}} \in \hat{\tilde{E}}^P$ 使得对任意 $\hat{a} \in \hat{E}$ 有

In particular, since we have $t'(a) = t(a)^{\tilde{c}}$ for any $a \in B$, the automorphism of $\hat{\tilde{B}}$ induced by $\hat{\tilde{c}}$ stabilizes $\hat{\tilde{E}}$; moreover, as far as the image of $\tilde{c}$ in $\tilde{B}/\pi\cdot\mathrm{Ker}(f)$ belongs to $Z(\tilde{B}/\pi\cdot\mathrm{Ker}(f))$, this automorphism of $\hat{\tilde{B}}$ acts trivially on $\hat{\tilde{E}}/\pi(\xi-1)\cdot\hat{\tilde{E}}$; thus, according to Proposition 14.5, there exists $\hat{\tilde{e}} \in \hat{\tilde{E}}^P$ such that, for any $\hat{a} \in \hat{E}$, we have

$$14.7.17 \qquad \hat{t}'(\hat{a}) = \hat{t}(\hat{a})^{\hat{\tilde{c}}} = \hat{t}(\hat{a})^{(1 \otimes \tilde{\imath} + \pi(\xi-1)\cdot\hat{\tilde{e}})}.$$

另一方面, 考虑下面被等式 $\hat{g}(1\otimes a) = 1\otimes(a\otimes_Q 1)$, 其中 $a\in E$, 定义的 Q-内 P-代数同构 (见 9.7)

On the other hand, consider the Q-interior P-algebra isomorphism defined by the equality $\hat{g}(1\otimes a) = 1\otimes(a\otimes_Q 1)$ for any $a\in E$ (see 9.7)

14.7.18 $$\hat{g} : \mathrm{Res}^R_Q(\hat{E}) \longrightarrow \hat{\tilde{E}} = \hat{\mathcal{O}} \otimes_{\mathcal{O}Z} \big(\mathrm{Res}^R_Q(E)\otimes_Q R\big)\,;$$

请注意, 对任意 $v\in R$ 有

note that, for any $v\in R$, we have

14.7.19 $$\hat{g}\big(1\otimes v{\cdot}i)\big) = 1\otimes(v{\cdot}i\otimes_Q 1) = \varphi(v)^{-1}\otimes(i\otimes_Q v)\,;$$

令 $\hat{s} = \hat{g}^{-1}\circ\hat{t}$ 与 $\hat{s}' = \hat{g}^{-1}\circ\hat{t}'$, 和 $\hat{e} = \hat{g}^{-1}(\hat{\tilde{e}})$; 这样, $\hat{s}$ 与 $\hat{s}'$ 是 $\mathrm{Res}^R_Q(\hat{E})$ 的 Q-内 P-代数自同构, 并且 $\hat{s}$ 在 $\hat{D}$ 上的限制与 $\hat{s}'$ 在 $\hat{D}'$ 上的限制都是恒等自同构; 进一步, 不难验证有

set $\hat{s} = \hat{g}^{-1}\circ\hat{t}$, $\hat{s}' = \hat{g}^{-1}\circ\hat{t}'$ and $\hat{e} = \hat{g}^{-1}(\hat{\tilde{e}})$; thus, $\hat{s}$ and $\hat{s}'$ are Q-interior P-algebra automorphisms of $\mathrm{Res}^R_Q(\hat{E})$, which respectively induce the identity automorphism on $\hat{D}$ and on $\hat{D}'$; furthermore, it is easily checked that we have

14.7.20 $$\hat{s}(v{\cdot}i) = \varphi(v){\cdot}(v{\cdot}i) = \hat{s}'(v{\cdot}i)$$

其中 $v\in R$; 特别是, 还有 $\hat{s}^p = \mathrm{id}_{\hat{E}} = \hat{s}'^p$.

for any $v\in R$; in particular, we still have $\hat{s}^p = \mathrm{id}_{\hat{E}} = \hat{s}'^p$.

而且, 由等式 14.7.17, 对任意 $\hat{a}\in\hat{E}$ 得到

Moreover, from equality 14.7.17, for any $\hat{a}\in\hat{E}$ we get

14.7.21 $$\hat{s}'(\hat{a}) = \hat{s}(\hat{a})^{(\hat{\imath} + \pi(\xi-1)\cdot\hat{e})}$$

其中 $\hat{\imath} = 1\otimes i$. 事实上, 由命题 14.3, 有

where we set $\hat{\imath} = 1\otimes i$. Actually, from Proposition 14.3, we have

14.7.22 $$\hat{E}^P = \hat{\mathcal{O}}C_R(P) + \mathrm{Ker}(\mathrm{Br}^{\hat{E}}_P)\,,$$

从而只要应用后面的命题 16.8, 就得到 $\hat{e}'\in\mathrm{Ker}(\mathrm{Br}^{\hat{E}}_P)$ 使得对任意 $\hat{a}\in\hat{E}$ 还有

and then, it suffices to apply Proposition 16.8 in section 16, to find $\hat{e}'\in\mathrm{Ker}(\mathrm{Br}^{\hat{E}}_P)$ such that, for any $\hat{a}\in\hat{E}$, we still have

14.7.23 $$\hat{s}'(\hat{a}) = \hat{s}(\hat{a})^{(\hat{\imath} + \pi(\xi-1)\cdot\hat{e}')}\,.$$

此时, 把后面的定理 16.7 应用到 $\hat{\mathcal{O}}$-代数 $\hat{E}^P$ 和 $\hat{s}$ 在 $\hat{E}^P$ 上的限制. 因为 P 在 R/Q 上平凡作用, 所以对任意 $v\in R$, P 稳定 $\hat{D}\otimes_Q v$, 从而不难验证

At this point, we apply Theorem 16.7 in section 16 to the $\hat{\mathcal{O}}$-algebra $\hat{E}^P$ and to the restriction of $\hat{s}$ to $\hat{E}^P$. Since P acts trivially on R/Q, P stabilizes $\hat{D}\otimes_Q v$ for any $v\in R$, and then it is not difficult to prove that

14.7.24 $$\mathrm{Ker}(\mathrm{Br}^{\hat{E}}_P) = \bigoplus_{w\in W}\mathrm{Ker}(\mathrm{Br}^{\hat{E}}_P)\cap(\hat{D}\otimes_Q w)\,,$$

其中 $W \subset R$ 是 R/Q 的代表集合; 因为 $\hat{E}^P/J(\hat{E}^P) \cong k$ 与 $\hat{E}(P) \neq \{0\}$ (见命题14.3), 所以有 $\mathrm{Ker}(\mathrm{Br}_P^{\hat{E}}) \subset J(\hat{E}^P)$; 进一步, 由等式 14.7.20, 对任意 $\hat{d} \in \hat{D}$ 有 $\hat{s}(\hat{d} \otimes_Q w) = \varphi(w)\hat{d} \otimes_Q w$, 从而得到

where $W \subset R$ is a set of representatives for R/Q; since we have $\hat{E}^P/J(\hat{E}^P) \cong k$ and $\hat{E}(P) \neq \{0\}$ (see proposition 14.3), we get $\mathrm{Ker}(\mathrm{Br}_P^{\hat{E}}) \subset J(\hat{E}^P)$; further, from equality 14.7.20, we have $\hat{s}(\hat{d} \otimes_Q w) = \varphi(w)\hat{d} \otimes_Q w$ for any $\hat{d} \in \hat{D}$, and therefore we obtain

14.7.25 $$(\hat{s} - \mathrm{id}_{\hat{E}})\big(\mathrm{Ker}(\mathrm{Br}_P^{\hat{E}})\big) \subset (\xi - 1)\cdot\mathrm{Ker}(\mathrm{Br}_P^{\hat{E}});$$

而且, 因为 $\hat{s}^p = \mathrm{id}_{\hat{E}} = \hat{s}'^p$, 所以从等式 14.7.23 不难验证

moreover, since we have $\hat{s}^p = \mathrm{id}_{\hat{E}} = \hat{s}'^p$, it is easy to check from equality 14.7.23 that

14.7.26 $$\prod_{h=0}^{p-1} \big(\hat{\imath} + \pi(\xi - 1)\cdot\hat{s}^h(\hat{e}')\big)^{-1} \in Z(\hat{E}).$$

现在, 由下面的定理 16.7, 存在 $\hat{m} \in \mathrm{Ker}(\mathrm{Br}_P^{\hat{E}}) \subset J(\hat{E}^P)$ 使得对任意 $\hat{a} \in \hat{E}$ 有

Now, it follows from Theorem 16.7 that there exists $\hat{m} \in \mathrm{Ker}(\mathrm{Br}_P^{\hat{E}}) \subset J(\hat{E}^P)$ such that , for any $\hat{a} \in \hat{E}$, we have

14.7.27 $$\hat{s}'(\hat{a})^{(\hat{\imath} + \pi\cdot\hat{m})} = \hat{s}(\hat{a}^{(\hat{\imath} + \pi\cdot\hat{m})});$$

特别是, 因为 $\hat{D}$ 与 $\hat{D}'$ 分别是被 $\hat{s}$ 与 $\hat{s}'$ 稳定的 $\hat{E}$ 的元素的集合, 所以还有

in particular, since $\hat{D}$ and $\hat{D}'$ are the sets of elements of $\hat{E}$ respectively fixed by $\hat{s}$ and $\hat{s}'$, we still have

14.7.28 $$\hat{D}' = (\hat{\imath} + \pi\cdot\hat{m})\hat{D}(\hat{\imath} + \pi\cdot\hat{m})^{-1}.$$

只要证明我们在 $J(E^P)$ 中能选择 $\hat{m}$. 令 Σ 记 $\hat{\mathcal{K}}$ 上 $\mathcal{K}$ 的 Galois 群, 并考虑 Σ 在 $\hat{B}$ 上的自然作用; 既然 Σ 显然稳定 $\hat{D}$ 与 $\hat{D}'$, 对任意 $\sigma \in \Sigma$ 还有

It remains to prove that we may choose $\hat{m}$ in $J(E^P)$. Denote by Σ the Galois group of $\hat{\mathcal{K}}$ over $\mathcal{K}$ and consider the natural action of Σ on $\hat{B}$; since Σ clearly stabilizes $\hat{D}$ and $\hat{D}'$, for any $\sigma \in \Sigma$ we also have

14.7.29 $$\hat{D}' = \big(\hat{\imath} + \pi\cdot\sigma(\hat{m})\big)\hat{D}\big(\hat{\imath} + \pi\cdot\sigma(\hat{m})\big)^{-1}.$$

那么, 记

Then, set

14.7.30 $$\hat{F} = \hat{\mathcal{O}}\cdot\hat{\imath} + \pi\cdot J(\hat{E}^P) = \hat{\mathcal{O}}\cdot\hat{\imath} + \pi\cdot\hat{E}^P$$

并令 $\hat{X}$ 记满足 $\hat{D}' = \hat{x}\hat{D}\hat{x}^{-1}$ 的 $\hat{x} \in \hat{F}$ 的集合; 这样 Σ 稳定 $\hat{X}$, 并且由于右乘以 $N_{\hat{F}^*}(\hat{D})$, 不难验证这个群 $N_{\hat{F}^*}(\hat{D})$ 在 $\hat{X}$ 上自

and denote by $\hat{X}$ the set of $\hat{x} \in \hat{F}$ fulfilling $\hat{D}' = \hat{x}\hat{D}\hat{x}^{-1}$; thus Σ stabilizes $\hat{X}$ and it is quite clear that the group $N_{\hat{F}^*}(\hat{D})$ acts regularly on $\hat{X}$ by the multiplication on

由可迁地作用; 特别是, $\hat{\imath}+\pi\cdot\hat{m}$ 在 $\hat{F}^*\rtimes\Sigma$ 里的稳定化子 $\hat{S}$ 也是 $\hat{F}^*$ 在 $\hat{F}^*\rtimes\Sigma$ 里的补.

the right; in particular, the stabilizer $\hat{S}$ of $\hat{\imath}+\pi\cdot\hat{m}$ in $\hat{F}^*\rtimes\Sigma$ also is a complement of $\hat{F}^*$ in $\hat{F}^*\rtimes\Sigma$.

此时, 考虑 $\mathcal{O}$-代数 $\hat{E}\Sigma$, 以及其子$\mathcal{O}$-代数 $\hat{D}$ 与 $\hat{F}\Sigma$; 包含映射 $\Sigma\to\hat{F}^*\rtimes\Sigma$ 与 $\hat{S}\cong\Sigma$ 的逆同构决定两个群同态

Then, consider the group algebra $\hat{E}\Sigma$ and the subalgebras $\hat{D}$ and $\hat{F}\Sigma$; the inclusion map $\Sigma\to\hat{F}^*\rtimes\Sigma$ and the inverse of the group isomorphism $\hat{S}\cong\Sigma$ induce two group homomorphisms

14.7.31 $$\iota\colon\Sigma\longrightarrow N_{(\hat{F}\Sigma)^*}(\hat{D})\,,\quad \varphi\colon\Sigma\longrightarrow N_{(\hat{F}\Sigma)^*}(\hat{D})\,.$$

而且, 不难验证

Moreover, it is quite clear that

14.7.32 $$\hat{S}\subset N_{(\hat{\imath}+\pi\cdot J(\hat{E}^P))\cdot\Sigma}(\hat{D})\,;$$

进一步, 只要记

furthermore, setting

14.7.33 $$\hat{T}=\hat{\imath}+\pi\cdot J\big(Z(\hat{E})^P\big)\,,\quad \hat{U}=\hat{\imath}+\pi\cdot J(\hat{D}^P)\,,$$

由推论 14.4 和命题 14.5, 就不难得到

from Corollary 14.4 and Proposition 14.5, it is easy to get

14.7.34 $$N_{\hat{\imath}+\pi\cdot J(\hat{E}^P)}(\hat{D})=\hat{T}\cdot\hat{U}\,.$$

现在, 考虑下面的商

Now, consider the following quotient

14.7.35 $$\tilde{N}=N_{(\hat{F}\Sigma)^*}(\hat{D})\big/J(\hat{D})^*\cap\hat{F}\Sigma$$

并且令 $\tilde{S}\,,\tilde{T}\,,\tilde{U}\,,\tilde{\Sigma}$ 分别记 $\hat{S}\,,\hat{T}\,,\hat{U}\,,\Sigma$ 在 $\tilde{N}$ 中的像; 它们在 $\hat{D}/J(\hat{D})$ 上都作用, 以及 $\hat{T}$ 与 Σ 的作用是平凡的. 另一方面, 显然有自然的群同态

and denote by $\tilde{S}\,,\tilde{T}\,,\tilde{U}$ and $\tilde{\Sigma}$ the respective images of $\hat{S}\,,\hat{T}\,,\hat{U}$ and Σ in $\tilde{N}$; all them act on $\hat{D}/J(\hat{D})$ and the actions of $\hat{T}$ and Σ are trivial. On the other hand, we clearly have a natural map

14.7.36 $$\tilde{U}\to\hat{D}^*/J(\hat{D})^*\cong\big(\hat{D}/J(\hat{D})\big)^*\,,$$

它既提升 $\hat{U}$ 的作用, 它的像也是幂幺的. 此时, 因为 $|\Sigma|$ 整除 $p-1$, 所以对适当的 $h\in\mathbb{N}$ 得到 (见包含式 14.7.32)

which lifts the action of $\tilde{U}$ and has a unipotent image. Consequently, since $|\Sigma|$ divides $p-1$, for a suitable $h\in\mathbb{N}$ we get (see the inclusion in 14.7.32)

14.7.37 $$\tilde{S}\subset(\tilde{T}\cdot\tilde{U}\cdot\tilde{\Sigma})^{p^h}\subset\tilde{T}\cdot\tilde{\Sigma}\,;$$

也就是说, 有 $\tilde{T}\cdot\tilde{\Sigma}=\tilde{T}\cdot\tilde{S}$, 并且, 由后面的推论 16.5, $|\Sigma|$-次幂映射在 $\tilde{T}$ 上是双射. 这样,

that is to say, we have $\tilde{T}\cdot\tilde{\Sigma}=\tilde{T}\cdot\tilde{S}$, and moreover, by Corollary 16.5 in section 16, the $|\Sigma|$-th power on $\tilde{T}$ defines a bijective map.

由下面的引理 14.10, 存在 $\tilde{t}\in\tilde{T}$ 使得 $\tilde{S}=\tilde{\Sigma}^{\tilde{t}}$, 从而也存在 $\hat{z}\in J\big(Z(\hat{E})^P\big)$ 满足 $\sigma^{\hat{\imath}+\pi\cdot\hat{z}}$ 与 $\varphi(\sigma)$ 在 $\tilde{N}$ 中的像是一样的, 其中 $\sigma\in\Sigma$.

Thus, by the Lemma 14.10 below, there is $\tilde{t}\in\tilde{T}$ fulfilling $\tilde{S}=\tilde{\Sigma}^{\tilde{t}}$ and therefore we find $\hat{z}\in J\big(Z(\hat{E})^P\big)$ such that, for any $\sigma\in\Sigma$, $\sigma^{\hat{\imath}+\pi\cdot\hat{z}}$ and $\varphi(\sigma)$ have the same image in $\tilde{N}$.

最后, 只要应用引理 12.5, 就得到 $\hat{d}\in J(\hat{D}^P)$ 使得

Finally, by applying Lemma 12.5, we find $\hat{d}\in J(\hat{D}^P)$ fulfilling

14.7.38 $$\hat{S}=\Sigma^{(\hat{\imath}+\pi\cdot\hat{z})(\hat{\imath}+\pi\cdot\hat{d})}\,;$$

此时, Σ 稳定 $\hat{X}$ 的元素

then, Σ stabilizes the element of $\hat{X}$

14.7.39 $$\hat{\imath}+\pi\cdot\hat{n}=(\hat{\imath}+\pi\cdot\hat{m})(\hat{\imath}+\pi\cdot\hat{d})^{-1}(\hat{\imath}+\pi\cdot\hat{z})^{-1}\,,$$

其中 $\hat{n}\in J(\hat{E}^P)$; 可是, 显然 $J(\hat{E}^P)^\Sigma=1\otimes J(E^P)$; 所以有 $\hat{n}=1\otimes n$, 其中 $n\in J(E^P)$, 并且从等式

for a suitable $\hat{n}\in J(\hat{E}^P)$; but, clearly $J(\hat{E}^P)^\Sigma=1\otimes J(E^P)$; consequently, we have $\hat{n}=1\otimes n$ for some $n\in J(E^P)$ and therefore, from the equality

14.7.40 $$\hat{\mathcal{O}}\otimes_{\mathcal{O}}D'=\hat{\mathcal{O}}\otimes_{\mathcal{O}}nDn^{-1}$$

可推出 $D'=nDn^{-1}$.

we get $D'=nDn^{-1}$.

引理 14.9. 设 B' 是一个 P-内代数并设 $g\colon B'\to B$ 是满足 $\mathrm{Ker}(g)\subset J(B')$ 的一个 P-内代数同态. 如果 $s\colon B\to B'$ 与 $t\colon B\to B'$ 是 P-内代数同态使得

Lemma 14.9. *Let B' be a P-interior algebra and $g\colon B'\to B$ a P-interior algebra homomorphism such that $\mathrm{Ker}(g)\subset J(B')$. If $s\colon B\to B'$ and $t\colon B\to B'$ are P-interior algebra homomorphisms fulfilling*

14.9.1 $$g\circ s=\mathrm{id}_B=g\circ t$$

那么存在 $c'\in(B'^P)^*$ 使得对任意 $a\in B$ 有 $t(a)=s(a)^{c'}$.

then there exists $c'\in(B'^P)^*$ such that we have $t(a)=s(a)^{c'}$ for any $a\in B$.

证明: 由 G-内代数 $\hat{B}=\mathrm{Ind}_P^G(B)$ 与 $\hat{B}'=\mathrm{Ind}_P^G(B')$ 的定义 (见注意 5.8), P-内代数同态 g, s, t 显然决定下面的 G-内代数同态

Proof: By the definition of $\hat{B}=\mathrm{Ind}_P^G(B)$ and $\hat{B}'=\mathrm{Ind}_P^G(B')$ (see Remark 5.8), the P-interior algebra homomorphisms g, s and t induce G-interior algebra homomorphisms

14.9.2 $$\hat{s}:\hat{B}\to\hat{B}',\quad \hat{t}:\hat{B}\to\hat{B}',\quad \hat{g}:\hat{B}'\to\hat{B}\,;$$

显然 $\mathrm{Ker}(\hat{g})\subset J(\hat{B}')$ 并且有

clearly $\mathrm{Ker}(\hat{g})\subset J(\hat{B}')$ and we have

14.9.3 $$\hat{g}\circ\hat{s}=\mathrm{id}_{\hat{B}}=\hat{g}\circ\hat{t}\,.$$

特别是, 对任意 G 的子群 H 有 $\hat{g}(\hat{B}'^H) = \hat{B}^H$, 从而 $\hat{g}$ 是一个严格覆盖同态 (见 4.13); 进一步, 因为 $\mathrm{id}_{\hat{B}}$ 显然是严格覆盖的, 所以 $\hat{s}$ 与 $\hat{t}$ 也是严格覆盖的 (见 4.14). 也就是说, $\hat{g}$ 从 $\hat{B}'$ 上的点群的集合到 $\hat{B}$ 上的点群的集合决定一个双射, 并且 $\hat{s}$ 与 $\hat{t}$ 决定其逆的映射. 特别是, P 在 B' 上有唯一点 γ' 并且它是局部的 (见 4.14).

In particular, for any subgroup H of G, we have $\hat{g}(\hat{B}'^H) = \hat{B}^H$ and therefore $\hat{g}$ is a strict covering (see 4.13); furthermore, since $\mathrm{id}_{\hat{B}}$ obviously is a strict covering homomorphism, $\hat{s}$ and $\hat{t}$ are strict covering homomorphisms too (see 4.14). That is to say, $\hat{g}$ induces a bijective map between the sets of pointed groups on $\hat{B}$ and on $\hat{B}'$ and the homomorphisms $\hat{s}$ and $\hat{t}$ induce the inverse map. In particular, P has a unique local point γ' on B' (see 4.14).

另一方面, 由定理 5.11, 存在完备诱导的 G-内代数 A 与 A' 和适当的 A 与 A' 上的 G-除子 ω, ω' 使得 $\hat{B} \cong A_\omega$, $\hat{B}' \cong A'_{\omega'}$; 此时, 由同构式 5.3.2 和 5.8.6, 得到

On the other hand, according to Theorem 5.11, there exist inductively complete G-interior algebras A and A', and suitable divisors ω and ω' of G on A and A' fulfilling $\hat{B} \cong A_\omega$ and $\hat{B}' \cong A'_{\omega'}$; then, from isomorphisms 5.3.2 and 5.8.6, we get

14.9.4 $$\omega = \mathrm{ind}_P^G(\gamma)\,,\quad \omega' = \mathrm{ind}_P^G(\gamma')\,,\quad B \cong A_\gamma\,,\quad B' \cong A'_{\gamma'}\,;$$

特别是, 既然在 $\mathcal{O}G$ 上我们有 $\{b\} = \alpha \subset \mathrm{ind}_P^G(\gamma)$, 存在 $\hat{B}^G$ 的本原幂等元 ℓ 使得

in particular, since over $\mathcal{O}G$ we have $\{b\} = \alpha \subset \mathrm{ind}_P^G(\gamma)$, there exists a primitive idempotent ℓ in $\hat{B}^G$ fulfilling

14.9.5 $$\mathcal{O}Gb \cong \ell\,\hat{B}\,\ell$$

(见 3.15 和 4.6.3), 并且 $\hat{s}(\ell)$ 与 $\hat{t}(\ell)$ 在 $\hat{B}'$ 上决定同样的 G-点; 这样, 存在 $e' \in (\hat{B}'^G)^*$ 使得 $\hat{t}(\ell) = \hat{s}(\ell)^{e'}$, 并对任意 $x \in G$ 有

(see 3.15 and 4.6.3), and moreover, $\hat{s}(\ell)$ and $\hat{t}(\ell)$ determine the same point of G on $\hat{B}'$; thus, there exists $e' \in (\hat{B}'^G)^*$ such that $\hat{t}(\ell) = \hat{s}(\ell)^{e'}$ and, for any $x \in G$, we have

14.9.6 $$\hat{s}(x{\cdot}\ell)^{e'} = x{\cdot}\hat{s}(\ell)^{e'} = \hat{t}(x{\cdot}\ell)\,;$$

所以, 由同构式 14.9.5, 还有 $\hat{t}(a) = \hat{s}(a)^{e'}$, 其中 $a \in \ell\hat{B}\ell$.

consequently, from isomorphism 14.9.5, we still have $\hat{t}(a) = \hat{s}(a)^{e'}$ for any $a \in \ell\hat{B}\ell$.

最后, 既然 $P_\gamma \subset G_\alpha$, 存在 $j \in \gamma$ 使得 $j\ell = j = \ell j$, 从而因为存在 $c \in (\hat{B}^P)^*$ 使得

Finally, as far as we have $P_\gamma \subset G_\alpha$, there exists $j \in \gamma$ fulfilling $j\ell = j = \ell j$ and therefore, since there is $c \in (\hat{B}^P)^*$ such that

$(1\otimes 1_B\otimes 1)^c = j$，所以对任意 $a\in B$ 得到

$(1\otimes 1_B\otimes 1)^c = j$, for any $a\in B$ we get

14.9.7 $$1\otimes t(a)\otimes 1 = \hat{t}\big((1\otimes a\otimes 1)^c\big)^{\hat{t}(c)^{-1}} = \hat{s}\big((1\otimes a\otimes 1)^c\big)^{e'\hat{t}(c)^{-1}} = \big(1\otimes s(a)\otimes 1\big)^{\hat{s}(c)\,e'\hat{t}(c)^{-1}};$$

特别是，$1\otimes 1_{B'}\otimes 1$ 与 $(\hat{B}'^P)^*$ 的元素 $\hat{s}(c)\,e'\hat{t}(c)^{-1}$ 可交换，从而

in particular, $1\otimes 1_{B'}\otimes 1$ and the element $\hat{s}(c)\,e'\hat{t}(c)^{-1}$ of $(\hat{B}'^P)^*$ commute, so that

14.9.8 $$c' = (1\otimes 1_{B'}\otimes 1)\big(\hat{s}(c)\,e'\,\hat{t}(c)^{-1}\big)$$

属于 $(B'^P)^*$，并且对任意 $a\in B$ 有 $t(a) = s(a)^{c'}$.

belongs to $(B'^P)^*$, and moreover we have $t(a) = s(a)^{c'}$ for any $a\in B$.

引理 14.10. **设 X 是一个群并设 Z 是一个 X 的规正交换子群. 假定商 $\bar{X} = X/Z$ 是有限的，再假定 $|\bar{X}|$-次幂映射在 Z 上是双射. 在 X 里存在唯一一个 Z 的补的 X-共轭类.**

Lemma 14.10. *Let X be a group and Z an abelian normal subgroup of X. Assume that the quotient $\bar{X} = X/Z$ is finite, and that the $|\bar{X}|$-th power in Z defines a bijective map. Then there exists a unique X-conjugacy class of complements of Z in X.*

证明: 对任意 $\bar{x}\in\bar{X}$ 取满足 $s(\bar{x})Z = \bar{x}$ 的 $s(\bar{x})\in X$；这样，有 $s(\bar{x})s(\bar{x}')Z = s(\bar{x}\bar{x}')Z$，其中 $\bar{x},\bar{x}'\in\bar{X}$，从而存在 Z 的元素 $z(\bar{x},\bar{x}')$ 使得

Proof: For any $\bar{x}\in\bar{X}$, pick $s(\bar{x})\in X$ such that $s(\bar{x})Z = \bar{x}$; thus, we have $s(\bar{x})s(\bar{x}')Z = s(\bar{x}\bar{x}')Z$ for any $\bar{x},\bar{x}'\in\bar{X}$, and therefore there exists an element $z(\bar{x},\bar{x}')$ of Z fulfilling

14.10.1 $$s(\bar{x})s(\bar{x}') = z(\bar{x},\bar{x}')s(\bar{x}\bar{x}');$$

只要使用结合律，就不难验证

from the associativity, it is easy to check that

14.10.2 $$z(\bar{x}',\bar{x}'')^{\bar{x}^{-1}} z(\bar{x}\bar{x}',\bar{x}'')^{-1} z(\bar{x},\bar{x}'\bar{x}'') z(\bar{x},\bar{x}')^{-1} = 1$$

其中 $\bar{x},\bar{x}',\bar{x}''\in\bar{X}$，而

for any $\bar{x},\bar{x}',\bar{x}''\in\bar{X}$, and where

14.10.3 $$z(\bar{x}',\bar{x}'')^{\bar{x}^{-1}} = s(\bar{x})z(\bar{x}',\bar{x}'')s(\bar{x})^{-1}$$

不依赖 $s(\bar{x})$ 的选择.

does not depend on the choice of $s(\bar{x})$.

特别是，如果 $\bar{x}''$ 跑遍 $\bar{X}$，那么这些等式的积产生

In particular, the sum of these equalities when $\bar{x}''$ runs over $\bar{X}$ gives

14.10.4 $$z(\bar{x},\bar{x}') = w(\bar{x})w(\bar{x}\bar{x}')^{-1}w(\bar{x}')^{\bar{x}^{-1}},$$

其中，对任意 $\bar{x}\in\bar{X}$，$w(\bar{x})$ 是在 Z 中唯一满足

where, for any $\bar{x}\in\bar{X}$, $w(\bar{x})$ is the unique element of Z which fulfills

14.10.5 $$w(\bar{x})^{|\bar{X}|} = \prod_{\bar{x}''\in\bar{X}} z(\bar{x},\bar{x}'')$$

的元素; 此时, 不难验证 $W = \{w(\bar{x})^{-1}s(\bar{x})\}_{\bar{x}\in\bar{X}}$ 是一个 X 的子群使得

then, it is quite easy to check that $W = \{w(\bar{x})^{-1}s(\bar{x})\}_{\bar{x}\in\bar{X}}$ is a subgroup of X such that

14.10.6 $$X = Z{\cdot}W\,, \quad Z\cap W = \{1\}\,.$$

类似地, 假定 W' 满足同样的条件, 并令 $t\colon\bar{X} \to W$ 与 $t'\colon\bar{X} \to W'$ 记对应的群同构; 那么, 不难验证满足

similarly, assume that W' fulfills the same condition, and denote by $t\colon\bar{X} \to W$ and $t'\colon\bar{X} \to W'$ the corresponding group isomorphisms; then, the element z of Z such that

14.10.7 $$z^{|\bar{X}|} = \prod_{\bar{x}\in\bar{X}} t(\bar{x})^{-1}t'(\bar{x})$$

的元素$z\in Z$,也满足$W'=W^z$.

fulfills $W' = W^z$.

15. 超聚焦子代数的存在性

15. Existence of the Hyperfocal Subalgebra

15.1. 如同在 §14 里一样，我们假定 $\mathcal{O}$ 的分式域 $\mathcal{K}$ 是特征为零的域，G 是一个有限群并 b 是一个 G-块；取 $\mathcal{O}Gb$ 上的极大的局部点群 $P_{\hat{\gamma}}$ 与 $i \in \hat{\gamma}$，并记

15.1. As in section 14, we assume that the quotient field $\mathcal{K}$ of $\mathcal{O}$ has characteristic zero, that G is a finite group and that b is a block of G; choose a maximal local pointed group $P_{\hat{\gamma}}$ on $\mathcal{O}Gb$ and $i \in \hat{\gamma}$, and set

15.1.1 $$B = i(\mathcal{O}G)i \cong (\mathcal{O}G)_{\hat{\gamma}}\,.$$

15.2. 设 Q 是 P 的正规子群使得它包含 $P_{\hat{\gamma}}$ 的超聚焦子群. 假定 k 是代数闭的；这一节里，我们要证明存在一个 B 的 P-稳定的幺子代数 D 使得

15.2. Let Q be a normal subgroup of P which contains the hyperfocal subgroup of $P_{\hat{\gamma}}$. Assume that k is algebraically closed; in this section, we will prove the existence of a P-stable unitary subalgebra D which fulfills

15.2.1 $$D \cap Pi = Qi\,, \quad B = D \otimes_Q P\,.$$

对 $|P/Q|$ 用归纳法；当然，能假定 $Q \neq P$，和考虑 P 的一个正规子群 R 使得 $Q \subset R$ 与 $|R/Q| = p$；那么，按归纳法存在 B 的一个 P-稳定幺子代数 E 使得

We argue by induction on $|P/Q|$; of course, we may assume that $Q \neq P$, and may consider a normal subgroup R of P such that $Q \subset R$ and $|R/Q| = p$; then, by the induction hypothesis, there exists a P-stable unitary subalgebra E of B fulfilling

15.2.2 $$E \cap Pi = Ri\,, \quad B = E \otimes_R P\,.$$

15.3. 如同在§14里一样记

15.3. As in section 14, set

15.3.1 $$\tilde{B} = \mathrm{Res}^R_Q(E) \otimes_Q P\,, \quad \tilde{E} = \mathrm{Res}^R_Q(E) \otimes_Q R\,;$$

不难验证对任意 $v \in R$，$\tilde{B}$ 的元素 $v^{-1}i \otimes_Q v$ 属于 $Z(\tilde{B}) \cap \tilde{E}^*$ 并且，这些元素的集合 Z 在这个群里就是阶为 p 的子群；请注意，因为 $Z \subset Z(\tilde{B}) \cap \tilde{E}^*$，所以 $\tilde{B}$ 与 $\tilde{E}$ 也有一个自然的 $\mathcal{O}Z$-代数结构；事实上，不难验证它们是自由 $\mathcal{O}Z$-模. 进一步，考虑自然的 P-内代数同态如下 (见 9.7.1)

it is easy to check that, for any $v \in R$, the element $v^{-1}i \otimes_Q v$ of $\tilde{B}$ belongs to $Z(\tilde{B}) \cap \tilde{E}^*$ and that the set Z of these elements is a subgroup of order p in this group; note that, since $Z \subset Z(\tilde{B}) \cap \tilde{E}^*$, $\tilde{B}$ and $\tilde{E}$ also have a natural $\mathcal{O}Z$-algebra structure; actually, it is not difficult to prove that they are free $\mathcal{O}Z$-modules. Further, consider the following natural P-interior algebra homomorphism (see 9.7.1)

15.3.2 $$\tilde{B} \xrightarrow{f} B = E \otimes_R P\,;$$

显然有 $f(v^{-1}i \otimes_Q v) = i$，其中 $v \in R$，并不难验证

we clearly have $f(v^{-1}i \otimes_Q v) = i$ for any $v \in R$, and it is easily checked that

15.3.3 $$\mathrm{Ker}(f) = \sum_{z \in Z} \tilde{B}(z - \tilde{\imath}) \,, \quad \mathcal{O} \otimes_{\mathcal{O}Z} \tilde{B} \cong B$$

其中 $\tilde{\imath} = i \otimes_Q 1$.

where we set $\tilde{\imath} = i \otimes_Q 1$.

15.4. 首先，我们要证明存在一个 P-内代数同态 $s: B \to \tilde{B}$ 使得 $f \circ s = \mathrm{id}_B$；这种 s 的存在性就是命题 15.7 的推论，并且从引理 14.10 可推出 s 的 $(\tilde{B}^P)^*$-共轭类是唯一的. 为了证明 s 的存在性需要下面的结果.

15.4. First of all, we will prove the existence of a P-interior algebra homomorphism $s: B \to \tilde{B}$ fulfilling $f \circ s = \mathrm{id}_B$; the existence of s is a corollary of Proposition 15.7, and Lemma 14.10 implies that the $(\tilde{B}^P)^*$-conjugacy class of s is unique. In order to prove the existence of s, we need the following result.

命题 15.5. 假定 k 是代数闭的. 对任意 B 上的局部点群 $S_{\hat{\varepsilon}}$ 与 $T_{\hat{\tau}}$ 有

Proposition 15.5. *Assume that k is algebraically closed. For any local pointed groups $S_{\hat{\varepsilon}}$ and $T_{\hat{\tau}}$ on B, we have*

15.5.1 $$融_B(S_{\hat{\varepsilon}}, T_{\hat{\tau}}) = \bigcup_{u \in P} 融_{\mathrm{Res}^R_Q(E)}\big((S_\varepsilon)^u, T_\tau\big) \circ \kappa^S_u$$

其中 ε 与 τ 分别是 E 上的 S-点与 T-点使得 $\varepsilon \otimes_R 1 \subset \hat{\varepsilon}$ 与 $\tau \otimes_R 1 \subset \hat{\tau}$.

where ε and τ respectively are the points of S and T on E which fulfill $\varepsilon \otimes_R 1 \subset \hat{\varepsilon}$ and $\tau \otimes_R 1 \subset \hat{\tau}$.

证明: 由定理 9.2，只要证明对任意 $\varphi \in 融_E(S_\varepsilon, T_\tau)$ 有 $\varphi = \psi \circ \kappa^S_u$，其中 $u \in P$ 与 $\psi \in 融_{\mathrm{Res}^R_Q(E)}\big((S_\varepsilon)^u, T_\tau\big)$，就得到等式 15.5.1; 当然，能假定 $\iota_T \circ \varphi \neq \iota_S$，其中 $\iota_S: S \to P$ 与 $\iota_T: T \to P$ 是包含映射; 由命题 8.13，也可假定 $\varphi(S) = T$.

Proof: According to Theorem 9.2, in order to prove equality 15.5.1, it suffices to show that, for any $\varphi \in 融_E(S_\varepsilon, T_\tau)$, we have $\varphi = \psi \circ \kappa^S_u$ for suitable $u \in P$ and $\psi \in 融_{\mathrm{Res}^R_Q(E)}\big((S_\varepsilon)^u, T_\tau\big)$; of course, we may assume that $\iota_T \circ \varphi \neq \iota_S$, where $\iota_S: S \to P$ and $\iota_T: T \to P$ are the inclusion maps; by Proposition 8.13, we may assume that $\varphi(S) = T$.

而且，由注意 11.11，有

Moreover, from Remark 11.11, we have

15.5.2 $$\begin{aligned} \iota_S &= \iota_0 \circ \nu_0 \\ \iota_{h-1} \circ \sigma_{h-1} \circ \nu_{h-1} &= \iota_h \circ \nu_h \\ \iota_n \circ \sigma_n \circ \nu_n &= \kappa^P_u \circ \iota_T \circ \varphi \end{aligned}$$

其中 $n = \|\tilde{\varphi}\|$，$0 \leq h \leq n$ 而，对 B 上的适当的自中心化子点群 $U^h_{\hat{\delta}^h}$，$\iota_h: U^h \to P$ 是包含映射

where $n = \|\tilde{\varphi}\|$, $0 \leq h \leq n$ and, for suitable self-centralizing pointed groups $U^h_{\hat{\delta}^h}$ on B, $\iota_h: U^h \to P$ denotes the inclusion map, σ_h is

与 σ_h 是 $\text{融}_B(U^h_{\hat{\delta}^h})$ 的 p'-元素与 $\nu_h \in \text{融}_B(S_{\hat{\varepsilon}}, U^h_{\hat{\delta}^h})$ 与 $u \in P$.

a p'-element of $\text{融}_B(U^h_{\hat{\delta}^h})$, $\nu_h \in \text{融}_B(S_{\hat{\varepsilon}}, U^h_{\hat{\delta}^h})$ and $u \in P$.

另一方面, 由定理 13.9, 存在唯一 E 上的局部 U^h-点 δ^h 使得 $\delta^h \otimes_R 1 \subset \hat{\delta}^h$; 此时, 由推论 9.3, σ_h 属于 $\text{融}_E(U^h_{\delta^h})$ 并且, 因为 $w^{-1}\sigma_h(w) \in \mathfrak{h}_G(P_{\hat{\gamma}}) \subset Q$, 其中 $w \in U^h$, 所以 σ_h 也属于 $\text{融}_{\mathrm{Res}^R_Q(E)}(U^h_{\delta^h})$.

On the other hand, it follows from Theorem 13.9 that there exists a unique local point δ^h of U^h on E fulfilling $\delta^h \otimes_R 1 \subset \hat{\delta}^h$; then, according to Corollary 9.3, σ_h belongs to $\text{融}_E(U^h_{\delta^h})$ and therefore, since we have $w^{-1}\sigma_h(w) \in \mathfrak{h}_G(P_{\hat{\gamma}}) \subset Q$ for any $w \in U^h$, σ_h also belongs to $\text{融}_{\mathrm{Res}^R_Q(E)}(U^h_{\delta^h})$.

别忘了, 由定理 10.19, σ_h, ν_h, φ 是被 G 的适当的元素 s_h, y_h, x 分别决定的; 特别是, 从等式 $\iota_n \circ \sigma_n \circ \nu_n = \kappa^P_u \circ \iota_T \circ \varphi$ 可推出

Recall that, by Theorem 10.19, σ_h, ν_h and φ are respectively determined by suitable elements s_h, y_h and x of G; in particular, the equality $\iota_n \circ \sigma_n \circ \nu_n = \kappa^P_u \circ \iota_T \circ \varphi$ implies that

15.5.3 $$(T_{\hat{\tau}})^u = (S_{\hat{\varepsilon}})^{xu} = (S_{\hat{\varepsilon}})^{y_n s_n} \subset (U^n_{\hat{\delta}^n})^{s_n} = U^n_{\hat{\delta}^n};$$

这样, 我们有 (见 3.14.1 和 5.2)

thus, we have (see 3.14.1 and 5.2)

15.5.4 $$\mathrm{m}^{\hat{\delta}^n}_{\tau^u} = \mathrm{res}^B_E\big(\mathrm{res}^{U^n}_{T^u}(\hat{\delta}^n)\big)(\tau^u) \neq 0$$

从而不难验证存在 E 上的 U^n-点 θ 使得 $\mathrm{m}^\theta_{\tau^u} \neq 0 \neq \mathrm{m}^{\hat{\delta}^n}_\theta$; 因为 δ^n 是唯一 E 上的 U^n-点使得 $\delta^n \otimes_R 1 \subset \hat{\delta}^n$, 所以有 $\theta = \delta^n$ 与 $T'_{\tau'} = (T_\tau)^u \subset U^n_{\delta^n}$.

and then it is quite clear that there exists a point θ of U^n on E fulfilling $\mathrm{m}^\theta_{\tau^u} \neq 0 \neq \mathrm{m}^{\hat{\delta}^n}_\theta$; since δ^n is the unique point of U^n on E such that $\delta^n \otimes_R 1 \subset \hat{\delta}^n$, we have $\theta = \delta^n$ and $T'_{\tau'} = (T_\tau)^u \subset U^n_{\delta^n}$.

令 $\iota' : T' \to U^n$ 记包含映射, 再令

Denote by $\iota' : T' \to U^n$ the inclusion map, and set

15.5.5 $$S^n = \nu_n(S) = (\sigma_n)^{-1}(T');$$

那么, 我们考虑从 τ' 到 δ^n 的 $\mathrm{Res}^R_Q(E)$-融合 $(\sigma_n)^{-1} \circ \iota'$; 由命题 8.13, 存在唯一 E 上的 S^n-点 ε^n 使得 $S^n_{\varepsilon^n} \subset U^n_{\delta^n}$ 与 $(\sigma_n)^{-1} \circ \iota' = \iota^n \circ \rho'_n$, 其中

then, we consider the fusion $(\sigma_n)^{-1} \circ \iota'$ from τ' to δ^n in $\mathrm{Res}^R_Q(E)$; according to Proposition 8.13, there exists a point ε^n of S^n on E fulfilling $S^n_{\varepsilon^n} \subset U^n_{\delta^n}$ and $(\sigma_n)^{-1} \circ \iota' = \iota^n \circ \rho'_n$ for a suitable

15.5.6 $$\rho'_n \in \text{融}_{\mathrm{Res}^R_Q(E)}(T'_{\tau'}, S^n_{\varepsilon^n})$$

与 $\iota^n : S^n \to U^n$ 是包含映射; 现在, 显然有

where $\iota^n : S^n \to U^n$ is the inclusion map; now, we clearly have

15.5.7 $$\nu_n = \iota^n \circ \rho'_n \circ \kappa^T_u \circ \varphi = \iota^n \circ \kappa^T_u \circ \rho_n \circ \varphi$$

其中$\rho_n = (\rho'_n)^{u^{-1}}$ (见 8.8.1), 以及$\rho_n \circ \varphi$是从S_ε到$(S^n_{\varepsilon^n})^{u^{-1}}$的$E$-融合. 因为$\iota_{S^n} = \iota_n \circ \iota^n$, 所以有 $\|\tilde{\rho}_n \circ \tilde{\varphi}\| = n-1$ (见 15.5.2); 按归纳法, 得到$\rho_n \circ \varphi = \psi \circ \kappa_v^S$, 其中$v \in P$与

where we set $\rho_n = (\rho'_n)^{u^{-1}}$ (see 8.8.1), and $\rho_n \circ \varphi$ is a fusion from S_ε to $(S^n_{\varepsilon^n})^{u^{-1}}$ in E. Since $\iota_{S^n} = \iota_n \circ \iota^n$, we have $\|\tilde{\rho}_n \circ \tilde{\varphi}\| = n-1$ (see 15.5.2); hence, by the induction hypothesis, we get $\rho_n \circ \varphi = \psi \circ \kappa_v^S$ for suitable $v \in P$ and

15.5.8 $$\psi \in 融_{\mathrm{Res}^R_Q(E)}\big((S_\varepsilon)^v, (S^n_{\varepsilon^n})^{u^{-1}}\big),$$

也得到$\varphi = (\rho_n)^{-1} \circ \psi \circ \kappa_v^S$.

and then we still get $\varphi = (\rho_n)^{-1} \circ \psi \circ \kappa_v^S$.

推论 15.6. 假定k是代数闭的. 那么$f\colon \tilde{B} \to B$是严格覆盖同态, 并且对任意$\tilde{B}$上的局部点群$S_{\tilde{\varepsilon}}$与$T_{\tilde{\tau}}$有

Corollary 15.6. *Assume that k is algebraically closed. Then $f\colon \tilde{B} \to B$ is a strict covering homomorphism and, for any local pointed groups $S_{\tilde{\varepsilon}}$ and $T_{\tilde{\tau}}$ on $\tilde{B}$, we have*

15.6.1 $$融_{\tilde{B}}(S_{\tilde{\varepsilon}}, T_{\tilde{\tau}}) = 融_B(S_{\hat{\varepsilon}}, T_{\hat{\tau}}),$$

其中$\hat{\varepsilon}$与$\hat{\tau}$分别是B上的S-点与T-点使得$f(\tilde{\varepsilon}) \subset \hat{\varepsilon}$与$f(\tilde{\tau}) \subset \hat{\tau}$.

where $\hat{\varepsilon}$ and $\hat{\tau}$ respectively are the points of S and T on B fulfilling $f(\tilde{\varepsilon}) \subset \hat{\varepsilon}$ and $f(\tilde{\tau}) \subset \hat{\tau}$.

证明: 由定理 9.5, 存在E上的局部S-点ε与局部T-点τ使得$\varepsilon \otimes_Q 1 \subset \tilde{\varepsilon}$与$\tau \otimes_Q 1 \subset \tilde{\tau}$, 以及$B$上的局部$S$-点$\hat{\varepsilon}$与局部$T$-点$\hat{\tau}$使得$\varepsilon \otimes_R 1 \subset \hat{\varepsilon}$与$\tau \otimes_R 1 \subset \hat{\tau}$; 特别是, 有$f(\tilde{\varepsilon}) \subset \hat{\varepsilon}$与$f(\tilde{\tau}) \subset \hat{\tau}$, 并且从定理 9.2 和命题 15.5 可推出

Proof: It follows from Theorem 9.5 that there exist respective local points ε and τ of S and T on E fulfilling $\varepsilon \otimes_Q 1 \subset \tilde{\varepsilon}$ and $\tau \otimes_Q 1 \subset \tilde{\tau}$, and respective local points $\hat{\varepsilon}$ and $\hat{\tau}$ of S and T on B such that $\varepsilon \otimes_R 1 \subset \hat{\varepsilon}$ and $\tau \otimes_R 1 \subset \hat{\tau}$; in particular, we have $f(\tilde{\varepsilon}) \subset \hat{\varepsilon}$ and $f(\tilde{\tau}) \subset \hat{\tau}$, and then Theorem 9.2 and Proposition 15.5 imply that

15.6.2 $$融_{\tilde{B}}(S_{\tilde{\varepsilon}}, T_{\tilde{\tau}}) = \bigcup_{u \in P} 融_{\mathrm{Res}^R_Q(E)}\big((S_\varepsilon)^u, T_\tau\big) \circ \kappa_u^S = 融_B(S_{\hat{\varepsilon}}, T_{\hat{\tau}});$$

只要应用推论 9.8 , 就证明完了.

Corollary 9.8 allows us to complete the proof.

命题 15.7. 假定k是代数闭的. 那么下面的G-内代数同态

Proposition 15.7. *Assume that k is algebraically closed. Then the G-interior algebra homomorphism*

15.7.1 $$\mathrm{Ind}_P^G(f)\colon \mathrm{Ind}_P^G(\tilde{B}) \longrightarrow \mathrm{Ind}_P^G(B)$$

是严格覆盖同态.

is a strict covering homomorphism.

证明: 由下面的引理 15.8, 对任意$\mathrm{Ind}_P^G(\tilde{B})$上的局部点群$S_{\tilde{\varepsilon}}$存在$x \in G$使得$S'_{\tilde{\varepsilon}'} = (S_{\tilde{\varepsilon}})^x$也

Proof: By Lemma 15.8 below, for any local pointed group $S_{\tilde{\varepsilon}}$ on $\mathrm{Ind}_P^G(\tilde{B})$ there exists $x \in G$ such that $S'_{\tilde{\varepsilon}'} = (S_{\tilde{\varepsilon}})^x$ is also a local

是 $\tilde{B}$ 上的局部点群 (见 4.8.1). 那么, 由推论 15.6, 存在 B 上的局部 S'-点 $\hat{\varepsilon}'$ 使得 $f(\tilde{\varepsilon}') \subset \hat{\varepsilon}'$, 并且 f 决定下面的 k-代数同构

pointed group on $\tilde{B}$ (see 4.8.1). Then, it follows from Corollary 15.6 that there is a local point ε' on S' fulfilling $f(\tilde{\varepsilon}') \subset \hat{\varepsilon}'$, and that f induces the following k-algebra isomorphism

15.7.2 $$\tilde{B}(S'_{\tilde{\varepsilon}'}) \cong B(S'_{\hat{\varepsilon}'});$$

记 $\hat{\varepsilon} = (\hat{\varepsilon}')^{x^{-1}}$; 这样, $\hat{\varepsilon}$ 是满足 $\mathrm{Ind}_P^G(f)(\tilde{\varepsilon}) \subset \hat{\varepsilon}$ 的 $\mathrm{Ind}_P^G(B)$ 上的局部 S-点; 特别是, 映射如下

set $\hat{\varepsilon} = (\hat{\varepsilon}')^{x^{-1}}$; thus, $\hat{\varepsilon}$ is a local point of S on $\mathrm{Ind}_P^G(B)$ such that $\mathrm{Ind}_P^G(f)(\tilde{\varepsilon}) \subset \hat{\varepsilon}$; in particular, the following map

15.7.3 $$\mathrm{res}_{\mathrm{Ind}_P^G(f)^S} : \mathcal{D}_{\mathrm{Ind}_P^G(\tilde{B})}(S) \longrightarrow \mathcal{D}_{\mathrm{Ind}_P^G(B)}(S)$$

决定下面的满射

induces a surjective map

15.7.4 $$\mathcal{LP}_{\mathrm{Ind}_P^G(\tilde{B})}(S) \longrightarrow \mathcal{LP}_{\mathrm{Ind}_P^G(B)}(S).$$

进一步, 如果 $T_{\tilde{\tau}}$ 与 $T_{\hat{\tau}}$ 分别是 $\mathrm{Ind}_P^G(\tilde{B})$ 与 $\mathrm{Ind}_P^G(B)$ 上的相互对应的局部点群, 并且对适当的 $y \in G$, $(T_{\tilde{\tau}})^y$ 也是 $\tilde{B}$ 上的局部点群, 那么从等式 15.6.1 可得到

Furthermore, if $T_{\tilde{\tau}}$ and $T_{\hat{\tau}}$ are mutually corresponding local pointed groups on $\mathrm{Ind}_P^G(\tilde{B})$ and $\mathrm{Ind}_P^G(B)$ respectively, and moreover, for a suitable $y \in G$, $(T_{\tilde{\tau}})^y$ is a local pointed group on $\tilde{B}$ too, then equality 15.6.1 implies that

15.7.5 $$融_{\tilde{B}}\big((S_{\tilde{\varepsilon}})^x, (T_{\tilde{\tau}})^y\big) = 融_B\big((S_{\hat{\varepsilon}})^x, (T_{\hat{\tau}})^y\big),$$

从而只要应用推论 8.19, 就有

and therefore, for Corollary 8.19, we get

15.7.6 $$\begin{aligned} 融_{\mathrm{Ind}_P^G(\tilde{B})}(S_{\tilde{\varepsilon}}, T_{\tilde{\tau}}) &= (\kappa_y^T)^{-1} \circ 融_{\mathrm{Ind}_P^G(\tilde{B})}\big((S_{\tilde{\varepsilon}})^x, (T_{\tilde{\tau}})^y\big) \circ \kappa_x^S \\ &= (\kappa_y^T)^{-1} \circ 融_{\tilde{B}}\big((S_{\tilde{\varepsilon}})^x, (T_{\tilde{\tau}})^y\big) \circ \kappa_x^S \\ &= (\kappa_y^T)^{-1} \circ 融_B\big((S_{\hat{\varepsilon}})^x, (T_{\hat{\tau}})^y\big) \circ \kappa_x^S \\ &= (\kappa_y^T)^{-1} \circ 融_{\mathrm{Ind}_P^G(B)}\big((S_{\hat{\varepsilon}})^x, (T_{\hat{\tau}})^y\big) \circ \kappa_x^S \\ &= 融_{\mathrm{Ind}_P^G(B)}(S_{\hat{\varepsilon}}, T_{\hat{\tau}}); \end{aligned}$$

特别是, 如果 $S_{\hat{\varepsilon}} = T_{\hat{\tau}}$ 那么 id_S 属于 $融_{\mathrm{Ind}_P^G(\tilde{B})}(S_{\tilde{\varepsilon}}, S_{\tilde{\tau}})$, 从而还有 $\tilde{\varepsilon} = \tilde{\tau}$ (见命题 8.10); 所以, 映射 15.7.4 是双射.

in particular, if $S_{\hat{\varepsilon}} = T_{\hat{\tau}}$ then id_S belongs to $融_{\mathrm{Ind}_P^G(\tilde{B})}(S_{\tilde{\varepsilon}}, S_{\tilde{\tau}})$ and therefore we still get $\tilde{\varepsilon} = \tilde{\tau}$ (see Proposition 8.10); consequently, the map 15.7.4 is bijective.

最后, 只要再次应用引理 15.8, 就从同构式 15.7.2 和等式 15.7.6 可得到

Finally, by applying Lemma 15.8 again, from isomorphism 15.7.2 and equality 15.7.6, we can conclude that

15.7.7 $$\dim_k\big(\mathrm{Ind}_P^G(\tilde{B})(S_{\tilde{\varepsilon}})\big) = \dim_k\big(\mathrm{Ind}_P^G(B)(S_{\hat{\varepsilon}})\big),$$

从而不难证明 $\mathrm{Ind}_P^G(f)$ 也决定 k-代数同构如下

and then it is not difficult to prove that $\mathrm{Ind}_P^G(f)$ induces a k-algebra isomorphism

15.7.8 $$\mathrm{Ind}_P^G(\tilde{B})(S_{\tilde{\varepsilon}}) \cong \mathrm{Ind}_P^G(B)(S_{\hat{\varepsilon}})\,.$$

此时, 由推论 6.19, $\mathrm{Ind}_P^G(f)$ 是严格覆盖同态.

Now, it follows from Corollary 6.19 that $\mathrm{Ind}_P^G(f)$ is a strict covering homomorphism.

引理 15.8. 设 H 是 G 的子群并设 C 是 H-内代数. 令 υ 与 θ 分别记由 $1_{\mathrm{Ind}_H^G(C)}$ 与 $1\otimes 1_C\otimes 1$ 决定的 $\mathrm{Ind}_H^G(C)$ 上的 G-除子与 H-除子. 对 $\mathrm{Ind}_H^G(C)$ 上的任意局部点群 U_δ 有

Lemma 15.8. *Let H be a subgroup of G and C an H-interior algebra. Denote respectively by υ and θ the divisors of G and H on $\mathrm{Ind}_H^G(C)$ determined by $1_{\mathrm{Ind}_H^G(C)}$ and $1\otimes 1_C\otimes 1$. For any local pointed group U_δ on $\mathrm{Ind}_H^G(C)$ we have*

15.8.1 $$\mathrm{m}_\delta^\upsilon = \sum_{\tilde{\varphi}\in E_G(U_\delta,H_\theta)} |C_G(U^{x(\tilde{\varphi})})/C_H(U^{x(\tilde{\varphi})})|\,\mathrm{m}_{\delta^{x(\tilde{\varphi})}}^\theta$$

其中 $x(\tilde{\varphi})\in G$ 是决定 $\tilde{\varphi}$ 的元素. 特别是, 存在 $x\in G$ 使得 $U^x\subset H$ 与 $\delta^x\subset\theta$.

where $x(\tilde{\varphi})\in G$ is an element which determines $\tilde{\varphi}$. In particular, there exists $x\in G$ such that $U^x\subset H$ and $\delta^x\subset\theta$.

证明: 因为 $\upsilon=\mathrm{ind}_H^G(\theta)$, 所以

Proof: Since $\upsilon=\mathrm{ind}_H^G(\theta)$, we have

15.8.2 $$\mathrm{res}_U^G(\upsilon)=\sum_{x\in X}\mathrm{ind}_{U\cap H^x}^U\big(\mathrm{res}_{U\cap H^x}^{H^x}(\theta^x)\big)$$

其中 $X\subset G$ 是一个 $H\backslash G/U$ 的代表集合; 可是, 既然 δ 是局部的, 对任意满足 $U\cap H^x\neq U$ 的 $x\in X$ 有

where $X\subset G$ is a set of representatives for $H\backslash G/U$; but, since δ is local, for any $x\in X$ such that $U\cap H^x\neq U$, we get

15.8.3 $$s_\delta\big(\mathrm{Tr}_{U\cap H^x}^U(x^{-1}\otimes 1_C\otimes x)\big)=0\,;$$

这样, 得到 $\mathrm{m}_\delta^\upsilon=\sum_{x\in Y}\mathrm{m}_\delta^{\theta^x}$, 其中 $Y\subset X$ 是满足 $U\subset H^x$ 的元素 $x\in X$ 的集合; 而且, 如果 $\mathrm{m}_\delta^{\theta^x}\neq 0$ 那么 $\delta\subset\mathrm{res}_U^{H^x}(\theta^x)$, 从而 x^{-1} 决定一个 $E_G(U_\delta,H_\theta)$ 的元素.

thus, we obtain $\mathrm{m}_\delta^\upsilon=\sum_{x\in Y}\mathrm{m}_\delta^{\theta^x}$, where $Y\subset X$ is the set of elements $x\in X$ fulfilling $U\subset H^x$; moreover, if $\mathrm{m}_\delta^{\theta^x}\neq 0$ then we certainly have $\delta\subset\mathrm{res}_U^{H^x}(\theta^x)$ and therefore x^{-1} determines an element of $E_G(U_\delta,H_\theta)$.

进一步, 因为对 $C_G(U)$ 的任意元素 z 有

Furthermore, since for any element z of $C_G(U)$ we have

15.8.4 $$\mathrm{res}_U^{H^{xz}}(\theta^{xz})=\mathrm{res}_U^{H^x}(\theta^x)\,,$$

所以还有 $\mathrm{m}_\delta^{\theta^{xz}}=\mathrm{m}_\delta^{\theta^x}$.另一方面,如果 $\mathrm{res}_U^{H^y}(\theta^y)=\mathrm{res}_U^{H^x}(\theta^x)$,其中 $y\in Y$,那么 x 与 y 在 $E_G(U_\delta,H_\theta)$ 中决定同一个元素当且仅当有 $x^{-1}y\in\tilde{C}_G(U)$. 所以, 得到等式 15.8.1; 特别是, 有 $E_G(U_\delta,H_\theta)\neq\emptyset$.

we also have $\mathrm{m}_\delta^{\theta^{xz}}=\mathrm{m}_\delta^{\theta^x}$. On the other hand, if we have $\mathrm{res}_U^{H^y}(\theta^y)=\mathrm{res}_U^{H^x}(\theta^x)$ for some $y\in Y$, then x and y determine the same element of $E_G(U_\delta,H_\theta)$ if and only if we have $x^{-1}y\in\tilde{C}_G(U)$. Consequently, we obtain equality 15.8.1 and, in particular, we have $E_G(U_\delta,H_\theta)\neq\emptyset$.

推论 15.9. 假定 k 是代数闭的. 存在唯一一个 P-内代数同态 $s\colon B\to\tilde{B}$ 的 $(\tilde{B}^P)^*$-共轭类使得 $f\circ s=\mathrm{id}_B$.

Corollary 15.9. *Assume that k is algebraically closed. There exists a unique $(\tilde{B}^P)^*$-conjugacy class of P-interior algebra homomorphisms $s\colon B\to\tilde{B}$ fulfilling $f\circ s=\mathrm{id}_B$.*

证明: 由定理 5.11, 存在完备诱导的 G-内代数 A 与 $\tilde{A}$ 和适当的 A 上的 G-点 $\hat{\alpha}$ 与 $\tilde{A}$ 上的 G-除子 $\tilde{\omega}$ 使得 $\mathcal{O}Gb\cong A_{\hat{\alpha}}$ 与 $\mathrm{Ind}_P^G(\tilde{B})\cong\tilde{A}_{\tilde{\omega}}$; 此时, 由同构式 5.3.2 和 5.8.6, 得到

Proof: According to Theorem 5.11, there exist two inductively complete G-interior algebras A and $\tilde{A}$, a suitable point $\hat{\alpha}$ of G on A, and a suitable divisor $\tilde{\omega}$ of G on $\tilde{A}$ such that $\mathcal{O}Gb\cong A_{\hat{\alpha}}$ and $\mathrm{Ind}_P^G(\tilde{B})\cong\tilde{A}_{\tilde{\omega}}$; so, from isomorphisms 5.3.2 and 5.8.6, we get

15.9.1 $\mathrm{Ind}_P^G(B)\cong A_{\mathrm{ind}_P^G(\hat{\gamma})}\,,\quad \tilde{\omega}=\mathrm{ind}_P^G(\tilde{\gamma})\,,\quad B\cong A_{\hat{\gamma}}\,,\quad \tilde{B}\cong\tilde{A}_{\tilde{\gamma}}\,,$

其中 $\tilde{\gamma}$ 是唯一 $\tilde{B}$ 上的 P-点(见推论 15.6); 所以 $\mathrm{Ind}_P^G(f)(\tilde{\gamma})\subset\hat{\gamma}$.

where $\tilde{\gamma}$ is the unique point of P on $\tilde{B}$ (see Corollary 15.6); hence $\mathrm{Ind}_P^G(f)(\tilde{\gamma})\subset\hat{\gamma}$.

由命题 15.8, 存在包含 $\mathrm{Ind}_P^G(f)(\tilde{\alpha})$ 的唯一一个 $\tilde{A}$ 上的 G-点 $\tilde{\alpha}$, 并且有 $P_{\tilde{\gamma}}\subset G_{\tilde{\alpha}}$ 和 $\mathrm{Ind}_P^G(f)$ 决定下面的 G-内代数同态

By Proposition 15.8, there exists a unique point $\tilde{\alpha}$ of G on $\tilde{A}$ which contains $\mathrm{Ind}_P^G(f)(\tilde{\alpha})$, we have $P_{\tilde{\gamma}}\subset G_{\tilde{\alpha}}$, and $\mathrm{Ind}_P^G(f)$ induces the following P-interior algebra homomorphism

15.9.2 $$g:\tilde{A}_{\tilde{\alpha}}\longrightarrow A_{\hat{\alpha}}\cong\mathcal{O}Gb\,,$$

它是严格覆盖同态 (见 4.13). 可是, 结构的同态 $\mathcal{O}G\to\tilde{A}_{\tilde{\alpha}}$ 显然决定一个 G-内代数同态 $t\colon\mathcal{O}Gb\to\tilde{A}_{\tilde{\alpha}}$ 使得 $g\circ t=\mathrm{id}_{\mathcal{O}Gb}$. 此时, 因为 $\mathrm{id}_{\mathcal{O}Gb}$ 是严格覆盖同态, 所以 t 也是严格覆盖同态.

which is a strict covering homomorphism (see 4.13). But, the structural homomorphism $\mathcal{O}G\to\tilde{A}_{\tilde{\alpha}}$ clearly determines a G-interior algebra homomorphism $t\colon\mathcal{O}Gb\to\tilde{A}_{\tilde{\alpha}}$ fulfilling $g\circ t=\mathrm{id}_{\mathcal{O}Gb}$. Hence, since $\mathrm{id}_{\mathcal{O}Gb}$ is a strict covering homomorphism, so is t .

这样, 既然 $\mathrm{Ind}_P^G(f)(\tilde{\gamma})\subset\hat{\gamma}$ 就有 $t(\hat{\gamma})\subset\tilde{\gamma}$, 而 t 决定一个 P-内代数同态 $s\colon B\to\tilde{B}$ 使得 $f\circ s\colon B\to B$ 是被 $\mathrm{id}_{\mathcal{O}Gb}$ 决定的; 也就是说, s 也是严格覆盖

Thus, since $\mathrm{Ind}_P^G(f)(\tilde{\gamma})\subset\hat{\gamma}$, we have $t(\hat{\gamma})\subset\tilde{\gamma}$ and therefore t induces a P-interior algebra homomorphism $s\colon B\to\tilde{B}$ such that $f\circ s\colon B\to B$ is induced by $\mathrm{id}_{\mathcal{O}Gb}$; that is to say, s is also a strict covering homomorphism

同态 (见 4.14) 并且 $f\circ s$ 是可嵌入同构; 此时, 因为它是 B 的自同构, 所以存在 $c\in (B^P)^*$ 使得有 $f\big(s(a)\big)^c = a$, 其中 $a\in B$. 特别是, 有 $f(\tilde B^P)=B^P$; 所以存在 $\tilde c\in(\tilde B^P)^*$ 使得 $f(\tilde c)=c$; 最后, 得到 $f\big(s(a)^{\tilde c}\big)=a$.

如果一个 P-内代数同态 $s'\colon B\to\tilde B$ 既是严格覆盖同态, 也满足 $f\circ s'=\mathrm{id}_B$, 那么只要应用引理 14.10, 就得到适当的 $\tilde c'\in(\tilde B^P)^*$ 使得 $s'(a)=s(a)^{\tilde c\tilde c'}$, 其中 $a\in B$.

(see 4.14) and $f\circ s$ is an *embeddable* isomorphism; consequently, since it is an automorphism of B, there exists $c\in (B^P)^*$ fulfilling $f\big(s(a)\big)^c = a$ for any $a\in B$. In particular, we have $f(\tilde B^P) = B^P$; hence, there exists $\tilde c\in(\tilde B^P)^*$ such that $f(\tilde c)=c$; finally, we get $f\big(s(a)^{\tilde c}\big)=a$.

If a P-interior algebra homomorphism $s'\colon B\to\tilde B$ is a strict covering homomorphism which fulfills $f\circ s'=\mathrm{id}_B$, then it follows from Lemma 14.10 that there exists $\tilde c'\in(\tilde B^P)^*$ such that $s'(a)=s(a)^{\tilde c\tilde c'}$ for any $a\in B$.

定理 15.10. **假定 k 是代数闭的. 存在 B 的 P-稳定的幺子代数 D 使得**

Theorem 15.10. *Assume that k is algebraically closed. There exists a P-stable unitary subalgebra D of B fulfilling*

15.10.1 $$D\cap Pi = Qi\,,\quad B = D\otimes_Q P\,.$$

证明: 设 $\hat{\mathcal K}$ 是被一个本原 p-次单位根 ξ 生成的 $\mathcal K$ 的扩张, 并令 $\hat{\mathcal O}$ 记 $\hat{\mathcal K}$ 中的 $\mathcal O$ 上的整元构成的环 (见 2.12); 设 $\varphi\colon R\to\hat{\mathcal O}^*$ 是一个群同态使得 $\mathrm{Ker}(\varphi)=Q$ 并考虑对应的同态 $\mathcal OZ\to\hat{\mathcal O}$; 令 (见 15.3.1)

Proof: Let $\hat{\mathcal K}$ be the extension of $\mathcal K$ generated by a primitive p-th root of unity ξ, and denote by $\hat{\mathcal O}$ the integral closure of $\mathcal O$ in $\hat{\mathcal K}$ (see 2.12); let $\varphi\colon R\to\hat{\mathcal O}^*$ be a group homomorphism such that $\mathrm{Ker}(\varphi)=Q$, and consider the corresponding homomorphism $\mathcal OZ\to\hat{\mathcal O}$; set (see 15.3.1)

15.10.2 $$\hat B = \hat{\mathcal O}\otimes_{\mathcal O} B\,,\quad \hat E = \hat{\mathcal O}\otimes_{\mathcal O} E\,,$$
$$\hat{\tilde B} = \hat{\mathcal O}\otimes_{\mathcal OZ}\tilde B\,,\quad \hat{\tilde E} = \hat{\mathcal O}\otimes_{\mathcal OZ}\tilde E\,.$$

由推论15.9, 存在一个 P-内代数同态 $s\colon B\to\tilde B$ 使得 $f\circ s=\mathrm{id}_B$; 那么只要考虑自然同态 $\tilde B\to\hat{\tilde B}$, 就得到一个 $\hat{\mathcal O}$ 上的 P-内代数同态 $\hat s\colon\hat B\to\hat{\tilde B}$.

事实上, 因为 $\tilde B$ 是自由 $\mathcal OZ$-模并且有 (见式子 15.3.3)

By Corollary 15.9, there exists a P-interior algebra homomorphism $s\colon B\to\tilde B$ fulfilling $f\circ s=\mathrm{id}_B$; then, considering the natural homomorphism $\tilde B\to\hat{\tilde B}$, we get a P-interior $\hat{\mathcal O}$-algebra homomorphism $\hat s\colon\hat B\to\hat{\tilde B}$.

Actually, since $\tilde B$ is a free $\mathcal OZ$-module and we have (see 15.3.3)

15.10.3 $$\tilde B = s(B)+\mathrm{Ker}(f) = s(B)+\sum_{z\in Z}\tilde B(z-\tilde\imath)$$
$$B\cong\mathcal O\otimes_{\mathcal OZ}\tilde B\,,$$

其中 $\tilde{\imath}=i\otimes_Q 1$, 所以 $\hat{s}$ 是满射并且有

where we set $\tilde{\imath}=i\otimes_Q 1$, $\hat{s}$ is surjective and we get

$$15.10.4 \qquad \mathrm{rank}_{\hat{\mathcal{O}}}(\hat{B})=\mathrm{rank}_{\mathcal{O}}(B)=\mathrm{rank}_{\mathcal{O}Z}(\tilde{B})=\mathrm{rank}_{\hat{\mathcal{O}}}(\hat{\tilde{B}}),$$

从而 $\hat{s}$ 是双射.

so that $\hat{s}$ is bijective.

进一步, 既然显然有

Furthermore, as far as we clearly have

$$15.10.5 \qquad \tilde{E}\cap P{\cdot}\tilde{\imath}=R{\cdot}\tilde{\imath}, \quad \tilde{B}=\tilde{E}\otimes_R P,$$

不难验证

it is easily checked that

$$15.10.6 \qquad \hat{\tilde{E}}\cap P{\cdot}(1\otimes\tilde{\imath})=R{\cdot}(1\otimes\tilde{\imath}), \quad \hat{\tilde{B}}=\hat{\tilde{E}}\otimes_R P.$$

那么, 在 $\hat{B}$ 里记 $\hat{\imath}=1\otimes i$ 与 $\hat{E}'=\hat{s}^{-1}(\hat{\tilde{E}})$; 这样得到

Then, setting $\hat{\imath}=1\otimes i$ and $\hat{E}'=\hat{s}^{-1}(\hat{\tilde{E}})$ in $\hat{B}$, we get

$$15.10.7 \qquad \begin{aligned}\hat{E}\cap P{\cdot}\hat{\imath}&=R{\cdot}\hat{\imath}, \quad \hat{B}=\hat{E}\otimes_R P,\\ \hat{E}'\cap P{\cdot}\hat{\imath}&=R{\cdot}\hat{\imath}, \quad \hat{B}=\hat{E}'\otimes_R P.\end{aligned}$$

现在, 我们断言

Now, we claim that

$$15.10.8 \qquad \hat{E}+(\xi-1){\cdot}\hat{B}=\hat{E}'+(\xi-1){\cdot}\hat{B}.$$

确实, 分别取 P/R 与 R/Q 的代表集合 V 与 W 使得它们包含 1; 请注意, $W{\cdot}V$ 是一个 P/Q 的代表集合. 而且, 对任意 $e\in E$ 与 $u\in W{\cdot}V$ 令 e_u 记 E 的元素使得

Indeed, choose respective sets V and W of representatives for P/R and R/Q, both containing 1; note that $W{\cdot}V$ is a set of representatives for P/Q. Moreover, for any $e\in E$ and any $u\in W{\cdot}V$ denote by e_u the element of E in such a way that we have

$$15.10.9 \qquad s(e)=\sum_{u\in W\cdot V} e_u\otimes_Q u;$$

因为 $f\circ s=\mathrm{id}_B$ 所以有

since $f\circ s=\mathrm{id}_B$, we have

$$15.10.10 \qquad \begin{aligned}e&=f\Big(\sum_{u\in W\cdot V} e_u\otimes_Q u\Big)=\sum_{u\in W\cdot V} e_u\otimes_R u\\ &=\sum_{v\in V}\Big(\sum_{w\in W} e_{wv}{\cdot}w\Big)\otimes_R v,\end{aligned}$$

从而得到 $e=\sum_{w\in W} e_w{\cdot}w$ 并还得到 $0=\sum_{w\in W} e_{wv}{\cdot}w$, 其中 $v\in V-\{1\}$.

and therefore we get $e=\sum_{w\in W} e_w{\cdot}w$ and $0=\sum_{w\in W} e_{wv}{\cdot}w$ for any $v\in V-\{1\}$.

令 $\hat{\hat{c}}$ 记任意 $c\in E$ 在 $\hat{\hat{E}}$ 中的像; 请注意, 对任意 $w\in W$ 有

Denote by $\hat{\hat{c}}$ the image of any $c \in E$ in $\hat{\hat{E}}$; note that, for any $w \in W$, we have

$$\begin{aligned} \varphi(w)\cdot\widehat{\widehat{c\cdot w}} &= \varphi(w) \otimes (c\cdot w \otimes_Q 1) \\ &= 1 \otimes \big((c\cdot w \otimes_Q 1)(w^{-1} i \otimes_Q w)\big) \\ &= 1 \otimes (c \otimes_Q w) = \hat{\hat{c}}\cdot w\,; \end{aligned} \tag{15.10.11}$$

特别是, 我们得到

in particular, we obtain

$$\begin{aligned} \hat{\hat{e}} &= \sum_{w\in W} \widehat{\widehat{e_w\cdot w}} = \sum_{w\in W} \varphi(w)^{-1}\cdot\hat{\hat{e}}_w\cdot w\,, \\ 0 &= \sum_{w\in W} \widehat{\widehat{e_{wv}\cdot w}} = \sum_{w\in W} \varphi(w)^{-1}\cdot\hat{\hat{e}}_{wv}\cdot w\,, \end{aligned} \tag{15.10.12}$$

其中 $v\in V-\{1\}$. 类似地, 令 $\hat{e}$ 记 e 在 $\hat{E}$ 的像; 那么有 $\hat{s}(\hat{e}) = \sum_{u\in W\cdot V} \hat{\hat{e}}_u \otimes_R u$. 最后, 得到

for any $v \in V - \{1\}$. Similarly, denoting by $\hat{e}$ the image of e in $\hat{E}$, we have $\hat{s}(\hat{e}) = \sum_{u\in W\cdot V} \hat{\hat{e}}_u \otimes_R u$. Finally, we obtain

$$\hat{s}(\hat{e}) - \hat{\hat{e}} \otimes_R 1 = \sum_{v\in V} \Big(\sum_{w\in W} (1-\varphi(w)^{-1})\cdot\hat{\hat{e}}_{wv}\cdot w\Big) \otimes_R v\,, \tag{15.10.13}$$

从而 $\hat{s}(\hat{e})-\hat{\hat{e}}\otimes_R 1$ 属于 $(\xi-1)\cdot\hat{\hat{B}}$, 并且也得到

hence $\hat{s}(\hat{e}) - \hat{\hat{e}} \otimes_R 1$ belongs to $(\xi-1)\cdot\hat{\hat{B}}$ and therefore we get

$$\hat{e} - (\hat{s})^{-1}(\hat{\hat{e}} \otimes_R 1) \in (\xi-1)\cdot\hat{B}\,,\quad (\hat{s})^{-1}(\hat{\hat{e}} \otimes_R 1) \in \hat{E}'\,. \tag{15.10.14}$$

现在, 等式 15.10.8 的证明完了.

So, the proof of equality 15.10.8 is complete.

这样, 由定理 14.7, 存在 $\hat{n}\in J(\hat{B}^P)$ 使得

Thus, it follows from Theorem 14.7 that there exists $\hat{n} \in J(\hat{B}^P)$ fulfilling

$$\hat{E}' = \hat{E}^{\hat{\imath}+(\xi-1)\cdot\hat{n}}\,; \tag{15.10.15}$$

此时, 下面的式子定义一个 $\hat{\mathcal{O}}$-代数自同构 $\hat{t}\colon \hat{E} \cong \hat{E}$

at this point, we can define the following $\hat{\mathcal{O}}$-algebra automorphism $\hat{t}\colon \hat{E} \cong \hat{E}$

$$\hat{t}(\hat{e}) = \big(\hat{\imath} + (\xi-1)\cdot\hat{n}\big)\hat{s}^{-1}(\hat{\hat{e}} \otimes_R 1)\big(\hat{\imath} + (\xi-1)\cdot\hat{n}\big)^{-1}\,, \tag{15.10.16}$$

其中 $e \in E$; 请注意, $\hat{t}$ 在商 $\hat{E}/(\xi-1)\cdot\hat{E}$ 上平凡作用; 进一步, 不难验证对任意 $u \in P$ 得到 $\hat{t}(\hat{e}^u) = \hat{t}(\hat{e})^u$ 并且, 由等式 15.10.11, 对任意 $v\in R$ 还得到 $\hat{t}(v\cdot\hat{\imath}) = \varphi(v)^{-1}\cdot(v\cdot\hat{\imath})$. 特别是, $\hat{t}^P$ 是 $\hat{E}$ 的 R-内 P-代数同构.

where $e \in E$; note that $\hat{t}$ acts trivially on the quotient $\hat{E}/(\xi-1)\cdot\hat{E}$; furthermore, it is easy to check that we have $\hat{t}(\hat{e}^u) = \hat{t}(\hat{e})^u$ for any $u \in P$, and moreover, according to equality 15.10.11, for any $v \in R$, we also have $\hat{t}(v\cdot\hat{\imath}) = \varphi(v)^{-1}\cdot(v\cdot\hat{\imath})$; in particular, $\hat{t}^P$ is an R-interior P-algebra isomorphism.

另一方面, 因为 (见 16.2.4) On the other hand, since (see 16.2.4)

15.10.17 $(\hat{t}-\mathrm{id}_{\hat{E}})(\hat{E}) \subset (\xi-1)\cdot\hat{E}\,,\quad (\xi-1)p\,\hat{\mathcal{O}} = (\xi-1)^p\hat{\mathcal{O}}\,,$

所以在 $\mathrm{End}_{\hat{\mathcal{O}}}(\hat{E})$ 中得到 in $\mathrm{End}_{\hat{\mathcal{O}}}(\hat{E})$ we obviously have

15.10.18 $$\hat{t}^{\,p} = \big(\mathrm{id}_{\hat{E}} + (\hat{t}-\mathrm{id}_{\hat{E}})\big)^p = \mathrm{id}_{\hat{E}} + \sum_{h=1}^{p}\binom{p}{h}(\hat{t}-\mathrm{id}_{\hat{E}})^h\,,$$

从而 $(\hat{t}^p-\mathrm{id}_{\hat{E}})(\hat{E})\subset(\xi-1)^p\cdot\hat{E}$; 也就是说, $\hat{t}^p$ 在 $\hat{E}/(\xi-1)^p\cdot\hat{E}$ 上平凡作用.

and therefore $(\hat{t}^p-\mathrm{id}_{\hat{E}})(\hat{E}) \subset (\xi-1)^p\cdot\hat{E}$; that is to say, $\hat{t}^p$ acts trivially on $\hat{E}/(\xi-1)^p\cdot\hat{E}$.

现在, 能应用命题 14.5; 而且, 既然 k 是代数闭的, 就有 $\hat{E}^P = \hat{\mathcal{O}}\cdot\hat{\imath} + J(\hat{E}^P)$; 这样, 对适当的 $\hat{m}\in J(\hat{E}^P)$ 我们得到

Now, we can apply Proposition 14.5; moreover, since k is algebraically closed, we have $\hat{E}^P = \hat{\mathcal{O}}\cdot\hat{\imath} + J(\hat{E}^P)$; thus, for a suitable $\hat{m}\in J(\hat{E}^P)$ we get

15.10.19 $$\hat{t^p}(\hat{e}) = \hat{e}^{\,\hat{\imath}+(\xi-1)^p\cdot\hat{m}}$$

其中 $\hat{e}\in\hat{E}$; 当然, $\hat{t}$ 与 $\hat{t^p}$ 相互交换, 从而不难验证

for any $\hat{e}\in\hat{E}$; since $\hat{t}$ and $\hat{t^p}$ obviously commute, it is not difficult to prove that

15.10.20 $$\big(\hat{\imath}+(\xi-1)^p\cdot\hat{m}\big)^{-1}\big(\hat{\imath}+(\xi-1)^p\cdot\hat{t}(\hat{m})\big)\in Z(\hat{E}^P)\,.$$

也就是说, 对适当的 $\hat{z}\in Z(\hat{E}^P)$ 得到

In other terms, for a suitable $\hat{z}\in Z(\hat{E}^P)$, we get

15.10.21 $$\hat{\imath}+(\xi-1)^p\cdot\hat{t}(\hat{m}) = \big(\hat{\imath}+(\xi-1)^p\cdot\hat{m}\big)\big(\hat{\imath}+(\xi-1)^p\cdot\hat{z}\big)\,,$$

最后得到 and finally we obtain

15.10.22 $$\hat{t}(\hat{m}) = \hat{m}+\hat{z}+(\xi-1)^p\cdot\hat{m}\hat{z}\,;$$

进一步, 既然 $(\xi-1)\cdot\hat{E}$ 包含 $(\hat{t}-\mathrm{id}_{\hat{E}})(\hat{E})$, 元素 $\hat{z}$ 显然属于 $(\xi-1)\cdot Z(\hat{E}^P)$.

furthermore, as far as $(\xi-1)\cdot\hat{E}$ contains $(\hat{t}-\mathrm{id}_{\hat{E}})(\hat{E})$, the element $\hat{z}$ clearly belongs to $(\xi-1)\cdot Z(\hat{E}^P)$.

所以, $\hat{t}$ 稳定下面的 $\hat{E}^P$ 的交换幺 $\hat{\mathcal{O}}$-子代数

Consequently, $\hat{t}$ stabilizes the following commutative $\hat{\mathcal{O}}$-subalgebra of $\hat{E}^P$

15.10.23 $$\hat{C} = Z(\hat{E}^P) + \sum_{h\in\mathbb{N}}\hat{\mathcal{O}}\cdot\hat{m}^h\,,$$

并且, $\hat{t}$ 在商 $\hat{C}/(\xi-1)\cdot\hat{C}$ 上平凡作用; 此时, 因为 $\hat{t^p}$ 在 $\hat{C}$ 上显然平凡作用, 所以只要应用下面的推论 16.6, 就得到

and moreover $\hat{t}$ acts trivially on the quotient $\hat{C}/(\xi-1)\cdot\hat{C}$; hence, since $\hat{t^p}$ acts trivially on $\hat{C}$, it suffices to apply Corollary 16.6 in section 16, to get $\hat{C} = \oplus_{h=0}^{p-1}\hat{C}_h$, where $\hat{C}_h$

$\hat{C} = \oplus_{h=0}^{p-1} \hat{C}_h$，其中$\hat{C}_h$是满足$\hat{t}(\hat{c}) = \xi^h{\cdot}\hat{c}$ 的 $\hat{c} \in \hat{C}$ 的集合；类似地，也得到

is the set of elements $\hat{c} \in \hat{C}$ which fulfill $\hat{t}(\hat{c}) = \xi^h{\cdot}\hat{c}$; similarly, we also get

15.10.24 $$Z(\hat{E}^P) = \bigoplus_{h=0}^{p-1} \hat{C}_h \cap Z(\hat{E}^P) .$$

现在，既然 $\hat{C}_h/\hat{C}_h \cap Z(\hat{E}^P)$ 是自由 $\hat{\mathcal{O}}$-模，$\hat{C}_h \cap Z(\hat{E}^P)$在$\hat{C}_h$里有一个补$\hat{M}_h$；最后，直和 $\hat{M} = \oplus_{h=0}^{p-1} \hat{M}_h$ 在 $\hat{\mathcal{O}}$-模 $\hat{C}$ 里是一个 $Z(\hat{E}^P)$ 的 $\hat{t}$-稳定补.

Now, as far as $\hat{C}_h/\hat{C}_h \cap Z(\hat{E}^P)$ is a free $\hat{\mathcal{O}}$-module, $\hat{C}_h \cap Z(\hat{E}^P)$ has a complement $\hat{M}_h$ in $\hat{C}_h$; finally, the direct sum $\hat{M} = \oplus_{h=0}^{p-1} \hat{M}_h$ is a $\hat{t}$-stable $\hat{\mathcal{O}}$-module complement of $Z(\hat{E}^P)$ in $\hat{C}$.

此时，由后面的命题 16.8，存在唯一 $\hat{m}' \in \hat{M}$ 并且唯一 $\hat{z}' \in Z(\hat{E}^P)$ 使得

At this point, by Proposition 16.8 in section 16, there exist a unique $\hat{m}' \in \hat{M}$ and a unique $\hat{z}' \in Z(\hat{E}^P)$ fulfilling

15.10.25 $$\hat{\imath} + (\xi - 1)^p{\cdot}\hat{m} = \big(\hat{\imath} + (\xi - 1)^p{\cdot}\hat{z}'\big)\big(\hat{\imath} + (\xi - 1)^p{\cdot}\hat{m}'\big) ;$$

特别是，我们有

in particular, we have

15.10.26 $$\hat{t}^p(\hat{e}) = \hat{e}^{\,\hat{\imath} + (\xi-1)^p{\cdot}\hat{m}'} , \quad \hat{t}(\hat{m}') = \hat{m}' ,$$

其中 $\hat{e} \in \hat{E}$；进一步，既然 $\hat{C} = \hat{\mathcal{O}}{\cdot}\hat{\imath} + J(\hat{C})$，也存在 $\hat{m}'' \in J(\hat{C})$ 使得我们有

for any $\hat{e} \in \hat{E}$; further, since $\hat{C} = \hat{\mathcal{O}}{\cdot}\hat{\imath} + J(\hat{C})$, we also can find $\hat{m}'' \in J(\hat{C})$ such that we have

15.10.27 $$\hat{t}^p(\hat{e}) = \hat{e}^{\,\hat{\imath} + (\xi-1)^p{\cdot}\hat{m}''} , \quad \hat{t}(\hat{m}'') = \hat{m}'' .$$

另一方面，由后面的推论 16.5，存在唯一 $\hat{c} \in \hat{\imath} + (\xi - 1){\cdot}J(\hat{C})$ 满足 $\hat{c}^p = \hat{\imath} + (\xi - 1)^p{\cdot}\hat{m}''$；特别是，有 $\hat{t}(\hat{c}) = \hat{c}$，并对任意 $\hat{e} \in \hat{E}$ 还有 $\hat{t}^p(\hat{e}) = \hat{e}^{\hat{c}^p}$；也就是说，我们得到

On the other hand, by Corollary 16.5 in section 16, there is a unique $\hat{c} \in \hat{\imath} + (\xi - 1){\cdot}J(\hat{C})$ fulfilling $\hat{c}^p = \hat{\imath} + (\xi - 1)^p{\cdot}\hat{m}''$; in particular, we have $\hat{t}(\hat{c}) = \hat{c}$ and, for any $\hat{e} \in \hat{E}$, we also have $\hat{t}^p(\hat{e}) = \hat{e}^{\hat{c}^p}$ for any $\hat{e} \in \hat{E}$; that is to say, we get

15.10.28 $$\big(\mathrm{in}_{\hat{E}}(\hat{c}^{-1}) \circ \hat{t}\big)^p = \mathrm{id}_{\hat{E}}$$

其中 $\mathrm{in}_{\hat{E}}(\hat{c}^{-1})$ 记由 $\hat{c}^{-1}$ 决定的 $\hat{E}$ 的 $\hat{\mathcal{O}}$-代数自同构.

where $\mathrm{in}_{\hat{E}}(\hat{c}^{-1})$ denotes the $\hat{\mathcal{O}}$-algebra automorphism of $\hat{E}$ induced by $\hat{c}^{-1}$.

最后，考虑 $\mathrm{in}_{\hat{E}}(\hat{c}^{-1}) \circ \hat{t} = \hat{t}'$；因为 $\hat{t}'$ 在 $\hat{E}/(\xi - 1){\cdot}\hat{E}$上平凡作用，所以只要再次应用推论16.6，就得到 $\hat{E} = \oplus_{h=0}^{p-1} \hat{E}_h$

Finally, consider $\hat{t}' = \mathrm{in}_{\hat{E}}(\hat{c}^{-1}) \circ \hat{t}$; since $\hat{t}'$ acts trivially on the quotient $\hat{E}/(\xi - 1){\cdot}\hat{E}$, applying Corollary 16.6 once again, we get $\hat{E} = \oplus_{h=0}^{p-1} \hat{E}_h$, where $\hat{E}_h$ is the set

其中$\hat{E}_h$是满足$\hat{t}'(\hat{e}) = \xi^h{\cdot}\hat{e}$的$\hat{e}\in\hat{E}$的集合.

of elements $\hat{e}\in\hat{E}$ which fulfill $\hat{t}'(\hat{e}) = \xi^h{\cdot}\hat{e}$.

特别是, $\hat{D} = \hat{E}_0$显然是一个$\hat{E}$的幺$\hat{\mathcal{O}}$-子代数; 而且, 因为$\hat{t}'$是$\hat{E}$的Q-内P-代数自同构使得对任意$v\in R$有$\hat{t}'(v{\cdot}\hat{\imath}) = \varphi(v){\cdot}(v{\cdot}\hat{\imath})$, 所以$P$稳定$\hat{D}$和$\hat{D}$包含$Q{\cdot}\hat{\imath}$, 并且对任意$v\in R$存在$h$使得$\hat{E}_h = \hat{D}{\cdot}v$; 这样, 有$\hat{E} = \oplus_{w\in W}\hat{D}{\cdot}w$. 最后, $\hat{D}$是P-稳定的幺子$\hat{\mathcal{O}}$-代数使得

In particular, $\hat{D} = \hat{E}_0$ clearly is a unitary $\hat{\mathcal{O}}$-subalgebra of $\hat{E}$; furthermore, since $\hat{t}'$ is a Q-interior P-algebra automorphism which fulfills $\hat{t}'(v{\cdot}\hat{\imath}) = \varphi(v){\cdot}(v{\cdot}\hat{\imath})$ for any $v\in R$, P stabilizes $\hat{D}$ and contains $Q{\cdot}\hat{\imath}$, and moreover, for any $v\in R$, exists h such that $\hat{E}_h = \hat{D}{\cdot}v$; thus, we have $\hat{E} = \oplus_{w\in W}\hat{D}{\cdot}w$. Finally, $\hat{D}$ is a P-stable unitary $\hat{\mathcal{O}}$-subalgebra fulfilling

15.10.29 $$\hat{D}\cap P{\cdot}\hat{\imath} = Q{\cdot}\hat{\imath}\,,\quad \hat{B} = \hat{D}\otimes_Q P\,.$$

令Σ记$\hat{\mathcal{K}}$上$\mathcal{K}$的Galois群, 并考虑Σ在$\hat{B}$上的自然作用; 对任意$\sigma\in\Sigma$显然有

Denote by Σ the Galois group of $\hat{\mathcal{K}}$ over $\mathcal{K}$ and consider the natural action of Σ on $\hat{B}$; for any $\sigma\in\Sigma$, we clearly have

15.10.30 $$\sigma(\hat{D})\cap P{\cdot}\hat{\imath} = Q{\cdot}\hat{\imath}\,,\quad \hat{B} = \sigma(\hat{D})\otimes_Q P\,;$$

所以, 下面的群

consequently, the following group

15.10.31 $$X = J(\hat{B}^P)^*\rtimes\Sigma$$

在满足条件15.10.29的P-稳定的幺子代数的集合上显然作用, 并且定理14.7推出子群$J(\hat{B}^P)^*$已经在这个集合上可迁作用.

clearly acts on the set of P-stable unitary subalgebras fulfilling condition 15.10.29, and then Theorem 14.7 implies that the subgroup $J(\hat{B}^P)^*$ already acts transitively on this set.

令Y记$\hat{D}$在X里的稳定化子; 那么, 从$J(\hat{B}^P)^*$的可迁性跟推论14.4可推出

Denote by Y the stabilizer of $\hat{D}$ in X; then, from the transitivity of $J(\hat{B}^P)^*$ and the Corollary 14.4, we get

15.10.32 $$X = J(\hat{B}^P)^*{\cdot}Y\,,\quad J(\hat{B}^P)^*\cap Y = J\big(Z(\hat{B})\big)^*{\cdot}J(\hat{D}^P)^*\,;$$

特别是, $Y/J(\hat{B}^P)^*\cap Y\cong\Sigma$; 进一步, 因为

in particular, we have $Y/J(\hat{B}^P)^*\cap Y\cong\Sigma$; consequently, since

15.10.33 $$J\big(Z(\hat{B})\big)^*\cap J(\hat{D}^P)^* = J\big(Z(\hat{D})^P\big)^*\,,$$

所以, 由推论16.5, $|\Sigma|$-次幂映射在$\bar{T} = J(Z(\hat{B}))^*/J(Z(\hat{D})^P)^*$上是双射, 从而只要把$\bar{Y} = Y/J(\hat{D}^P)^*$应用引理14.10, 就得到$\bar{T}$在$\bar{Y}$里有一个补$\bar{S}$.

it follows from Corollary 16.5 that the $|\Sigma|$-th power induces a bijective map on $\bar{T} = J(Z(\hat{B}))^*/J\big(Z(\hat{D})^P\big)^*$ and then, by Lemma 14.10 applied to $\bar{Y} = Y/J(\hat{D}^P)^*$, we get the existence of a complement $\bar{S}$ of $\bar{T}$ in $\bar{Y}$.

令 S 记 $\bar{S}$ 在Y里的原像, 再令 $\hat{F}$ 记被 S 生成的 $(\hat{B}^P)\Sigma$ 的 $\mathcal{O}$-子代数 (见 4.1); 此时, 我们考虑 $\mathcal{O}$-代数 $(\hat{B}^P)\Sigma$ 的 $\mathcal{O}$-子代数 $\hat{D}^P$ 与 $\hat{F}$; 因为包含映射 $S \to \hat{F}^*$ 显然决定群同态如下

Denote by S the converse image of $\bar{S}$ in Y, and by $\hat{F}$ the $\mathcal{O}$-subalgebra of $(\hat{B}^P)\Sigma$ generated by S (see 4.1); then, we consider the $\mathcal{O}$-subalgebras $\hat{D}^P$ and $\hat{F}$ of $(\hat{B}^P)\Sigma$; since the inclusion map $S \to \hat{F}^*$ clearly induces a group homomorphism

15.10.34 $$\bar{\rho} : \bar{S} \longrightarrow N_{\hat{F}^*}(\hat{D}^P)/J(\hat{D}^P)^* ,$$

所以, 由引理 12.5, 存在 $\bar{\rho}$ 的提升 $\rho: \bar{S} \to S \subset Y$.

it follows from Lemma 12.5 that $\bar{\rho}$ can be lifted to a homomorphism $\rho: \bar{S} \to S \subset Y$.

类似地, 既然包含映射 $X \to \big((\hat{B}^P)\Sigma\big)^*$ 显然决定下面的群同态

Similarly, as far as the inclusion map $X \to \big((\hat{B}^P)\Sigma\big)^*$ clearly induces the following group homomorphism

15.10.35 $$\Sigma \longrightarrow N_{((\hat{B}^P)\Sigma)^*}(\hat{B}^P)/J(\hat{B}^P)^* ,$$

只要把引理 12.5 应用到$\mathcal{O}$-代数 $(\hat{B}^P)\Sigma$ 的 $\mathcal{O}$-子代数 $(\hat{B}^P)\Sigma$ 与 $\hat{B}^P$ 就得到存在 $\hat{n}' \in J(\hat{B}^P)$ 使得在X里有 $\rho(\bar{S})^{\hat{\imath}+\hat{n}'} = \Sigma$. 最后, Σ 稳定 $\hat{D}^{\hat{\imath}+\hat{n}'}$, 所以从式子 15.10.29 可得到

it suffices to apply Lemma 12.5 to the $\mathcal{O}$-algebra $(\hat{B}^P)\Sigma$ and to its $\mathcal{O}$-subalgebras $(\hat{B}^P)\Sigma$ and $\hat{B}^P$, to find $\hat{n}' \in J(\hat{B}^P)$ fulfilling $\rho(\bar{S})^{\hat{\imath}+\hat{n}'} = \Sigma$ in X. Finally, Σ stabilizes $\hat{D}^{\hat{\imath}+\hat{n}'}$ and therefore, from 15.10.19, we get

15.10.36 $$(\hat{D}^{\hat{\imath}+\hat{n}'})^\Sigma \cap P{\cdot}\hat{\imath} = Q{\cdot}\hat{\imath}\,, \quad B = (\hat{D}^{\hat{\imath}+\hat{n}'})^\Sigma \otimes_Q P\,.$$

16. 论在$\mathcal{O}$-代数上的指数函数和对数函数

16. On the Exponential and Logarithmic Functions in $\mathcal{O}$-algebras

16.1. 这里，我们假定 $\mathcal{O}$ 的分式域 $\mathcal{K}$ 是特征为零的域. 如果 $\mathcal{O}$ 包含一个本原 p-次单位根 ξ，那么在交换 $\mathcal{O}$-代数的适当的理想上能考虑所谓指数函数. 同样的经常，这个函数可以给出一个从加法的结构到乘法的结构的一个同态; 特别是，乘以 $n \in \mathbb{N}$ 成为 n-次幂，从而这个函数可帮证明存在适当的元素的 n-次根. 在 §14 与 §15 里，我们需要这种结果; 这一节里，要表示它们.

16.1. Here, we assume that the quotient field $\mathcal{K}$ of $\mathcal{O}$ has characteristic zero. If $\mathcal{O}$ contains a primitive p-th root of unity ξ then, in suitable ideals of the commutative $\mathcal{O}$-algebras, we can consider the so-called *exponential function*. As usual, this function is a homomorphism from the additive structure to the multiplicative one; in particular, the multiplication by $n \in \mathbb{N}$ becomes the n-th power, and thus this function is helpful in proving the existence of the n-th root of some elements. In §14 and §15, we need this kind of results; in this section, we will prove them.

16.2. 首先请注意，如果 $1 \le h, \ell \le p-1$ 那么 $\mathcal{O}^*$ 包含 $(\xi^h - 1)/(\xi^\ell - 1)$; 确实，因为 ξ^ℓ 生成 $\langle \xi \rangle$ 所以对适当的 n 有 $\xi^h = (\xi^\ell)^n$，从而还有

16.2. First of all, note that whenever $1 \le h, \ell \le p-1$, $\mathcal{O}^*$ contains $(\xi^h-1)/(\xi^\ell-1)$; indeed, since ξ^ℓ generates $\langle \xi \rangle$, we have $\xi^h = (\xi^\ell)^n$ for a suitable n and therefore we also have

16.2.1
$$(\xi^h - 1)/(\xi^\ell - 1) = \sum_{m=0}^{n-1} \xi^{\ell m} .$$

类似地，既然在多项式环 $\mathcal{K}[X]$ 中有

Similarly, since in the polynomial ring $\mathcal{K}[X]$ we have

16.2.2
$$X^p - 1 = \prod_{h=0}^{p-1} (X - \xi^h) ,$$

显然得到

we certainly obtain

16.2.3
$$\sum_{h=0}^{p-1} X^h = \prod_{h=1}^{p-1} (X - \xi^h) ;$$

特别是，只要考虑满足 $f(X)=1$ 的 $\mathcal{K}$-代数同态 $f\colon \mathcal{K}[X] \to \mathcal{K}$ 就有

in particular, considering the $\mathcal{K}$-algebra homomorphism $f\colon \mathcal{K}[X] \to \mathcal{K}$ such that $f(X) = 1$, we get

16.2.4
$$p\,\mathcal{O} = \prod_{h=1}^{p-1} (1 - \xi^h)\mathcal{O} = (\xi - 1)^{p-1}\mathcal{O} ;$$

那个函数的存在性依赖于这个等式. 固定一个 $\mathcal{O}$-代数 A, 再固定满足 $J(A)^e \subset J \subset J(A)$ 的 A 的理想 J 其中 e 是某个整数; 考虑加法群 J 与乘法群 1_A+J.

the existence of that function depends on this equality. Fix an $\mathcal{O}$-algebra A and an ideal J of A, which fulfills $J(A)^e \subset J \subset J(A)$ for some positive integer e; consider the additive group J and the multiplicative group 1_A+J.

命题 16.3. 假定 $\mathcal{O}$ 包含一个本原 p-次单位根 ξ 和 A 是交换的. 存在相互逆群同构

Proposition 16.3. *Assume that $\mathcal{O}$ has a primitive p-th root of unity ξ and that A is commutative. There exist mutually inverse group isomorphisms*

16.3.1
$$\mathfrak{exp}_J : (\xi-1)\cdot J \cong 1_A + (\xi-1)\cdot J$$
$$\mathfrak{log}_J : 1_A + (\xi-1)\cdot J \cong (\xi-1)\cdot J$$

使得对任意 $m \in (\xi-1)\cdot J$ 有

which, for any $m \in (\xi-1)\cdot J$, fulfill

16.3.2
$$\mathfrak{exp}_J(m) = 1_A + \sum_{h\ge 1} m^h/h!\,,$$
$$\mathfrak{log}_J(1_A - m) = -\sum_{h\ge 1} m^h/h\,.$$

而且, 设 A' 是一个交换 $\mathcal{O}$-代数并设 J' 是一个 A' 的理想使得 $J' \subset J(A')$. 如果 $f\colon A \to A'$ 是满足 $f(J) \subset J'$ 的一个 $\mathcal{O}$-代数幺同态, 那么有

Moreover, let A' be a commutative $\mathcal{O}$-algebra and J' an ideal of A' such that $J' \subset J(A')$. If $f\colon A \to A'$ is a unitary $\mathcal{O}$-algebra homomorphism such that $f(J) \subset J'$ then we have

16.3.3
$$f\big(\mathfrak{exp}_J(m)\big) = \mathfrak{exp}_{J'}\big(f(m)\big)\,,$$
$$f\big(\mathfrak{log}_J(1_A - m)\big) = \mathfrak{log}_{J'}\big(1_{A'} - f(m)\big)\,.$$

证明: 对任意 $h \in \mathbb{N}$ 令

Proof: For any $h \in \mathbb{N}$, we set

16.3.4
$$h = \sum_\ell r_\ell(h)\,p^\ell\,,\quad 0 \le r_\ell(h) \le p-1$$

其中 ℓ 跑遍 $\mathbb{N}$; 当然, 对某个 n, 如果 $\ell \ge n$ 那么 $r_\ell(h) = 0$; 首先, 要证明下面的等式

where ℓ runs over $\mathbb{N}$; obviously, for some n, we have $r_\ell(h) = 0$ whenever $\ell \ge n$; first of all, we have to prove the following equality

16.3.5
$$h!\,\mathcal{O} = p^{\sum_{j\ge 1}\sum_{\ell\ge j} r_\ell(h)\,p^{\ell-j}}\mathcal{O} = (\xi-1)^{h-\sum_{\ell\in\mathbb{N}} r_\ell(h)}\mathcal{O}\,.$$

只要对 h 用归纳法, 就能假定 $h \ge 1$ 与

We argue by induction on h, and may assume that $h \ge 1$ and that

16.3.6
$$(h-1)!\,\mathcal{O} = p^{\sum_{j\ge 1}\sum_{\ell\ge j} r_\ell(h-1)\,p^{\ell-j}}\mathcal{O}\,;$$

如果 p 不整除 h，那么有 $h!\,\mathcal{O} = (h-1)!\,\mathcal{O}$，并且对任意 $\ell \geq 1$ 有 $r_\ell(h) = r_\ell(h-1)$.

if p does not divide h then we have $h!\,\mathcal{O} = (h-1)!\,\mathcal{O}$, and for any $\ell \geq 1$, we clearly have $r_\ell(h) = r_\ell(h-1)$.

假定 p 整除 h 和设 n 是极大的整数使得 p^n 整除 h；那么显然有

Assume that p divides h and let n be the maximal integer such that p^n divides h; then, we clearly obtain

16.3.7 $$h!\,\mathcal{O} = (h-1)!\,p^n\mathcal{O} = p^{n+\sum_{j\geq 1}\sum_{\ell\geq j} r_\ell(h-1)\,p^{\ell-j}}\,\mathcal{O}\,;$$

进一步, 因为 $h=\sum_{\ell\geq n} r_\ell(h)\,p^\ell$ 所以

moreover, since $h = \sum_{\ell\geq n} r_\ell(h)\,p^\ell$, we get

16.3.8 $$\begin{aligned} h-1 &= (p^n-1) + \big(r_n(h)-1\big)\,p^n + \sum_{\ell\geq n+1} r_\ell(h)\,p^\ell \\ &= \sum_{\ell=0}^{n-1}(p-1)\,p^\ell + \big(r_n(h)-1\big)\,p^n + \sum_{\ell\geq n+1} r_\ell(h)\,p^\ell\,, \end{aligned}$$

从而显然 $r_\ell(h-1) = p-1$ 或 $r_n(h)-1$ 或 $r_\ell(h)$, 当 $\ell \leq n-1$ 或 $\ell = n$ 或 $\ell \geq n+1$. 这样, 有

and therefore we clearly have $r_\ell(h-1) = p-1$ or $r_n(h)-1$ or $r_\ell(h)$, according to $\ell \leq n-1$ or $\ell = n$ or $\ell \geq n+1$. Thus, we get

16.3.9 $$\begin{aligned} &\sum_{\ell\geq 1}\sum_{j=1}^{\ell} r_\ell(h-1)\,p^{\ell-j} \\ &= \sum_{\ell=1}^{n-1}\sum_{j=1}^{\ell}(p-1)\,p^{\ell-j} - \sum_{j=1}^{n} p^{n-j} + \sum_{\ell\geq n}\sum_{j=1}^{\ell} r_\ell(h)\,p^{\ell-j} \\ &= \sum_{\ell=1}^{n-1}(p^\ell-1) - \sum_{j=1}^{n} p^{n-j} + \sum_{\ell\geq n}\sum_{j=1}^{\ell} r_\ell(h)\,p^{\ell-j} \\ &= -n + \sum_{j\geq 1}\sum_{\ell\geq j} r_\ell(h)\,p^{\ell-j}\,, \end{aligned}$$

从而得到

and therefore we obtain

16.3.10 $$h!\,\mathcal{O} = p^{\sum_{j\geq 1}\sum_{\ell\geq j} r_\ell(h)\,p^{\ell-j}}\,\mathcal{O}\,.$$

最后, 因为显然有

Finally, since we obviously have

16.3.11 $$\begin{aligned} (p-1)\sum_{\ell\geq 1}\sum_{j=1}^{\ell} r_\ell(h)\,p^{\ell-j} &= \sum_{\ell\geq 1} r_\ell(h)(p^\ell-1) \\ &= h - \sum_{\ell\geq 1} r_\ell(h)\,, \end{aligned}$$

所以, 由等式 16.2.4, 得到

from equality 16.2.4, we get

16.3.12
$$\begin{aligned} h!\mathcal{O} &= (\xi-1)^{(p-1)\sum_{j\geq 1}\sum_{\ell\geq j} r_\ell(h)\,p^{\ell-j}}\mathcal{O} \\ &= (\xi-1)^{h-\sum_{\ell\geq 1} r_\ell(h)}\mathcal{O}\,. \end{aligned}$$

特别是, 对任意 $n\in J$ 有

In particular, for any $n\in J$, we have

16.3.13
$$(\xi-1)^h\cdot n^h/h! \in (\xi-1)^{\sum_{\ell\geq 1} r_\ell(h)}\cdot J^h$$

从而, 对任意 $m\in(\xi-1)\cdot J$ 与 $h\geq 1$, 元素 $m^h/h!$ 与 m^h/h 属于 $(\xi-1)\cdot J^h$; 也就是说, A 的序列 $\left\{\sum_{h=1}^r m^h/h!\right\}_{r\in\mathbb{N}}$ 与 $\left\{\sum_{h=1}^r m^h/h\right\}_{r\in\mathbb{N}}$ 有极限 (见 2.14); 所以可令

and therefore, for any $m\in(\xi-1)\cdot J$ and any $h\geq 1$, the elements $m^h/h!$ and m^h/h belong to $(\xi-1)\cdot J^h$; that is to say, the sequences $\left\{\sum_{h=1}^r m^h/h!\right\}_{r\in\mathbb{N}}$ and $\left\{\sum_{h=1}^r m^h/h\right\}_{r\in\mathbb{N}}$ of A have a limit (see 2.14); consequently, we are able to set

16.3.14
$$\begin{aligned} \mathfrak{exp}_J(m) &= 1_A + \lim_{n\to\infty}\Big\{\sum_{h=1}^r m^h/h!\Big\} = 1_A + \sum_{h\geq 1} m^h/h!\,, \\ \mathfrak{log}_J(1_A-m) &= -\lim_{n\to\infty}\Big\{\sum_{h=1}^r m^h/h\Big\} = -\sum_{h\geq 1} m^h/h\,. \end{aligned}$$

而且, 设 A' 是一个交换 $\mathcal{O}$-代数并设 J' 是一个 A' 的理想使得 $J'\subset J(A')$; 如果 $f\colon A\to A'$ 是满足 $f(J)\subset J'$ 的 $\mathcal{O}$-代数幺同态, 那么显然有

Moreover, let A' be a commutative $\mathcal{O}$-algebra and J' an ideal of A' such that $J'\subset J(A')$; if $f\colon A\to A'$ is a unitary $\mathcal{O}$-algebra homomorphism fulfilling $f(J)\subset J'$ then we clearly have

16.3.15
$$\begin{aligned} f\Big(\sum_{h=1}^r m^h/h!\Big) &= \sum_{h=1}^r f(m)^h/h! \\ f\Big(\sum_{h=1}^r m^h/h\Big) &= \sum_{h=1}^r f(m)^h/h \end{aligned}$$

并且 $f(m)^h/h!$ 与 $f(m)^h/h$ 属于 $(\xi-1)\cdot J'^h$, 从而有 (见 2.11)

and the elements $f(m)^h/h!$ and $f(m)^h/h$ belong to $(\xi-1)\cdot J'^h$; hence, we have (see 2.11)

16.3.16
$$\begin{aligned} f\big(\mathfrak{exp}_J(m)\big) &= \mathfrak{exp}_{J'}\big(f(m)\big)\,, \\ f\big(\mathfrak{log}_J(1_A-m)\big) &= \mathfrak{log}_{J'}\big(1_{A'}-f(m)\big)\,. \end{aligned}$$

这两个函数都是单射; 确实, 如果 $m,n\in(\xi-1)\cdot J$ 与 $m\neq n$, 那么存在 $\ell\in\mathbb{N}$ 使得 $m-n\in(\xi-1)\cdot(J^\ell-J^{\ell+1})$;

These two functions are injective; indeed, if $m,n\in(\xi-1)\cdot J$ and $m\neq n$, then we have $m-n\in(\xi-1)\cdot(J^\ell-J^{\ell+1})$ for a suitable $\ell\in\mathbb{N}$; consequently, setting

此时，记 $a_{h,j} = m^{h-j}n^{j-1}/h!$ 与 $b_{h,j} = m^{h-j}n^{j-1}/h$；因为有

$a_{h,j} = m^{h-j}n^{j-1}/h!$ and $m^{h-j}n^{j-1}/h = b_{h,j}$, since we have

16.3.17
$$\mathfrak{exp}_J(m) - \mathfrak{exp}_J(n) = (m-n)\big(1_A + \sum_{h\geq 2}\sum_{j=1}^{h} a_{h,j}\big)$$
$$\mathfrak{log}_J(1_A - m) - \mathfrak{log}_J(1_A - n) = (n-m)\big(1_A + \sum_{h\geq 2}\sum_{j=1}^{h} b_{h,j}\big)$$

并且元素 $a_{h,j}$ 与 $b_{h,j}$ 属于

and the elements $a_{h,j}$ and $b_{h,j}$ belong to

16.3.18
$$(\xi - 1)^{-1+\sum_{\ell\geq 1} r_\ell(h)} \cdot J^{h-1} \subset J\,,$$

所以还有

we get

16.3.19
$$\mathfrak{exp}_J(m) \neq \mathfrak{exp}_J(n)\,, \quad \mathfrak{log}_J(1_A - m) \neq \mathfrak{log}_J(1_A - n)\,.$$

进一步，既然对任意 $r \in \mathbb{N}$ 与 $m, n \in (\xi - 1)\cdot J$ 显然

Moreover, as far as, for any $r \in \mathbb{N}$ and any $m, n \in (\xi - 1)\cdot J$, we clearly have

16.3.20
$$\begin{aligned}\big(1_A + \sum_{h=1}^{r} m^h/h!\big)\big(1_A + \sum_{h=1}^{r} n^h/h!\big) + (\xi-1)^{r+1}\cdot J^{r+1} \\ = 1_A + \sum_{h=1}^{r}\sum_{j=1}^{h}\binom{h}{j} a_{h,j} + (\xi-1)^{r+1}\cdot J^{r+1} \\ = 1_A + \sum_{h=1}^{r}(m+n)^h/h! + (\xi-1)^{r+1}\cdot J^{r+1}\,,\end{aligned}$$

就不难验证

it is not difficult to prove that

16.3.21
$$\mathfrak{exp}_J(m)\mathfrak{exp}_J(n) = \mathfrak{exp}_J(m+n)$$

也就是说，$\mathfrak{exp}$ 是群同态.

that is to say, $\mathfrak{exp}$ is a group homomorphism.

另一方面，为了证明

On the other hand, in order to prove

16.3.22
$$\mathfrak{log} \circ \mathfrak{exp} = \mathrm{id}_{(\xi-1)\cdot J}$$

考虑多项式环 $\mathcal{K}[X]$ 和对应的导映射 $\mathfrak{d}\colon \mathcal{K}[X] \to \mathcal{K}[X]$；也就是说，如果 $\sum_{h=0}^{r} \lambda_h X^h$ 是一个多项式，那么导出多项就是

consider the polynomial ring $\mathcal{K}[X]$ and the corresponding derivate map $\mathfrak{d}\colon \mathcal{K}[X] \to \mathcal{K}[X]$; in other words, if $\sum_{h=0}^{r} \lambda_h X^h$ is a polynomial then the derivate polynomial is

16.3.23
$$\mathfrak{d}\big(\sum_{h=0}^{r} \lambda_h X^h\big) = \sum_{h=1}^{r} \lambda_h h X^{h-1}\,;$$

别忘了, $\mathrm{Ker}(\eth) = \mathcal{K}$; 还别忘了, 如果 $P = \sum_{h=0}^{r} \lambda_h X^h$ 与 Q 是两个多项式, 那么合成多项式是

recall that $\mathrm{Ker}(\eth) = \mathcal{K}$, and that, if Q and $P = \sum_{h=0}^{r} \lambda_h X^h$ are two polynomials, then the *composed* polynomial is

16.3.24 $$P \circ Q = \sum_{h=0}^{r} \lambda_h Q^h$$

并且不难验证

and moreover, it is easily checked that

16.3.25 $$\eth(P \circ Q) = \big(\eth(P) \circ Q\big)\eth(Q)\,.$$

特别是, 对任意 $r \in \mathbb{N}$ 考虑下面的多项式

In particular, for any $r \in \mathbb{N}$ consider the following polynomials

16.3.26 $$\mathrm{E}_r = \sum_{h=0}^{r} X^h/h! = 1 - F_r\,, \quad \mathrm{L}_r = -\sum_{h=1}^{r}(1-X)^h/h\,;$$

对任意 $r \geq 1$ 显然有

for any $r \geq 1$, we clearly have

16.3.27 $$\eth(\mathrm{E}_r) = \mathrm{E}_{r-1}\,, \quad \eth(\mathrm{L}_r) = \sum_{h=0}^{r-1}(1-X)^h$$

从而得到

and therefore we get

16.3.28 $$\begin{aligned}\eth(\mathrm{L}_r \circ \mathrm{E}_r) &= \sum_{h=0}^{r-1}(\mathrm{F}_r)^h(1-\mathrm{F}_{r-1}) \\ &= 1 - (F_r)^r + \sum_{h=0}^{r-1}(\mathrm{F}_r)^h(X^r/r!)\,;\end{aligned}$$

而且, 因为有

moreover, since we have

16.3.29 $$\frac{X^s}{s!}\,\frac{X^t}{t!} = \binom{s+t}{s}\frac{X^{s+t}}{(s+t)!}$$

其中 $s, t \in \mathbb{N}$, 所以对适当的 $z_{r,h} \in \mathbb{Z}$ 其中 $r \leq h \leq r^2$, 有

for any $s, t \in \mathbb{N}$, for suitable $z_{r,h} \in \mathbb{Z}$ where $r \leq h \leq r^2$, we have

16.3.30 $$\eth(\mathrm{L}_r \circ \mathrm{E}_r) = 1 + \sum_{h=r}^{r^2} z_{r,h} X^h/h!\,;$$

这样既然 $\mathrm{L}_r \circ \mathrm{E}_r \in X\mathcal{K}[X]$, 就有

thus, as far as $\mathrm{L}_r \circ \mathrm{E}_r \in X\mathcal{K}[X]$, we get

16.3.31 $$\mathrm{L}_r \circ \mathrm{E}_r = X + \sum_{h=r}^{r^2} z_{r,h} X^{h+1}/(h+1)!\,.$$

所以，对任意 $r\in\mathbb{N}-\{0\}$ 存在一个 $\mathbb{Z}$-序列 $\{z_{r,h}\}_{r\le h\le r^2}$ 使得对任意 $m\in(\xi-1)\cdot J$ 有

Consequently, for any $r\in\mathbb{N}-\{0\}$, there exists a sequence $\{z_{r,h}\}_{r\le h\le r^2}$ in $\mathbb{Z}$ such that, for any $m\in(\xi-1)\cdot J$, we have

16.3.32 $$\mathrm{L}_r\big(\mathrm{E}_r(m)\big)=m+\sum_{h=r}^{r^2} z_{r,h}m^{h+1}/(h+1)!\,;$$

可是，既然 $m^h/h!\in(\xi-1)\cdot J^h$ 其中 $h\ge 1$，就有

but, as far as $m^h/h!\in(\xi-1)\cdot J^h$ for any $h\ge 1$, we get

16.3.33 $$\mathfrak{exp}(m)-\mathrm{E}_r(m)\in(\xi-1)\cdot J^{r+1}$$

从而，由等式 16.3.17，还有

and then, from equality 16.3.17, we still get

16.3.34 $$\mathfrak{log}_J\big(\mathfrak{exp}_J(m)\big)-\mathfrak{log}_J\big(\mathrm{E}_r(m)\big)\in(\xi-1)\cdot J^{r+1}\,.$$

类似地，因为有

Similarly, since we have

16.3.35 $$\mathrm{E}_r(m)\in 1_A+(\xi-1)\cdot J\,,$$

所以还有

we get

16.3.36 $$\mathfrak{log}_J\big(\mathrm{E}_r(m)\big)-\mathrm{L}_r\big(\mathrm{E}_r(m)\big)\in(\xi-1)\cdot J^{r+1}\,.$$

最后，对任意 $r\ge 1$ 得到

Finally, for any $r\ge 1$ we obtain

16.3.37 $$\mathfrak{log}_J\big(\mathfrak{exp}_J(m)\big)-m\in(\xi-1)\cdot J^{r+1}\,,$$

从而也得到

and therefore we also obtain

16.3.38 $$\mathfrak{log}_J\big(\mathfrak{exp}_J(m)\big)=m\,.$$

特别是，$\mathfrak{log}_J$ 是满射，以及它是单射；所以 $\mathfrak{log}_J$ 与 $\mathfrak{exp}_J$ 都是双射，并且它们是相互逆的，从而 $\mathfrak{log}_J$ 也是群同构.

In particular, $\mathfrak{log}_J$ is surjective, and we already know that it is injective; hence, $\mathfrak{log}_J$ and $\mathfrak{exp}_J$ are both bijective, inverse of each other, so that $\mathfrak{log}_J$ is an isomorphism too.

16.4. 对任意 $r\in\mathbb{N}-p\mathbb{N}$ 令

16.4. For any $r\in\mathbb{N}-p\mathbb{N}$, we denote by

16.4.1 $$\mathfrak{p}_r : 1_A+J\longrightarrow 1_A+J$$

记满足 $\mathfrak{p}_r(1_A+n)=(1_A+n)^r$，的幂函数，其中 $n\in J$. 而且如果 $\mathcal{O}$ 包含一个本原 p-次单位根 ξ 那么再令

the *power function* defined by $\mathfrak{p}_r(1_A+n)=(1_A+n)^r$ for any $n\in J$. Moreover, if $\mathcal{O}$ contains a p-th root of unity ξ, we still denote by

16.4.2 $$\mathfrak{p}_p : 1_A+(\xi-1)\cdot J\longrightarrow 1_A+(\xi-1)^p\cdot J$$

记对应的p-次幂函数; 确实从等式 16.2.4 不难验证, 对任意 $n\in J$, $(1_A+(\xi-1)\cdot n)^p$ 属于 $1_A+(\xi-1)^p\cdot J$

the corresponding *p-th power function*; indeed, from equality 16.2.4 it is not difficult to prove that $(1_A+(\xi-1)\cdot n)^p$ belongs to $1_A+(\xi-1)^p\cdot J$ for any $n\in J$.

推论 16.5. 如果p不整除$r\in\mathbb{N}$那么幂函数 $\mathfrak{p}_r$ 是双射. 而且, 如果 $\mathcal{O}$ 包含一个本原 p-次单位根 ξ 并 A 是交换的, 那么幂函数 $\mathfrak{p}_p$ 是群同构.

Corollary 16.5. *If p does not divide $r\in\mathbb{N}$ then the power function $\mathfrak{p}_r$ is bijective. Moreover, if $\mathcal{O}$ contains a primitive p-th root of unity ξ and A is commutative, then the power function $\mathfrak{p}_p$ is an isomorphism.*

证明: 我们能假定 $r\neq 1$; 如果 $m,n\in J$ 与 $m\neq n$, 那么存在 $\ell\in\mathbb{N}$ 使得 $m-n\in J^\ell-J^{\ell+1}$; 此时, 因为有

Proof: We may assume that $r\neq 1$; if $m,n\in J$ and $m\neq n$ then there exists $\ell\in\mathbb{N}$ such that $m-n\in J^\ell-J^{\ell+1}$; consequently, since we have

$$16.5.1\quad (1_A+m)^r-(1_A+n)^r=(m-n)\big(r\cdot 1_A+\sum_{h=2}^{r}\binom{r}{h}\big(\sum_{j=1}^{h}m^{h-j}n^{j-1}\big)\big)\,,$$

所以它属于 $r\cdot(m-n)+J^{\ell+1}$ 从而 $(1_A+m)^r\neq(1_A+n)^r$; 也就是说, $\mathfrak{p}_r$ 是单射.

this difference belongs to $r\cdot(m-n)+J^{\ell+1}$ and therefore $(1_A+m)^r\neq(1_A+n)^r$; that is to say, $\mathfrak{p}_r$ is injective.

为了证明它是满射, 令 $\hat{\mathcal{K}}$ 记包含一个本原 p-次单位根 ξ 的 $\mathcal{K}$ 的扩张, 再令 $\hat{\mathcal{O}}$ 记 $\hat{\mathcal{K}}$ 中的 $\mathcal{O}$ 上的整元构成的环 (见 2.12); 如果 $n\in J$ 那么在 $\hat{A}=\hat{\mathcal{O}}\otimes_{\mathcal{O}}A$ 里考虑下面的交换 $\hat{\mathcal{O}}$-子代数

In order to prove that it is surjective, denote by $\hat{\mathcal{K}}$ an extension of $\mathcal{K}$ which contains a primitive p-th root of unity, and by $\hat{\mathcal{O}}$ the integral closure of $\mathcal{O}$ in $\hat{\mathcal{K}}$ (see 2.12); if $n\in J$ then consider the following commutative $\hat{\mathcal{O}}$-subalgebra of $\hat{A}=\hat{\mathcal{O}}\otimes_{\mathcal{O}}A$

$$16.5.2\qquad \hat{A}'=\hat{\mathcal{O}}\otimes 1_A+\sum_{h\geq 1}\hat{\mathcal{O}}\otimes n^h\,.$$

存在 $\ell\in\mathbb{N}$ 使得 $1\otimes n^{p^\ell}$ 属于 $(\xi-1)\cdot\hat{J}'$, 其中 $\hat{J}'$ 是 $\hat{A}'$ 的根; 此时, 下面的元素

There exists $\ell\in\mathbb{N}$ such that $1\otimes n^{p^\ell}$ belongs to $(\xi-1)\cdot\hat{J}'$, where $\hat{J}'$ denotes the radical of $\hat{A}'$; then, the following element

$$16.5.3\quad (1_{\hat{A}}+1\otimes n)^{p^\ell}=1_{\hat{A}}+p\sum_{h=1}^{p^\ell-1}\frac{p^{\ell-1}}{h}\binom{p^\ell-1}{h-1}\otimes n^h+1\otimes n^{p^\ell}$$

属于$1_{\hat{A}}+(\xi-1)\cdot\hat{J}'$; 这样, 只要选择$s\in\mathbb{N}$使得$sr\equiv 1\,(\mathrm{mod}\,p^\ell)$

belongs to $1_{\hat{A}}+(\xi-1)\cdot\hat{J}'$; thus, choosing $s\in\mathbb{N}$ fulfilling $sr\equiv 1\,(\mathrm{mod}\,p^\ell)$, the element

就得到 $(1_{\hat{A}}+1\otimes n)^{sr-1}$ 属于 $1_{\hat{A}}+(\xi-1)\cdot\hat{J}'$；所以下面的元素

$(1_{\hat{A}}+1\otimes n)^{sr-1}$ also belongs to $1_{\hat{A}}+(\xi-1)\cdot\hat{J}'$; consequently, the following element

16.5.4 $$\hat{a}=(1_{\hat{A}}+1\otimes n)^{s}\mathfrak{exp}_{\hat{J}'}\Big(r^{-1}\cdot\mathfrak{log}_{\hat{J}'}\big((1_{\hat{A}}+1\otimes n)^{sr-1}\big)\Big)^{-1}$$

属于 $1_{\hat{A}}+\hat{\mathcal{O}}\otimes J$ 并且不难验证 $\mathfrak{p}_r(\hat{a})=1_{\hat{A}}+1\otimes n$；进一步，因为 $\hat{a}$ 是唯一的，所以 $\hat{\mathcal{K}}$ 在 $\mathcal{K}$ 上的 Galois 群稳定它；从而，这个元素属于 $1\otimes(1_A+J)$.

belongs to $1_{\hat{A}}+\hat{\mathcal{O}}\otimes J$ and it is not difficult to check that $\mathfrak{p}_r(\hat{a})=1_{\hat{A}}+1\otimes n$; moreover, since such an element $\hat{a}$ is unique, the Galois group of $\hat{\mathcal{K}}$ over $\mathcal{K}$ stabilizes it and therefore this element belongs to $1\otimes(1_A+J)$.

假定$\mathcal{O}=\hat{\mathcal{O}}$以及$A$是交换的；因为对任意 $m\in(\xi-1)\cdot J$ 有

Assume that $\mathcal{O}=\hat{\mathcal{O}}$ and that A is commutative; since for any $m\in(\xi-1)\cdot J$ we have

16.5.5 $$\mathfrak{log}_J\big((1_A+m)^p\big)=p\cdot\mathfrak{log}_J(1_A+m)\in(\xi-1)p\cdot J,$$

所以，由等式16.2.4和命题16.3，还有

it follows from equality 16.2.4 and Proposition 16.3 that we still have

16.5.6 $$\begin{aligned}(1_A+m)^p&=\mathfrak{exp}_J\big(p\cdot\mathfrak{log}_J(1_A+m)\big)\\&=\mathfrak{exp}_{p\cdot J}\big(p\cdot\mathfrak{log}_J(1_A+m)\big)\in 1_A+(\xi-1)^p\cdot J.\end{aligned}$$

也就是说，只要令

That is to say, denoting by

16.5.7 $$\mathfrak{m}_p\colon(\xi-1)\cdot J\to(\xi-1)^p\cdot J$$

记乘以p (见等式 16.2.4)，幂函数 $\mathfrak{p}_r$ 就是三个群同构 $\mathfrak{exp}_{p\cdot J}$, $\mathfrak{m}_p$, $\mathfrak{log}_J$ 的合成.

the multiplication by p (see equality 16.2.4), $\mathfrak{p}_r$ is equal to the composition of the three group homomorphisms $\mathfrak{exp}_{p\cdot J}$, $\mathfrak{m}_p$ and $\mathfrak{log}_J$.

推论 16.6. 假定 k 是代数闭的与 $\mathcal{O}$ 包含一个本原 p-次单位根 ξ，还假定 A 是交换的. 那么 $1_A+(\xi-1)\cdot A$ 的挠子群是直积 $\prod_{i\in I}\langle\xi\cdot i\rangle$，其中 I 是 A 的本原幂等元的集合. 特别是，设 G 是有限群并设 M 是不可分解 $\mathcal{O}G$-模使得

Corollary 16.6. *Assume that k is algebraically closed, that $\mathcal{O}$ contains a primitive p-th root of unity ξ, and that A is commutative. Then the torsion subgroup of $1_A+(\xi-1)\cdot A$ is the direct product $\prod_{i\in I}\langle\xi\cdot i\rangle$, where I is the set of all the primitive idempotents of A. In particular, let G be a finite group and M an indecomposable $\mathcal{O}G$-module fulfilling*

16.6.1 $$G\cdot\mathrm{id}_M\subset\mathrm{id}_M+(\xi-1)\cdot\mathrm{End}_{\mathcal{O}}(M).$$

那么 $M\cong\mathcal{O}$ 并且 $\langle\xi\rangle$ 包含 G 的像.

Then $M\cong\mathcal{O}$ and $\langle\xi\rangle$ contains the image of G.

证明: 记 $J=J(A)$，$\delta=\xi-1$；由命题 16.3，$1_A+\delta\cdot J$ 与 $\delta\cdot J$ 是相互同构的；从而 $\{1_A\}$ 就是

Proof: Set $J=J(A)$ and $\delta=\xi-1$; according to Proposition 16.3, $1_A+\delta\cdot J$ and $\delta\cdot J$ are mutually isomorphic; hence, the tor-

$1_A + \delta{\cdot}J$ 的挠子群. 另一方面, 显然有 $\delta{\cdot}A \subset J$, 并且从 A 到 $1_A + \delta{\cdot}A$ 的自然映射诱导下面的群同构

sion group $1_A + \delta{\cdot}J$ is just $\{1_A\}$. On the other hand, we clearly have $\delta{\cdot}A \subset J$ and the natural map from A to $1_A + \delta{\cdot}A$ induces the following group isomorphism

16.6.2 $$A/J \cong \big(1_A + \delta{\cdot}A\big)/\big(1_A + \delta{\cdot}J\big)\,;$$

所以, $1_A + \delta{\cdot}A$ 的挠子群的指数是 p.

hence, the torsion subgroup of $1_A + \delta{\cdot}A$ has exponent p.

取 $i \in I$ 并考虑 $a \in A$ 使得 $(1_A + \delta{\cdot}a)^p = 1_A$; 因为 k 是代数闭, 所以有 $Ai = \mathcal{O}{\cdot}i + Ji$, 从而不难验证对适当的 $n \in Ji$ 与 $\lambda \in 1 + \delta\mathcal{O}$ 还有 $i + \delta{\cdot}ai = \lambda{\cdot}(i + \delta{\cdot}n)$; 这样, 得到 $i = \lambda^p{\cdot}(i + \delta{\cdot}n)^p$, 从而 $\lambda^p{\cdot}i$ 属于交如下 (见 16.2.4)

Choose $i \in I$ and consider $a \in A$ such that $(1_A + \delta{\cdot}a)^p = 1_A$; since k is algebraically closed, we have $Ai = \mathcal{O}{\cdot}i + Ji$, and then it is easy to check that, for suitable $n \in Ji$ and $\lambda \in \delta\mathcal{O}$, we still have $i + \delta{\cdot}ai = \lambda{\cdot}(i + \delta{\cdot}n)$; thus, we get $i = \lambda^p{\cdot}(i + \delta{\cdot}n)^p$ and therefore $\lambda^p{\cdot}i$ belongs to the intersection (see 16.2.4)

16.6.3 $$(i + \delta^p\mathcal{O}){\cdot}i \cap (i + \delta^p{\cdot}Ji) = \big(1 + \delta^p{\cdot}J(\mathcal{O})\big){\cdot}i\,;$$

所以只要应用推论 16.5, 就得到满足 $(\mu{\cdot}i)^p = (i + \delta{\cdot}n)^p$ 的元素 $\mu \in 1 + \delta J(\mathcal{O})$, 从而仍由推论 16.5, 还有 $\mu{\cdot}i = i + \delta{\cdot}n$; 最后, 得到 $i + \delta{\cdot}ai = \lambda\mu{\cdot}i$ 属于 $\langle \xi{\cdot}i \rangle$.

consequently, applying Corollary 16.5, we find $\mu \in 1 + \delta J(\mathcal{O})$ which fulfills $(\mu{\cdot}i)^p = (i + \delta{\cdot}n)^p$ and therefore, always applying Corollary 16.5, we have $\mu{\cdot}i = i + \delta{\cdot}n$; finally, we get that $i + \delta{\cdot}ai = \lambda\mu{\cdot}i$ belongs to $\langle \xi{\cdot}i \rangle$.

为了证明最后一论断, 能假定 $G \subset \mathrm{id}_M + \delta{\cdot}\mathrm{End}_{\mathcal{O}}(M)$; 既然 $J(\mathcal{O})$ 包含 $\delta\mathcal{O}$, 只要应用推论 16.5, 就得到 G 是 p-群. 另一方面, 如果 $f \in \mathrm{End}_{\mathcal{O}}(M)$ 满足 $\mathrm{id}_M + \delta{\cdot}f \in Z(G)$, 那么考虑 $\mathrm{End}_{\mathcal{O}}(M)$ 的交换 $\mathcal{O}$-子代数

In order to prove the last statement, we may assume that $G \subset \mathrm{id}_M + \delta{\cdot}\mathrm{End}_{\mathcal{O}}(M)$; since $J(\mathcal{O})$ contains $\delta\mathcal{O}$, it follows from Corollary 16.5 that G is a p-group. On the other hand, if $f \in \mathrm{End}_{\mathcal{O}}(M)$ fulfills $\mathrm{id}_M + \delta{\cdot}f \in Z(G)$, then consider the commutative $\mathcal{O}$-subalgebra of $\mathrm{End}_{\mathcal{O}}(M)$

16.6.4 $$F = \mathcal{O}{\cdot}\mathrm{id}_M + \sum_{h\ge 1} \mathcal{O}{\cdot}f^h\,;$$

因为 $F \subset \mathrm{End}_{\mathcal{O}G}(M)$ 所以 id_M 在 F 里是本原的; 此时, 由上述的论断, 有 $\mathrm{id}_M + \delta{\cdot}f \in \langle \xi{\cdot}\mathrm{id}_M \rangle$; 也就是说, $Z(G) \subset \langle \xi{\cdot}\mathrm{id}_M \rangle$. 特别是, 既然 G 是一个 p-群, 不难证明

since $F \subset \mathrm{End}_{\mathcal{O}G}(M)$, id_M is primitive in F; then, according to the argument above, we have $\mathrm{id}_M + \delta{\cdot}f \in \langle \xi{\cdot}\mathrm{id}_M \rangle$; in other terms, we get $Z(G) \subset \langle \xi{\cdot}\mathrm{id}_M \rangle$. In particular, as far as G is a p-group, it is not difficult to see that

16.6.5 $$G \cap \big(\mathrm{id}_M + \delta^2{\cdot}\mathrm{End}_{\mathcal{O}}(M)\big) = \{\mathrm{id}_M\}\,;$$

可是, 类似于 16.6.2, 对任意 $\mathcal{O}$-代数 B 显然有 $\delta\cdot B \subset J(B)$ 并且得到下面的群同构

but, similarly to 16.6.2, for any $\mathcal{O}$-algebra B we clearly have $\delta\cdot B \subset J(B)$ and thus we get the following group isomorphism

16.6.6
$$B/\delta\cdot B \cong (1_B + \delta\cdot B)/(1_A + \delta^2\cdot B);$$

从而$G=Z(G)$.证明是完整的.

therefore $G = Z(G)$, and we are done.

定理 16.7. 假定 $\mathcal{O}$ 包含一个本原 p-次单位根 ξ. 设 s 是满足 $s^p = \mathrm{id}_A$ 的 A 的 $\mathcal{O}$-代数自同构. 假定 s 稳定 A 的理想 J 并有 $(s-\mathrm{id}_A)(J) \subset (\xi-1)\cdot J$. 如果 $n \in J$ 满足

Theorem 16.7. *Assume that $\mathcal{O}$ contains a primitive p-th root of unity ξ. Let s be an $\mathcal{O}$-algebra automorphism of A such that $s^p = \mathrm{id}_A$. Assume that s stabilizes an ideal J of A and that $(s-\mathrm{id}_A)(J) \subset (\xi-1)\cdot J$. If $n \in J$ fulfills*

16.7.1
$$\prod_{h=0}^{p-1}\big(1_A + (\xi-1)\cdot s^h(n)\big) \in Z(A),$$

那么存在 $m \in J$ 与 $z \in Z(A)^*$ 使得

then there exist $m \in J$ and $z \in Z(A)^*$ such that

16.7.2
$$\big(1_A + (\xi-1)\cdot n\big)z = \big(1_A + s(m)\big)\big(1_A+m\big)^{-1}.$$

证明: 取 $n \in J$ 和记 $\delta = \xi - 1$; 如果 $s(n) = n$ 那么条件 16.7.1 成为 $\big(1_A+\delta\cdot n\big)^p \in Z(A)$; 此时, 考虑 A 的交换 $\mathcal{O}$-子代数

Proof: Pick $n \in J$ and set $\delta = \xi - 1$; if $s(n) = n$ then condition 16.7.1 becomes $\big(1_A+\delta\cdot n\big)^p \in Z(A)$; in this case, consider the commutative $\mathcal{O}$-subalgebra of A

16.7.3
$$B = Z(A) + \sum_{h\geq 1}\mathcal{O}\cdot n^h$$

和B的理想$J\cap B$;由推论16.5, 存在$J\cap B$的唯一元素 m 使得

and the ideal $J\cap B$ of B; by Corollary 16.5, there is a unique element m of $J\cap B$ fulfilling

16.7.4
$$(1_A + \delta\cdot n)^p = 1_A + \delta^p\cdot m,$$

并不难验证 m 属于 $J\cap Z(A)$; 可是, 只要把推论 16.5 应用到 $Z(A)$ 与 $J\cap Z(A)$, 就得到 $z \in J\cap Z(A)$ 使得

and it is quite clear that m belongs to $J\cap Z(A)$; but, by applying Corollary 16.5 to $Z(A)$ and $J\cap Z(A)$, we find $z \in J\cap Z(A)$ such that

16.7.5
$$\big(1_A + \delta\cdot z\big)^p = 1_A + \delta^p\cdot m$$

仍由推论 16.5, 最后有 $n = z$, 它属于 $Z(A)$.

and, always applying Corollary 16.5, we finally get that $n = z$ belongs to $Z(A)$.

假定 $s(n) \neq n$ 和对任意 $\ell \in \mathbb{N} - \{0\}$ 考虑 $\mathcal{O}\langle s\rangle$-模 J^ℓ；不难验证对任意 $\ell \geq 2$ 也得到 $(s - \mathrm{id}_A)(J^\ell) \subset \delta \cdot J^\ell$；这样，由推论16.6 就得到 $\mathcal{O}\langle s\rangle$-模分解 $J^\ell = \oplus_{h=0}^{p-1}(J^\ell)_h$ 使得对任意 $1 \leq h \leq p-1$ 与 $m \in (J^\ell)_h$ 有 $s(m) = \xi^h \cdot m$；特别是，有 $n = \sum_{h=0}^{p-1} n_h$，其中 $n_h \in J_h$。现在，对 ℓ 用归纳法；就要证明存在 J 的序列 $\{m_\ell\}_{\ell \in \mathbb{N}}$ 使得对任意 $\ell \in \mathbb{N}$, $J^{\ell+1}$ 包含 $m_{\ell+1} - m_\ell$ 和对适当的 $r_\ell \in J^{\ell+1}$ 有

Assume that $s(n) \neq n$ and, for any $\ell \in \mathbb{N} - \{0\}$, consider the $\mathcal{O}\langle s\rangle$-module J^ℓ; it is easy to check that, for any $\ell \geq 2$, we still have $(s - \mathrm{id}_A)(J^\ell) \subset \delta \cdot J^\ell$; thus, according to Corollary 16.6, we get an $\mathcal{O}\langle s\rangle$-module decomposition $J^\ell = \oplus_{h=0}^{p-1}(J^\ell)_h$ such that, for any $1 \leq h \leq p-1$ and any $m \in (J^\ell)_h$, we have $s(m) = \xi^h \cdot m$; in particular, we get $n = \sum_{h=0}^{p-1} n_h$, where $n_h \in J_h$. Now, arguing by induction on ℓ, we will prove the existence of a sequence $\{m_\ell\}_{\ell \in \mathbb{N}}$ in J such that, for any $\ell \in \mathbb{N}$, $J^{\ell+1}$ contains $m_{\ell+1} - m_\ell$ and, for a suitable $r_\ell \in J^{\ell+1}$, we have

16.7.6 $$\big(1_A + s(m_\ell)\big)^{-1}\big(1_A + \delta \cdot n\big)\big(1_A + m_\ell\big) = 1_A + \delta \cdot (n_0 + r_\ell).$$

令 $m_0 = 0$，固定 $\ell \geq 1$，而假定 $\{m_j\}_{j \leq \ell-1}$ 已经得到了；特别是，我们假定对适当的 $r_{\ell-1} \in J^\ell$ 有

Set $m_0 = 0$, fix $\ell \geq 1$, and assume that we already know $\{m_j\}_{j \leq \ell-1}$; in particular, we assume that, for a suitable $r_{\ell-1} \in J^\ell$, we have

16.7.7 $$\big(1_A + s(m_{\ell-1})\big)^{-1}\big(1_A + \delta \cdot n\big)\big(1_A + m_{\ell-1}\big) = 1_A + \delta \cdot (n_0 + r_{\ell-1});$$

此时，一定有

at this point, we certainly have

16.7.8 $$r_{\ell-1} = \sum_{h=0}^{p-1} r_{\ell-1,h}$$

其中 $r_{\ell-1,h} \in (J^\ell)_h$，$0 \leq h < p$。

for suitable $r_{\ell-1,h} \in (J^\ell)_h$, $0 \leq h < p$.

考虑元素如下 (见 16.2.4)

Consider the element (see 16.2.4)

16.7.9 $$t_\ell = \sum_{h=1}^{p-1} \big((\xi - 1)/(\xi^h - 1)\big) \cdot r_{\ell-1,h} \in J^\ell;$$

因为 $(1_A + s(t_\ell))^{-1} - (1_A - s(t_\ell))$ 属于 $J^{2\ell}$ (见 2.11)，所以得到

since $(1_A + s(t_\ell))^{-1} - (1_A - s(t_\ell))$ belongs to $J^{2\ell}$ (see 2.11), we get

16.7.10 $$\begin{aligned}\big(1_A + s(t_\ell)\big)^{-1}\big(1_A + \delta \cdot (n_0 + r_{\ell-1})\big)\big(1_A + t_\ell\big) + \delta . J^{\ell+1} &= \\ = 1_A - \big(s(t_\ell) - t_\ell\big) + \delta \cdot (n_0 + r_{\ell-1}) + \delta \cdot J^{\ell+1} & \\ = 1_A + \delta \cdot n_0 + \delta \cdot J^{\ell+1};&\end{aligned}$$

也就是说，下面的元素   that is to say, the following element

16.7.11 $\qquad m_\ell = (1_A + m_{\ell-1})(1_A + t_\ell) - 1_A$

满足上述条件 16.7.6.   fulfills the condition 16.7.6 above.

令 $m = \lim_{\ell\to\infty}\{m_\ell\}$ （见 2.11）；它显然属于 J，并且因为 $J(A)^e \subset J$ 与，对任意 $\ell \geq 1$，$m - m_\ell \in J(A)^{\ell+1}$，所以从等式 16.7.6 可推出

Set $m = \lim_{\ell\to\infty}\{m_\ell\}$ (see 2.11); clearly, it belongs to J, and moreover, since $J(A)^e \subset J$ and $m - m_\ell \in J(A)^{\ell+1}$ for any $\ell \geq 1$, it follows from equality 16.7.6 that

$$\big(1_A + s(m)\big)^{-1}\big(1_A + \delta\cdot n\big)\big(1_A + m\big) + J^{\ell+1} =$$

16.7.12
$$= \big(1_A + s(m_{e\ell})\big)^{-1}\big(1_A + \delta\cdot n\big)\big(1_A + m_{e\ell}\big) + J^{\ell+1}$$
$$= 1_A + \delta\cdot n_0 + J^{\ell+1};$$

这样，得到   thus, we get

16.7.13 $\qquad \big(1_A + s(m)\big)^{-1}\big(1_A + \delta\cdot n\big)\big(1_A + m\big) = 1_A + \delta\cdot n_0;$

所以对任意 $h \in \mathbb{N}$ 也得到   consequently, for any $h \in \mathbb{N}$ we still get

16.7.14 $\quad \big(1_A + \delta\cdot s^h(n)\big) = \big(1_A + s^{h+1}(m)\big)\big(1_A + \delta\cdot n_0\big)\big(1_A + s^h(m)\big)^{-1},$

从而条件 16.7.1 成为   and therefore condition 16.7.1 becomes

16.7.15 $\qquad \big(1_A + m\big)\big(1_A + \delta\cdot n_0\big)^p\big(1_A + m\big)^{-1} \in Z(A);$

这样，$\big(1_A + \delta\cdot n_0\big)^p$ 属于 $Z(A)$；进一步，既然 $s(n_0) = n_0$，上述的结果推出 n_0 也属于 $Z(A)$；最后，得到

thus, $\big(1_A + \delta\cdot n_0\big)^p$ belongs to $Z(A)$; moreover, since $s(n_0) = n_0$, the result above implies that n_0 also belongs to $Z(A)$; finally, we get

16.7.16 $\qquad \big(1_A + \delta\cdot n\big)\big(1_A + \delta\cdot n_0\big)^{-1} = \big(1_A + s(m)\big)\big(1_A + m\big)^{-1}.$

命题 16.8. 设 B 是一个 A 的幺 $\mathcal{O}$-子代数并设 M 是一个 A 的 $\mathcal{O}$-子模使得 $A = B + M$. 对任意 $\pi \in J(\mathcal{O})$ 有

Proposition 16.8. *Let B be a unitary $\mathcal{O}$-subalgebra of A and M an $\mathcal{O}$-submodule of A such that $A = B + M$. Then, for any $\pi \in J(\mathcal{O})$, we have*

16.8.1 $\qquad 1_A + \pi\cdot A = (1_A + \pi\cdot M)(1_A + \pi\cdot B).$

而且，如果 $B \cap M = \{0\}$ 那么由乘法决定的如下映射是双射

Moreover, if $B \cap M = \{0\}$ then the following map, determined by the product, is bijective.

16.8.2 $\qquad \mathfrak{m}: (1_A + \pi\cdot M) \times (1_A + \pi\cdot B) \longrightarrow 1_A + \pi\cdot A.$

证明: 显然有

Proof: We clearly have

16.8.3 $$(1_A + \pi\cdot M)\cdot(1_A + \pi\cdot B) \subset 1_A + \pi\cdot A\,.$$

固定 $a \in A$ 和对任意 A 的序列 $\{a_\ell\}_{\ell\in\mathbb{N}}$ 记

Fix $a \in A$ and, for any sequence $\alpha = \{a_\ell\}_{\ell\in\mathbb{N}}$ in A, we set

16.8.4 $$\Sigma_\ell(\alpha) = \sum_{h=1}^{\ell} \pi^h\cdot a_h\,;$$

首先要证明分别存在 B 与 M 的序列 $\beta = \{b_\ell\}_{\ell\in\mathbb{N}}$ 与 $\mu = \{m_\ell\}_{\ell\in\mathbb{N}}$ 使得对任意 $\ell \in \mathbb{N}$ 和对适当的 $a_\ell \in A$ 有

first of all, we will prove the existence of respective sequences $\beta = \{b_\ell\}_{\ell\in\mathbb{N}}$ and $\mu = \{m_\ell\}_{\ell\in\mathbb{N}}$ in B and M such that, for any $\ell \in \mathbb{N}$ and for a suitable $a_\ell \in A$, we have

16.8.5 $$1_A + \pi\cdot a = \big(1_A + \Sigma_\ell(\mu) + \pi^{\ell+1}\cdot a_\ell\big)\big(1_A + \Sigma_\ell(\beta)\big)\,.$$

对 ℓ 用归纳法; 如果 $\ell = 0$ 那么选择 $b_0 = 0 = m_0$ 和 $a_0 = a$ 显然满足这个等式. 假定 $\ell \geq 1$; 那么, 既然 $A = B + M$, 存在 $b'_\ell \in B$ 与 $m_\ell \in M$ 使得 $a_{\ell-1} = m_\ell + b'_\ell$, 并且不难验证

We argue by induction on ℓ; if $\ell = 0$ then the choice $b_0 = 0 = m_0$ and $a_0 = a$ clearly fulfills this equality. Assume that $\ell \geq 1$; then, since $A = B + M$, there exist $b'_\ell \in B$ and $m_\ell \in M$ such that $a_{\ell-1} = m_\ell + b'_\ell$, and it is easily checked that

16.8.6 $$\begin{aligned} 1_A+\pi\cdot a &= \\ &= \big(1_A+\Sigma_{\ell-1}(\mu)+\pi^\ell\cdot a_{\ell-1}\big)\big(1_A+\pi^\ell\cdot b'_\ell\big)^{-1}\big(1_A+\pi^\ell\cdot b'_\ell\big)\big(1_A+\Sigma_{\ell-1}(\beta)\big) \\ &= \big(1_A+\Sigma_\ell(\mu)+\pi^{\ell+1}\cdot b'_\ell\big)\big(1_A+\pi^\ell\cdot b'_\ell\big)^{-1}\big(1_A+\Sigma_{\ell-1}(\beta)+\pi^\ell\cdot b_\ell\big) \\ &= \big(1_A+\Sigma_\ell(\mu)+\pi^{\ell+1}\cdot a_\ell\big)\big(1_A+\Sigma_\ell(\beta)\big) \end{aligned}$$

其中

where we set

16.8.7 $$b_\ell = b'_\ell\big(1_A + \Sigma_{\ell-1}(\beta)\big)\,, \quad a_\ell = -b'_\ell(1_A + \pi^\ell\cdot b'_\ell)^{-1}\Sigma_\ell(\mu)\,;$$

这样, 等式 16.8.5 获证. 此时, 在 B 与 M 中分别令

thus, equality 16.8.5 is proved. At that point, in B and M, we respectively set

16.8.8 $$b = \lim_{\ell\to\infty}\{\sum_{h=0}^{\ell}\pi^h\cdot b_{h+1}\} = \sum_{h\in\mathbb{N}}\pi^h\cdot b_{h+1}\,, \quad m = \sum_{h\in\mathbb{N}}\pi^h\cdot m_{h+1}\,;$$

这样, 对任意 $\ell \in \mathbb{N}$ 有

thus, for any $\ell \in \mathbb{N}$ we have

16.8.9 $$\begin{aligned} (1_A + \pi\cdot m)(1_A + \pi\cdot b) + \pi^{\ell+1}\cdot A &= \\ &= \big(1_A + \Sigma_\ell(\mu)\big)\big(1_A + \Sigma_\ell(\beta)\big) + \pi^{\ell+1}\cdot A \\ &= 1_A + \pi\cdot a + \pi^{\ell+1}\cdot A\,, \end{aligned}$$

从而还有 and therefore we still have

16.8.10 $$(1_A + \pi{\cdot}m)(1_A + \pi{\cdot}b) = 1_A + \pi{\cdot}a\,;$$

特别是, $\mathfrak{m}$ 是满射. in particular, $\mathfrak{m}$ is surjective.

最后, 假定存在 $m' \in M$ 与 $b' \in B - \{b\}$ 使得 Finally, assume that there are $m' \in M$ and $b' \in B - \{b\}$ such that

16.8.11 $$(1_A + \pi{\cdot}m)(1_A + \pi{\cdot}b) = (1_A + \pi{\cdot}m')(1_A + \pi{\cdot}b')\,;$$

因为有 since we have

16.8.12 $$1_A \neq (1_A + \pi{\cdot}b')(1_A + \pi{\cdot}b)^{-1} \in 1_A + \pi{\cdot}B\,,$$

所以存在 $\ell \geq 1$ 与 $c \in B - \pi{\cdot}B$ 使得 there exist $\ell \geq 1$ and $c \in B - \pi{\cdot}B$ such that

16.8.13 $$\begin{aligned} 1_A + \pi{\cdot}m &= (1_A + \pi{\cdot}m')(1_A + \pi^{\ell}{\cdot}c) \\ &= 1_A + \pi{\cdot}m' + \pi^{\ell}{\cdot}c + \pi^{\ell+1}{\cdot}m'c\,; \end{aligned}$$

可是, 存在 $d \in B$ 与 $n \in M$ 使得 $m'c = d + n$; 所以, 有 but, there exist $d \in B$ and $n \in M$ fulfilling $m'c = d + n$; hence, we have

16.8.14 $$m - m' - \pi^{\ell}{\cdot}n = \pi^{\ell-1}{\cdot}(c + \pi{\cdot}d)$$

并且, 既然 $c \notin \pi{\cdot}B$, 可推出 $c + \pi{\cdot}d \neq 0$; 这样, 有 and moreover, since $c \notin \pi{\cdot}B$, we clearly get $c + \pi{\cdot}d \neq 0$; thus, we have

16.8.15 $$0 \neq \pi^{\ell-1}{\cdot}(c + \pi{\cdot}d) \in M \cap B\,.$$

参考文献 References

[1] Laurence BARKER — *Induction, restriction and G-algebras* Comm. in Algebra, 22(1994), 6349-83

[2] Michel BROUÉ & Lluís PUIG — *A Frobenius Theorem for Blocks* Inventiones math., 56(1980), 117-28

[3] Yun FAN — *The source algebras of nilpotent blocks over arbitrary ground-fields* J. of Algebra, 165(1994), 606-32

[4] Yun FAN — *Relative local control and the block source algebra* Science in China (Series A) 40(1997), 785-98

[5] Yun FAN & Lluís PUIG — *On blocks with nilpotent coefficient extensions* Algebras and Rep. Theory 2(1999) Err., 27-73

[6] Lluís PUIG — *Sur un Théorème de Green* Math. Zeit., 166(1979), 117-29

[7] Lluís PUIG — *Pointed Groups and Construction of Characters* Math. Zeit., 176(1981), 265-92

[8] Lluís PUIG — *Local fusions in block source algebras* J. of Algebra, 104(1986), 358-369

[9] Lluís PUIG — *Nilpotent blocks and their source algebras* Inventiones math., 93(1988), 77-116

[10] Lluís PUIG — *The hyperfocal subalgebra of a block* Inventiones math., 141(2000), 365-97

[11] Lluís PUIG — *Source algebras of p-central group extensions* Journal of Algebra, 235(2001), 359-98

[12] Jean-Pierre SERRE — *"Linear Representations of Finite Groups"* Graduate Texts in Mathematics, 42(1977) Springer-Verlag, New York

[13] Jean-Pierre SERRE — *"Local Fields"* Graduate Texts in Mathematics, 67(1979) Springer-Verlag, New York

[14] Jacques THÉVENAZ — *"G-Algebras and Modular Representation Theory"* Oxford Mathematical Monographs, 1995 Clarendon Press, OXFORD

名词索引 Index

Springer Monographs in Mathematics

This series publishes advanced monographs giving well-written presentations of the "state-of-the-art" in fields of mathematical research that have acquired the maturity needed for such a treatment. They are sufficiently self-contained to be accessible to more than just the intimate specialists of the subject, and sufficiently comprehensive to remain valuable references for many years. Besides the current state of knowledge in its field, an SMM volume should also describe its relevance to and interaction with neighbouring fields of mathematics, and give pointers to future directions of research.

Abhyankar, S.S. **Resolution of Singularities of Embedded Algebraic Surfaces** 2nd enlarged ed. 1998
Andrievskii, V.V.; Blatt, H.-P. **Discrepancy of Signed Measures and Polynomial Approximation** 2002
Armitage, D.H.; Gardiner, S.J. **Classical Potential Theory** 2001
Arnold, L. **Random Dynamical Systems** 1998
Aubin, T. **Some Nonlinear Problems in Riemannian Geometry** 1998
Bang-Jensen, J.; Gutin, G. **Digraphs** 2001
Baues, H.-J. **Combinatorial Foundation of Homology and Homotopy** 1999
Brown, K.S. **Buildings** 3rd printing 2000 (1st ed. 1998)
Cherry, W.; Ye, Z. **Nevanlinna's Theory of Value Distribution** 2001
Ching, W.K. **Iterative Methods for Queuing and Manufacturing Systems** 2001
Crabb, M.C.; James, I.M. **Fibrewise Homotopy Theory** 1998
Dineen, S. **Complex Analysis on Infinite Dimensional Spaces** 1999
Elstrodt, J.; Grunewald, F. Mennicke, J. **Groups Acting on Hyperbolic Space** 1998
Fadell, E.R.; Husseini, S.Y. **Geometry and Topology of Configuration Spaces** 2001
Fedorov, Y.N.; Kozlov, V.V. **A Memoir on Integrable Systems** 2001
Flenner, H.; O'Carroll, L. Vogel, W. **Joins and Intersections** 1999
Gelfand, S.I.; Manin, Y.I. **Methods of Homological Algebra** 2nd ed. 2002
Griess, R.L.Jr. **Twelve Sporadic Groups** 1998
Ivrii, V. **Microlocal Analysis and Precise Spectral Asymptotics** 1998
Jorgenson, J.; Lang, S. **Spherical Inversion on SLn (R)** 2001
Kozlov, V.; Maz'ya, V. **Differential Equations with Operator Coefficients** 1999
Landsman, N.P. **Mathematical Topics between Classical & Quantum Mechanics** 1998
Lemmermeyer, F. **Reciprocity Laws: From Euler to Eisenstein** 2000
Malle, G.; Matzat, B.H. **Inverse Galois Theory** 1999
Mardesic, S. **Strong Shape and Homology** 2000
Narkiewicz, W. **The Development of Prime Number Theory** 2000
Parker, C.; Rowley, P. **Symplectic Amalgams** 2002
Prestel, A.; Delzell, C.N. **Positive Polynomials** 2001
Ranicki, A. **High-dimensional Knot Theory** 1998
Ribenboim, P. **The Theory of Classical Valuations** 1999
Rowe, E.G.P. **Geometrical Physics in Minkowski Spacetime** 2001
Rudyak, Y.B. **On Thom Spectra, Orientability and Cobordism** 1998
Ryan, R.A. **Introduction to Tensor Products of Banach Spaces** 2002
Saranen, J,; Vainikko, G. **Periodic Integral and Pseudodifferential Equations with Numerical Approximation** 2002
Schneider, P. **Nonarchimedean Functional Analysis** 2002
Serre, J-P. **Complex Semisimple Lie Algebras** 2001 (reprint of first ed. 1987)
Serre, J.-P. **Galois Cohomology** corr. 2nd printing 2002 (1st ed. 1997)
Serre, J-P. **Local Algebra** 2000
Springer, T.A.; Veldkamp, F.D. **Octonions, Jordan Algebras, and Exceptional Groups** 2000
Sznitman, A.-S. **Brownian Motion, Obstacles and Random Media** 1998
Tits, J.; Weiss, R.M. **Moufang Polygons** 2002
Uchiyama, A. **Hardy Spaces on the Euclidean Space** 2001
Üstünel, A.-S.; Zakai, M. **Transformation of Measure on Wiener Space** 2000
Yang, Y. **Solitons in Field Theory and Nonlinear Analysis** 2001